新編
作物学用語集

日本作物学会編

養賢堂

日本作物学会用語委員会

平成12年度

　　　　委員長　今井　勝　明治大学農学部
　　　　委　員（五十音順）
　　　　　　　加藤盛夫　筑波大学農林学系（幹事）
　　　　　　　高野　泰　東京大学大学院農学生命科学研究科
　　　　　　　田代　亨　三重大学生物資源学部
　　　　　　　森田茂紀　東京大学大学院農学生命科学研究科

専門委員・協力者（五十音順）

秋山　侃・安藤　豊・石井龍一・伊藤　治・伊藤大雄・伊東睦泰・稲永　忍・井上吉雄・岩間和人・内田直次・及川武久・大杉　立・岡野邦夫・小田雅行・片山忠夫・兼子　真・刈屋国男・川上直人・窪田文武・後藤雄佐・小林和彦・近藤始彦・三枝正彦・齋藤和幸・齊藤邦行・斎藤　均・佐々木修・塩澤宏康・塩谷哲夫・篠原　温・芝山秀次郎・大門弘幸・高橋　清・高橋　幹・田中耕司・谷山鉄郎・長南信雄・辻本　壽・寺島一郎・中川博視・野内　勇・箱山　晋・花田毅一・林　久喜・平沢　正・平田昌彦・堀江　武・松井重雄・松浦朝奈・松江勇次・松田智明・道山弘康・三宅　博・矢島正晴・山内　章・山岸順子・山岸　徹・山崎耕宇・山本由徳・吉田智彦・葭田隆治・和田富吉・和田道宏・渡辺利通・Cervantes, Emy

平成21年度

　　　　委員長　松田智明　茨城大学
　　　　委　員（五十音順）
　　　　　　　新田洋司　茨城大学農学部（幹事）
　　　　　　　後藤雄佐　東北大学大学院農学研究科
　　　　　　　高野　泰　東京大学大学院農学生命科学研究科
　　　　　　　平沢　正　東京農工大学大学院共生科学技術研究院
　　　　　　　山本由徳　高知大学農学部
　　　　　　　吉田智彦　宇都宮大学農学部

専門委員・協力者（五十音順）

　池田　武・井上吉雄・今井　勝・鴻田一絵・杉本秀樹・辻　博之・中村　聡・林　久喜

新編の刊行に当たって

　今日，作物に関する研究分野は古典的なものから最先端のものまで非常に多岐にわたっており，分野を異にする研究者間での意志の疎通が困難なことさえある．したがって，論文を読み書きする際に用語の共通理解が必要であることは言うまでもない．また，用語には長い歴史の間に使用頻度が減じたり用法が変わり意味が不明確になったものや，他分野から導入されたばかりで未だ定着していないものなどが常に存在する．これらの問題に対応すべく従来から日本作物学会では用語集を編纂してきたが，今回は新編と銘打ってかなり大幅な改訂を行った．

　「作物学用語集」（編集委員長：村田吉男，編集委員：花田毅一・北條良夫・星川清親・大久保隆弘・山崎耕宇）は，本学会設立50周年を記念して昭和52年に日本作物学会紀事40巻1号（昭和46年）綴じ込みの「作物学用語集」（問題用語小委員会編）を土台として誕生した．編集の基本方針は，使いやすさを重視し作物学関連用語を広く採用すること，および作物・雑草名を充実することであった．幸い，学会員内外に好評で広く行き渡り活用されたが，その後の学問の急速な発展に適合させるため，用語の再検討と新規用語の追加をともなう「改訂作物学用語集」（編集委員長：山崎耕宇，編集委員：石井龍一・石原邦・川口数美・星川清親・森脇　勉・渡邉和之，幹事：森田茂紀）が10年後の昭和62年に刊行された．これも好評を博したが，さらに時代の変遷に対応すべく平成6年に河野恭廣用語委員会委員長（当時）の下で改訂の検討が始まり，平成8年に至って本委員会が本格的に改訂作業に着手した．まず，「改訂作物学用語集」に採録された用語を，56の分野に重複させつつ分類した原案を用意し，多数の専門委員を委嘱して加除訂正をお願いした．その後，用語委員会の各委員による分野毎の整理・選択，および検討会議を重ねてようやく和英8,190，英和9,160の見出し語からなる「新編作物学用語集」の刊行に至ったのである．今回の編集方針も，これまでの用語集編集の精神を継承することであったが，作物学における研究領域の深化・拡大を願って，環境，バイオテクノロジー，統計，情報処理などに関する用語を充実させたことが特記されよう．幸いコンピュータを利用したので作業時間は大幅に短縮されたが，用語の採否や配列順序などを巡っては紆余曲折したり躊躇することも多かった．用語の取捨選択や順位づけに当たっては，日本農学

会をはじめとする国内関連学協会の用語集や各種の辞典類を参考にした．とりわけ，植物の学名に関しては「朝日百科植物の世界」(朝日新聞社)と「世界有用植物事典」(平凡社)を主な拠り所とした．

おわりに，用語の検討に際して多大のご協力を頂いた専門委員・協力者各位，ならびに参照させて頂いた出版物の編集者・出版社各位には衷心より御礼申し上げる．また，「作物学用語集」以来長年にわたり日本作物学会にご協力を頂いている株式会社養賢堂には深く謝意を表する次第である．今回の改訂に当たっても及川 清社長の快諾を得，矢野勝也，故大津弘一の両氏には種々ご指導頂いた．重ねて衷心より御礼申し上げる．

平成12年3月

日本作物学会用語委員会
委員長 今 井 　 勝

訂正版の発行に当たって

「新編 作物学用語集」は，発行後9年が経過し，見直しが必要と思われる用語や，重要な新出用語が見受けられるようになった．

本用語集の訂正には，新出用語を加えたデータベース上での整理と検討作業が必要である．加えて，現在，本学会では作物学用語を解説する専門書の編集が進められている．また，新出用語にたいする学会員の意見を集約する時間も必要である．

このようなことから今回の訂正版では，学会員からの意見が多かった用語を中心に見直し，修正した．また，農林水産省等の試験研究機関は，新旧の名称の対照表を追補した．さらに，「籾」は通常使われている字体に変更した．

さいごに，本用語集の発行に長年ご協力いただいている株式会社 養賢堂，ならびに同社 田中 道雄 氏に心より感謝申し上げる．

平成21年9月

日本作物学会用語委員会
委員長 松 田 智 明

    i. 略号あるいは正式名
   例：adenosine triphosphate (ATP)　アデノシン三リン酸
       ATP (adenosine triphosphate)　アデノシン三リン酸
   ii. 一部の単語のみ異なる場合(( )およびその中は無視して配列)
   例：autumn sowing (seeding)　秋播き
   iii. 難読漢字の読み
   例：axillary bud　腋芽(えきが)
 (2) [ ]省略可([ ]は無視して配列)
   例：artificial [farmyard] manure　速成堆肥

付記　生長と成長については，「成長」とする．

付表　国際単位系(SI単位)および単位の換算表を巻末に付した．

ギリシア文字の読み方

大	小	名	称	大	小	名	称
(A)	α	alpha	アルファ	(N)	ν	nu	ニュー
(B)	β	beta	ベータ	Ξ	ξ	xi	クサイ，クシー
Γ	γ	gamma	ガンマ	(O)	o	omicron	オミクロン
Δ	δ	delta	デルタ	Π	π	pi	パイ
(E)	ε	epsilon	イプシロン	(P)	ρ	rho	ロー
(Z)	ζ	zeta	ゼータ	Σ	σ	sigma	シグマ
(H)	η	eta	イータ，エータ	(T)	τ	tau	タウ
Θ	θ	theta	テータ，シータ	Υ	υ	upsilon	ユプシロン
(I)	ι	iota	イオタ	Φ	ϕ	phi	ファイ
(K)	κ	kappa	カッパ	(X)	χ	chi	カイ
Λ	λ	lambda	ラムダ	Ψ	ψ	psi	プサイ，プシー
(M)	μ	mu	ミュー	Ω	ω	omega	オメガ

( )で囲んだ文字はローマ字と区別できないので科学論文には普通使えない．

和英の部

Japanese – English

[あ]

アイ (藍) Chinese indigo, *Persicaria tinctoria* (Aiton) H. Gross (= *Polygonum tinctorium* Lour.)

アイシーピーぶんこうぶんせき ICP分光分析, 誘導結合高周波プラズマ分光分析 inductively coupled plasma spectrometry

アイソザイム, イソ酵素 isozyme

アイソトープ, 同位元素, 同位体 isotope

あいは 合葉【タバコ】 cutters

アウス【イネの生態型】 aus

あえん 亜鉛 zinc (Zn)

アオイモドキ→エノキアオイ

アオウキクサ *Lemna aoukikusa* Beppu et Murata (= *L. paucicostata* Hegelm.)

あおがり 青刈り soiling

あおがりきゅうよ 青刈り給与, 青刈り利用 [fresh] forage feeding, [green] soiling

あおがりさくもつ 青刈り作物 soiling crop

あおがりしりょう 青刈り飼料, 生草 fresh forage, green forage, soilage, green chop

あおがりトウモロコシ 青刈りトウモロコシ soiling corn (maize)

あおがりりよう 青刈り利用, 青刈り給与 [fresh] forage feeding, [green] soiling

あおがれびょう 青枯病 bacterial wilt

あおしにまい 青死に米 green dead-rice kernel

あおだち 青立ち【イネ】 straight head, straighthead

アオノリュウゼツラン century plant, *Agave americana* L.

アオバナルーピン (青花ルーピン) blue lupine, *Lupinus angustifolius* L.

アオビユ slender amaranth, green amaranth, *Amaranthus viridis* L.

あおまい 青米 green rice kernel

アオミドロ *Spirogyra arcla* Kutz.

あか 亜科 subfamily

アカウキクサ *Azolla pinnata* R. Br. ssp. *asiatica* R. M. K. Saunders et K. Fowler

あかかびびょう 赤かび病【ムギ類】 scab, Fusarium blight

あかがれ [びょう] 赤枯れ [病]【イネの生理病】 stifle disease

アカクローバ red clover, *Trifolium pratense* L.

アカザ *Chenopodium album* L. var. *centrorubrum* Makino

あかさびびょう 赤さび (錆) 病【コムギ】 leaf rust, brown rust

あかだねウンダイ 赤種ウンダイ (蕓薹) →在来種ナタネ

アカツメクサ→アカクローバ

あかまい 赤米 red [kerneled] rice

アカミノキ→ログウッド

アカヤジオウ (赤矢地黄) *Rehmannia glutinosa* (Gaertn.) Libosh. ex Fisch. et C. A. Mey. var. *purpurea* (Makino)

アガロース agarose

あかんたい 亜寒帯 subarctic zone, subpolar zone

あきうえ 秋植え autumn (fall) planting

あきおち 秋落ち "akiochi"

あきおちすいでん 秋落水田 "akiochi" paddy field

あきおちていこうせい 秋落抵抗性【イネ】 resistance to "akiochi"

あきがた 秋型【ソバ】 autumn ecotype, late-summer ecotype

あきごえ 秋肥 autumn manuring, fall dressing

あきざき 秋咲き fall flowering

あきさく 秋作 1) autumn (fall) cropping 2) autumn (fall) crop

あきさくもつ　秋作物　autumn (fall) crop
あきさめ　秋雨　autumnal rain, Akisame
あきせいし　秋整枝【チャ】　autumn skiffing
あきソバ　秋ソバ　autumn buckwheat, late-summer buckwheat
あきダイズ　秋ダイズ　autumn soybean, late-summer soybean
アギナシ　*Sagittaria aginashi* Makino
アキノウナギツカミ　*Persicaria sieboldii* (Meisn.) Ohki (= *Polygonum sagittatum* L. var. *sieboldii* (Meisn.) Maxim.)
アキノエノコログサ　giant foxtail, *Setaria faberi* Herrm.
アキノノゲシ　*Lactuca indica* L. var. *laciniata* (O. Kuntze) Hara
あきまき　秋播き　autumn sowing (seeding), fall sowing (seeding)
あきまきがた　秋播き型　winter type
あきまきせい　秋播き性　winter habit
あきまきせいていど　秋播き性程度　degree of winter habit
あきまきひんしゅ　秋播き品種　winter variety
あきめ　秋芽【チャ】　autumn shoot
アキメヒシバ　violet crabgrass, *Digitaria violascens* Link
アクチン　actin
アクチンフィラメント　actin filament
アグリビジネス　agribusiness
アグロバクテリウム・ツメファシエンス【細菌】　*Agrobacterium tumefacience*
アグロバクテリウム・リゾゲネス【細菌】　*Agrobacterium rhizogenes*
アグロフォレストリー　agroforestry
あげつぎ　揚げ接ぎ　bench-grafting, indoor-grafting
あげどこ　揚げ床　raised bed, raised seedbed
アサ 1)→タイマ　2) (麻)【麻類】 hemps
あさうえ　浅植え　shallow planting
あさうね　浅うね(畝), 低うね(畝)　low ridge (bed)
アサガオ (朝顔)　Japanese morning glory, *Ipomoea nil* (L.) Roth (= *Pharbitis nil* (L.) Choisy)
あさがり　浅刈り【チャ】　light trimming of canopy
あさぎり　浅剪り　light pruning
アサクサノリ　laver, *Porphyra tenera* Kjellman
アサびき　麻挽き　scraping hemp fiber
あさまき　浅播き　shallow sowing (seeding)
あさみず　浅水　shallow flooding
アサみゆ　麻実油, 大麻子油　hemp seed oil
アシ (葦)→ヨシ (葭)
あじ　味　taste
アジアイネ→イネ
アジアかいはつぎんこう　アジア開発銀行　Asian Development Bank (ADB)
アジアこうかだいがく　アジア工科大学　Asian Institute of Technology (AIT)
アジアせいさんせいきこう　アジア生産性機構　Asian Productivity Organization (APO)
アジアそさいけんきゅうかいはつセンター　アジア蔬菜研究開発センター　Asian Vegetable Research and Development Center (AVRDC)
アジアメン (アジア棉)　Asiatic cotton, *Gossypium arboreum* L. および *G. herbaceum* L.
アシカキ　*Leersia japonica* Makino
あしにざっそう　脚荷雑草　ballast weed
あじぶっしつ　味物質　taste substance
あしぶみだっこくき　足踏み脱穀機　foot thresher
あしゅ　亜種　subspecies
あしょうさんイオン　亜硝酸イオン

nitrite ion (NO_2^-)
あしょうさんえん　亜硝酸塩　nitrite
アズキ(小豆)　adzuki (azuki) bean, small red bean, *Vigna angularis* (Willd.) Ohwi et Ohashi (= *Phaseolus angularis* L.)
アスコルビンさん　アスコルビン酸　ascorbic acid (AsA)
アスパラギン　asparagine (Asn)
アスパラギンさん　アスパラギン酸　aspartic acid (Asp)
あぜ(畦)　levee
アゼガヤ　Chinese sprangletop, *Leptochloa chinensis* (L.) Nees
アゼガヤツリ　globe sedge, *Pycreus flavidus* (Retz.) T. Koyama (= *Cyperus globosus* All.)
あぜさく　あぜ(畦)作　levee planting
アセチレンかんげんほう　アセチレン還元法　acetylene reduction method
あぜづくり　あぜ(畦)作り　levee building
アゼトウガラシ　*Lindernia angustifolia* (Benth.) Wettst.
アセトカーミン　acetocarmine
アセトシリンゴン　acetosyringone
アセトン　acetone
アゼナ　common false pimpernel, *Lindernia procumbens* (Krock.) Philcox (= *L. pyxidaria* L.)
あぜぬり　あぜ(畦)塗り，くろ(畔)塗り　levee coating
あぜマメ　あぜ(畦)豆　soybean grown on levee
あぜみち　あぜ(畦)道　footpath between paddy fields
アゼムシロ　lobelia, *Lobelia chinensis* Lour.
アセンヤクノキ(阿仙薬樹)　catechu, cutch, black catechu, *Acacia catechu* (L.f.) Willd.
アゾトバクター　azotobacter
あたたかさのしすう　暖かさの指数，温量指数　warmth index
アッサムゴム→インドゴム
あっぺん　圧扁　rolling
あっぺんおおむぎ　圧扁大麦　rolled barley
あっぺんこむぎ　圧扁小麦　rolled wheat
あつポテンシャル　圧ポテンシャル　pressure potential
あつボンベほう　圧ボンベ法，プレッシャーチャンバー法　pressure chamber method
あつまき　厚播き，密播　dense sowing (seeding), thick sowing (seeding)
あつりゅうせつ　圧流説　pressure flow theory
アデニン　adenine (A)
アデノシンさんリンさん　アデノシン三リン酸　adenosine triphosphate (ATP)
アデノシンにリンさん　アデノシン二リン酸　adenosine diphosphate (ADP)
あとさく　後作　succeeding cropping
あとさく[もつ]　後作[物]　succeeding crop
あなうえ　穴植え　dibble planting
あなまき　穴播き　dibbling
アナローグ　類似体　analogue
アナログ【情報】　analog
アニス　anise, *Pimpinella anisum* L.
アニリンブルー　aniline blue
アヌウ　anu, jicamas, *Tropaeolum tuberosum* Ruiz. et Pav.
あねったい　亜熱帯　subtropical zone, subtropics
アバカ→マニラアサ
アーバスキュラーきんこん　アーバスキュラー菌根　arbuscular mycorrhiza
アビシニアバショウ　Abyssinian banana, *Ensete ventricosum* (Welw.) Cheesman
アブシジンさん　アブシジン酸　abscisic acid (ABA)
アブノメ　dopatrium, *Dopatrium*

junceum (Roxb.) Buch.-Ham. ex Benth.
あぶらかす　油粕　oil meal
アブラギリ (油桐)　tung, Japanese tung-oil tree, *Aleurites cordata* (Thunb.) R. Br. ex Steud.
あぶらしょうじ　油障子　oil paper sash
アブラナ (油菜) →ナタネ
アブラムシ, アリマキ　aphid
アブラヤシ (油椰子)　oil palm, African oil palm, *Elaeis guineensis* Jacq.
アフリカイネ→グラベリマイネ
アフリカヒゲシバ→ローズグラス
アベナくっきょくしけんほう　アベナ屈曲試験法　*Avena* curvature test
アベナしんちょうしけんほう　アベナ伸長試験法　*Avena* straight-growth test
アボカド　avocado, alligator pear, *Persea americana* Mill.
アポトーシス　apoptosis
アポプラスト　apoplast
アポミクシス　無配偶生殖　apomixis
アマ (亜麻)　flax, *Linum usitatissimum* L.
アマウイキョウ (甘茴香) →イタリアウイキョウ
アマチャ (甘茶)　*Hydrangea macrophylla* (Thunb. ex Murray) Ser. var. *amacha* Makino
アマにかす　亜麻仁粕　linseed meal
アマにゆ　亜麻仁油　linseed oil
アマハステビア, ステビア　stevia, kaa he-e, *Stevia rebaudiana* (Bertoni) Hemsl.
アマン【イネの生態型】　aman
アミガサユリ→バイモ
アミノさん　アミノ酸　amino acid
1-アミノシクロプロパン-1-カルボンさん　1-アミノシクロプロパン-1-カルボン酸　1-aminocyclopropane-1-carboxylic acid (ACC)
アミラーゼ　amylase
アミロース　amylose
アミログラム　amylogram
アミログラムとくせい　アミログラム特性　amylographic characteristics
アミロプラスト　amyloplast
アミロペクチン　amylopectin
アメダス, 地域気象観測システム　AMeDAS (Automated Meteorological Data Acquisition System)
アメリカアゼナ　low false-pimpernel, *Lindernia dubia* (L.) Penn.
アメリカアリタソウ　wormseed goosefoot, *Chenopodium ambrosioides* L. var. *anthelminticum* (L.) A. Gray
アメリカキンゴジカ　prickly sida, *Sida spinosa* L.
アメリカサトイモ　yautia, tannia, *Xanthosoma sagittifolium* (L.) Schott
アメリカスズメノヒエ→バヒアグラス
アメリカセンダングサ　devil's beggarticks, *Bidens frondosa* L.
アメリカチョウセンアサガオ　sacred datura, *Datura meteloides* Dunal. (= *D. inoxa* Mill.)
アメリカヒゲシバ→ローズグラス
アメリカホドイモ　potato bean, groundnut, *Apios americana* Medik.
アメリカマコモ　wildrice, Indian rice, *Zizania aquatica* L. および *Z. palustris* L.
アメリカミズキンバイ→ヒレタゴボウ
アメリカヤマゴボウ→ヨウシュヤマゴボウ
アーモンド (扁桃)　almond, *Prunus amygdalus* Batsch
あらいそ　洗い麻　bleached hemp stem
あらおこし　荒起し　coarse plowing, first plowing
あらこ　荒粉　1) dried corm slice【コンニャク】　2) meal【麦・豆など】
あらしろ　荒代　coarse puddling, first

puddling
アラニン　alanine (Ala)
あらぬか　荒糠　coarse bran
アラビアコーヒー [ノキ]　Arabian coffee, *Coffea arabica* L.
アラビアゴム　gum arabic, gum Senegal, *Acacia senegal* Willd.
アラミナ→オオバボンテンカ
あられ (霰)　snow pellets, graupel
アリマキ, アブラムシ　aphid
ありゅうさんガス　亜硫酸ガス⇒二酸化イオウ　sulfur dioxide (SO_2)
アールアイプラスミド　Riプラスミド　Ri plasmid
アールエヌエーポリメラーゼ　RNA ポリメラーゼ　RNA polymerase
アルカリせいどじょう　アルカリ性土壌　alkaline soil
アルカリどじょう　アルカリ土壌　alkali soil
アルカリほうかいど　アルカリ崩壊度【イネ】　alkali solubility
アルカロイド　alkaloid
アルギニン　arginine (Arg)
アルゴリズム　algorithm
アルコール　alcohol
アルコールおんどけい　アルコール温度計　alcohol thermometer
アルコールはっこう　アルコール発酵　alcohol fermentation
アルコール [りょう] さくもつ　アルコール [料] 作物　alcoholic crop
アルサイククローバ　alsike clover, *Trifolium hybridum* L.
アルビノ　albino
アルファナフチルアミン　α-ナフチルアミン　α-naphthylamine
アルファルファ　alfalfa, lucerne, *Medicago sativa* L.【*M.* × *media* Pers. を含む】
アルブミン　albumin
アルベド　albedo
アルボウイルス　arbovirus

アルミニウム　aluminum (Al)
アルメニアコムギ　wild timopheevi wheat, *Triticum araraticum* Jakubz.
アレイさくつけ　アレイ作付け, 灌木間作　alley cropping
アレチノギク　hairy fleabane, *Conyza bonariensis* (L.) Cronq. (= *Erigeron bonariensis* L.)
アレニウスプロット　Arrhenius plot
アレロパシー　他感作用　allelopathy
アロールート　arrowroot, West Indian arrowroot, *Maranta arundinacea* L.
アロステリックこうか　アロステリック効果　allosteric effect
アロフェン　allophane
アロメトリー　allometry, 相対成長　relative growth
アワ (粟)　Italian millet, foxtail millet, *Setaria italica* (L.) P. Beauv.
あわせつぎ　合わせ接き　splice grafting
あんき　暗期　dark period
あんきょ　暗きょ (渠)　underdrain, conduit
あんきょはいすい　暗きょ (渠) 排水　underdrainage, tile drainage, pipe drainage
あんこきゅう　暗呼吸　dark respiration
アンシミドール　ancymidol
あんしやけんびきょう　暗視野顕微鏡　dark-field microscope
アンズ (杏)　apricot, *Prunus armeniaca* L.
あんぜんしようきじゅん　安全使用基準【農薬】　direction for safe use
アンチオーキシン　抗オーキシン　antiauxin
アンチセンスアールエヌエー　アンチセンス RNA　antisense RNA
アンチセンスディーエヌエー　アンチセンス DNA　antisense DNA
あんていせい　安定性　stability
あんていどういげんそ　安定同位元素,

安定同位体　stable isotope
あんていどういたい　安定同位体，安定同位元素　stable isotope
アンテナしきそ　アンテナ色素　antenna pigment, 集光性色素　light harvesting pigment
アントシアン　anthocyan
アンナンウルシ→トンキンウルシ
あんつつが　暗発芽　dark germination
あんはつがしゅし　暗発芽種子　dark germinater, negative photoblastic seed
あんはんのう　暗反応　dark reaction
アンペラソウ　Chinese mat rush, *Lepironia articulata* (Retz.) Domin (= *L. mucronata* Rich.)
アンミ　tooth pick, *Ammi visnaga* (L.) Lam.
アンモニア　ammonia
アンモニアかせい[さよう]　アンモニア化成[作用]　ammonification
アンモニアしょくぶつ　アンモニア植物　ammonia plant
アンモニアたいちっそ　アンモニア態窒素　ammonium nitrogen
アンローディング　unloading

[い]

イ(藺)→イグサ
イオウ(硫黄)　sulfur (S)
いおう[びょう]　萎黄[病], 黄化病　yellows
イオウさんかぶつ　イオウ(硫黄)酸化物　sulfur oxides (SO_x)
イオンこうかんじゅし　イオン交換樹脂　ion exchange resin
イオンでんきょく　イオン電極　ion electrode
イオンポンプ　ionic pump
いか[さよう]　異化[作用]　catabolism, dissimilation
イガガヤツリ　*Pycreus polystachyos* (Rottb.) P. Beauv. (= *Cyperus polystachyos* Rottb.)
いかちゅうか　異花柱花　heterostyled flower
いかんそく　維管束　vascular bundle
いかんそくかんけいせいそう　維管束間形成層　interfascicular cambium
いかんそくしょう　維管束鞘　vascular bundle sheath
いかんそくしょうえんちょうぶ　維管束鞘延長部　bundle sheath extention
いかんそくしょくぶつ　維管束植物　vascular plant, Tracheophyta
いかんそくないけいせいそう　維管束内形成層　fascicular cambium, intrafascicular cambium
いきあおまい　活青米　live green-kerneled rice
いきち　いき(閾)値　threshold value
イグサ(藺草)　[mat] rush, *Juncus effusus* L. var. *decipiens* Buchenau (= *J. decipiens* Nakai)
いくしゅ　育種　breeding
いくしゅか　育種家　breeder
いくしゅがく　育種学　thremmatology, breeding science
いくしゅかしゅし　育種家種子　breeder ['s] seed
いくしゅけいか　育種経過　breeding process
いくしゅけいかく　育種計画　breeding program
いくしゅざいりょう　育種材料, 育種素材　breeding material
いくしゅしけんち　育種試験地　breeding station
いくしゅそざい　育種素材, 育種材料　breeding material
いくしゅ[ほう]ほう　育種[方]法　breeding method
いくしゅほ[じょう]　育種圃[場]　breeding field, breeding nursery
いくしゅもくひょう　育種目標

breeding objective
いくせい　育成　raising
いくびょう　育苗　rearing of seedling, raising seedling
いくびょうおんど　育苗温度　nursery temperature
いくびょうき　育苗器, 出芽器　nursery chamber
いくびょうセンター　育苗センター　nursery center
いくびょうばこ　育苗箱　nursery box
いくびょうポット　育苗ポット　seedling pot
いけい[かく]ぶんれつ　異型[核]分裂　heterotypic division
いけいかちゅう　異型花柱　heterostyle
いけいかちゅうせい　異型花柱性　heterostyly
いけいこうはい　異系交配　outbreeding, exogamy
いけいさいぼう　異型細胞, 異形細胞　idioblast
いけいしょくぶつ　異型植物　off-type plant
いけいずいげんしょう　異型ずい(蕊)現象　heterostylism
いけいせつごうせい　異型接合性　heterozygosity
いけいせつごうたい　異型接合体　heterozygote
いけいようせい　異形葉性　heterophylly
いこうせいじょそうざい　移行性除草剤　translocating herbicide
いじけいとう　維持系統　maintainer
いしけっていしえんシステム　意思決定支援システム　decision support system
いじこきゅう　維持呼吸　maintenance respiration
いしつせい　異質性　heterogeneity
いしつばいすうたい　異質倍数体　allopolyploid, alloploid
イシミカワ　Persicaria perfoliata (L.) H. Gross (= Polygonum perfoliatum L.)
いしゅう　異臭　off-flavor, offensive smell
いしゅくびょう　萎縮病　1) dwarf, stunt【イネ, ダイズ】2) rosette, green mosaic【ムギ類】
いしゅこくりゅう　異種穀粒　foreign grain
いじょうきしょう　異常気象　abnormal (unusual) weather
いじょうち　異常値　abnormal value
いじょうはっせい　異常発生　abnormal occurrence
いしょく　移植　transplanting, transplantation
いしょくき　移植期　transplanting time
いしょくき　移植機　transplanter
いしょくさいばい　移植栽培　transplanting culture
いしょくどこ　移植床　transplanting bed
いしょせい　異所性　allopatry
いしょせいの　異所性の　allopatric
いすうせい　異数性　heteroploidy, aneuploidy
いすうたい　異数体　heteroploid
いせいたい　異性体　isomer
いそうさけんびきょう　位相差顕微鏡　phase-contrast microscope
イソこうそ　イソ酵素, アイソザイム　isozyme
イソロイシン　isoleucine (Ile)
いたくさいしゅ　委託採種　trusted seed production
イタドリ　Japanese knotweed, *Reynoutria japonica* Houtt. (= *Polygonum cuspidatum* Sieb. et Zucc.)
イタリアウイキョウ　Florence fennel, sweet anise, *Foeniculum vulgare* Mill. var. *dulce* (Mill.) Batt. et Trab. (= *F. dulce* Mill.)

イタリアンライグラス　Italian ryegrass, *Lolium multiflorum* Lam.
いちゲノムしゅ　一ゲノム種，一基種　monogenomic species
いちげんせいの　一元性の　monophyletic
いちげんぶんるい　一元分類　one-way classification
イチゴ(苺)　strawberry, *Fragaria × ananassa* Duch.
いちこたいいちれつけんてい　一個体一列検定，一母本一列検定　plant-to-row test
いちじいかんそく　一次維管束　primary vascular bundle
イチジク(無花果)　fig, *Ficus carica* L.
いちじこん　一次根　primary root
いちじ[さいぼう]へき　一次[細胞]壁　primary [cell] wall
いちじしこう　一次枝梗　primary rachis-branch of panicle, primary branch of panicle
いちじしぶ　一次篩部　primary phloem
いちじせいちょう　一次成長　primary growth
いちじせんい　一次遷移　primary succession
いちじそしき　一次組織　primary tissue
いちじてきかんきょうへんい　一時的環境変異　temporal environmental variation (modification)
いちじひだい[せいちょう]　一次肥大[成長]　primary thickening [growth]
いちじぶんげつ　一次分げつ　primary tiller
いちじぶんれつそしき　一次分裂組織　primary meristem
いちじへきこういき　一次壁孔域　primary pit-field
いちじもくぶ　一次木部　primary xylem

いちじょうほう　位置情報　positional information
いちせんしょくたいしょくぶつ　一染色体植物　monosomic plant
いちだいざっしゅ　一代雑種　F_1 hybrid
いちだいざっしゅひんしゅ　一代雑種品種　hybrid variety
いちにちせっしゅきょようりょう　一日摂取許容量　acceptable daily intake (ADI)
いちねんご　一年子【コンニャク】　first year crop
いちねんせい[の]　一年生[の]　annual
いちねんせいさくもつ　一年生作物　annual crop
いちねんせいざっそう　一年生雑草　annual weed
いちねんそう　一年草　annual grass, annual herb, annual
いちばんこ　一番粉【ソバ】　No.1 flour
いちばんじょそう　一番除草　first weeding
いちばんそう　一番草【牧草】　first crop, primary canopy
いちばんちゃ　一番茶　first crop of tea
いちばんちゃき　一番茶期　tea season of first crop
いちばんなり　一番成り　first setting [of fruit]
イチビ→ボウマ
いちぼほんいちれつけんてい　一母本一列検定，一個体一列検定　plant-to-row test
いちもうさく　一毛作　single cropping
いちもうさくでん　一毛作田　single-cropped paddy field
いちょう(萎凋)⇒しお(萎)れ　wilting
イチョウウキゴケ　*Ricciocarpos natans* (L.) Corda
いちょうけいすう　いちょう(萎凋)係数⇒しお(萎)れ係数　wilting coeffcient

いちようせいしけん 一様性試験, 均一性試験【圃場】 uniformity trial, homogeneity trial
いちようぶんぷ 一様分布 uniform distribution
いちようらんすう 一様乱数 uniform random number
いちりゅうけいコムギ 一粒系コムギ einkorn wheat
いちりゅうざや 一粒莢 one-seeded pod
いつえき いつ(溢)液⇒排水 guttation
いっかいおや 一回親, 非反復親 nonrecurrent parent
いっかいせんばつ 一回選抜 single selection
いっかせいはつげん 一過性発現 transient expression
いっかせんしょくたい 一価染色体 univalent [chromosome]
いつぎ 居接き field grafting
いっきさく 一期作 first cropping
いっきしゅ 一基種, 一ゲノム種 monogenomic species
いっけいく 一畦区 rod-row plot
いっさんかちっそ 一酸化窒素 nitrogen monoxide (NO)
いっさんかにちっそ 一酸化二窒素 dinitrogen monoxide (N_2O)
いっしゅつしゅ 逸出種 escaped species
いっしゅつしょくぶつ 逸出植物 escaped plant
いっすい(ひとほ)じゅう 一穂重 weight of a head
いっすい(ひとほ)りゅうすう 一穂粒数 grain number per head
いっすいいちれつけんてい 一穂一列検定 ear-to-row test, head-to-row test
いっぱんくみあわせのうりょく 一般組合せ能力 general combining ability
いっぴつ いっ(溢)泌⇒出液【現象】 bleeding, exudation
いっぴつえき いっ(溢)泌液⇒出液【物質】 bleeding sap, exudate
いっぽんうえ 一本植え single planting
いっぽんだて 一本立て singling
いつりゅうかんがい いつ(溢)流灌漑 overflow irrigation, runoff irrigation
いでん 遺伝 heredity, inheritance
いでんあんごう 遺伝暗号 genetic code
いでんがく 遺伝学 genetics
いでんかくとくりょう 遺伝獲得量 genetic gain
いでんし 遺伝子 gene
いでんしいにゅう 遺伝子移入 introgression
いでんしがた 遺伝子型 genotype
いでんしがた-かんきょうこうごさよう 遺伝子型-環境交互作用 genotype-environment interaction
いでんしがたせんばつ 遺伝子型選抜 genotypic selection
いでんしがたぶんさん 遺伝子型分散 genotypic variance
いでんしくみかえ 遺伝子組換え gene recombination, recombination of genes
いでんしクローニング 遺伝子クローニング gene cloning
いでんしげん 遺伝資源 genetic resources
いでんしこうがく 遺伝子工学 genetic engineering
いでんしざ 遺伝子座 locus (*pl.* loci)
いでんしじゅう 遺伝子銃, パーティクルガン particle gun
いでんしそうさ 遺伝子操作 gene manipulation
いでんしちず 遺伝子地図 gene map
いでんしとつぜんへんい 遺伝子突然変異 gene mutation
いでんしはつげん 遺伝子発現 gene

expression
いでんしひんど 遺伝子頻度 gene frequency
いでんしライブラリー 遺伝子ライブラリー gene library
いでんそうかん 遺伝相関 genetic correlation
いでんてきアルゴリズム 遺伝的アルゴリズム genetic algorithm
いでんてきくみかえ 遺伝の組換え genetic recombination
いでんてきぜいじゃくせい 遺伝的ぜい(脆)弱性 genetic vulnerability
いでんてきたようせい 遺伝的多様性 genetic diversity
いでんてきはいけい 遺伝の背景 genetic background
いでんひょうしき 遺伝標識 genetic marker
いでんぶんさん 遺伝分散 genetic variance
いでんぶんせき 遺伝分析 genetic analysis
いでんへんい 遺伝変異 genetic variation, hereditary variation
いでんようしき 遺伝様式 mode of inheritance
いでんりつ 遺伝率 heritability
いど 緯度 latitude
イドイヤシ(イドイ椰子)→クジャクヤシ
いどうか 易動化 mobilization
いどうき 移動期, ディアキネシス期【減数分裂】 diakinesis stage
いどうこうさく 移動耕作, 焼畑農耕 shifting cultivation, slash-and-burn agriculture
いどうデンプン 移動デンプン transitory starch
いどうへいきん 移動平均 moving average
いなかぶ 稲株 rice stubble
イナゴマメ carob, locust bean, *Ceratonia siliqua* L.
いなさく 稲作 rice cropping, rice cultivation
いなだ 稲田 paddy field, ricefield, riceland
いなたば 稲束 rice bundle
いなわら 稲わら(藁) rice straw
いにゅうこうざつ 移入交雑, 浸透交雑, 導入交雑 introgressive hybridization, introgression
イヌガラシ Indian field cress, *Rorippa indica* (L.) Hochr.
イヌサフラン colchicum, autumn crocus, meadow saffron, *Colchicum autumnale* L.
イヌタデ tafted knotweed, *Persicaria longiseta* (De Bruyn) Kitag. (= *Polygonum longisetum* De Bruyn)
イヌノヒゲ *Eriocaulon miquelianum* Koern.
イヌビエ barnyard grass, *Echinochloa crus-galli* (L.) Beauv. var. *crus-galli*
イヌビユ livid amaranth, *Amaranthus lividus* L.
イヌホオズキ black nightshade, *Solanum nigrum* L.
イヌホタルイ *Schoenoplectus juncoides* (Roxb.) Palla ssp. *juncoides* (= *Scirpus juncoides* Roxb. ssp. *juncoides* Roxb.)
イヌムギ→レスキュグラス
イヌリン inulin
イネ(稲) rice, *Oryza sativa* L.【ssp. *japonica* 日本型イネ, ssp. *indica* インド型イネ, ssp. *javanica* ジャワ型イネ】
イネか イネ科 Gramineae, Poaceae
イネかさくもつ イネ科作物 gramineous crop
イネかざっそう イネ科雑草 gramineous weed, grass weed
イネかしょくぶつ イネ科植物 gramineous plant, grass

いねかぼくそう　イネ科牧草　forage grass
いねかり　稲刈り　rice harvest, rice reaping
いねこき　稲こ(扱)き　rice threshing
イネ・ムギにもうさく　イネ・ムギ二毛作　rice-barley double cropping, rice-wheat double cropping
イノコズチ　*Achyranthes bidentata* Blume var. *bidentata* (= *A. japonica* (Miq.) Nakai)
イノンド, ディル　dill, *Anethum graveolens* L.
イピルイピル→ギンゴウカン
いぶつ　異物, 夾雑物(きょうざつぶつ)　foreign matter, impurity
いぶんかいせいゆうきぶつ　易分解性有機物　easily decomposable organic matter
いへんいでんし　易変遺伝子　mutable gene
いへんしきそたい　易変色素体　mutable plastid
イボクサ　marsh dayflower, *Murdannia keisak* (Hassk.) Hand.-Mazz. (= *Aneilema keisak* Hassk.)
イムノアッセイ, 免疫検定法　immunoassay
いも　tuber, corm
イモゼリ　aracacha, *Arracacia xanthorrhiza* Bancr. (= *A. esculenta* DC.)
いもちびょう　いもち(稲熱)病【イネ】　blast, neck rot
いもるい　いも類　root and tuber crops
いやち　忌地　1) sick soil【土壌】 2) soil sickness【現象】
イランイラン　ylang-ylang, ilang-ilang, *Canangium odoratum* Hook. f. et Thoms. (= *Cananga odorata* Baill.)
いりあいけん　入会権　common
いりあいち　入会地　commons, common land

いリッぺし　イリッぺ脂　illipe butter
イロガワリコンニャク　*Amorphophallus variabilis* Bl.
いろしゅうさ　色収差　chromatic aberration
いんイオン　陰イオン　anion
いんイオンこうかんようりょう　陰イオン交換容量　anion exchange capacity (AEC)
いんかしょくぶつ　陰花植物　cryptogam
インゲンマメ(隠元豆)　kidney bean, French bean, *Phaseolus vulgaris* L.
いんし　因子, 要因　factor
いんしじっけん　因子実験, 要因実験　factorial experiment
いんしせっけい　因子設計, 要因設計　factorial design
インシチュ　in situ
インシチュ・ハイブリダイゼーション　in situ hybridization
いんじゅ　陰樹　shade [tolerant] tree
いんせいしょくぶつ　陰生植物　shade plant
インターネット　internet
インターフェース　interface
インターミーディエイト・ホィートグラス　intermediate wheatgrass, *Agropyron intermedium* Beauv., *A. trichophorum* (Link) K. Richt. および *A. pulcherrimum* Grossh.
インドールさくさん　インドール酢酸　indoleacetic acid (IAA)
インドがた　インド型【イネ】　indica type
インドクワズイモ　giant taro, *Aloccasia macrorrhiza* (L.) Schott
インドゴム　Indian rubber, Assam rubber, *Ficus elastica* Roxb.
インドジャボク(インド蛇木)　snakewood, Java devilpepper, *Rauvolfia serpentina* Benth. ex Kurtz
インドビエ　Indian barnyard millet,

billion dollar grass, *Echinochloa frumentacea* (Roxb.) Link
イントロン intron, 介在配列 intervening sequence
インドわいせいコムギ インド矮性コムギ Indian dwarf wheat, *Triticum sphaerococcum* Perc.
インビトロ, 生体外で in vitro
インビボ, 生体内で in vivo
インベルターゼ invertase
いんよう 陰葉 shade leaf
いんりょう 飲料 beverage

[う]

ウイキョウ (茴香) fennel, *Foeniculum vulgare* Mill.
ウィーピングラブグラス weeping lovegrass, *Eragrostis curvula* (Schrad.) Nees
ウィメラライグラス wimmera ryegrass, *Lolium rigidium* Gaud.
ウイルス virus
ウイルスびょう ウイルス病 virus disease
ウイロイド viroid
ウィンターベッチ→ヘアリベッチ
ウインドマシン, 防霜ファン【チャ】 wind machine
うえあな 植え穴 planting hole, planting pit
うえいたみ 植え傷み transplanting injury
うえかえ 植替え transplanting
うえかえどこ 植替え床【タバコ】 secondary bed
うえきばち 植木鉢, ポット, 鉢 pot
うえしろ 植代 final puddling
ウェスタン・ウィートグラス western wheatgrass, *Pascopyrum smithii* (Rydb.) A. Löve
ウェスタンブロットほう ウェスタンブロット法 western blot technique, western blotting
うえつけ 植付け planting
うえつけき 植付け機 planter
うえつけみつど 植付密度⇒栽植密度 planting density, planting rate
うえなおし 植え直し, 改植 replanting
うえなわ 植え縄 planting cord
うえみぞ 植え溝 planting furrow, planting trench
うがい 雨害 rain damage
うき 雨季 rainy season
ウキアゼナ disc waterhyssop, *Bacopa rotundifolia* (Michx.) Wettst.
うきイネ 浮稲 floating rice
ウキクサ giant duckweed, *Spirodela polyrhiza* (L.) Schleid.
うきさく 雨季作, 雨期作 rainy season crop[ping], wet-season crop[ping]
うきなえ 浮苗 floating seedling
うきなえどこ 浮苗床 floating nursery
ウキヤガラ river bulrush, *Bolboschoenus fluviatilis* (Torr.) T. Koyama ssp. *yagara* (Ohwi) T. Koyama (= *Scirpus yagara* Ohwi)
ウコン (欝金) turmeric, *Curcuma longa* L. (= *C. domestica* Valet.)
うしけっせいアルブミン 牛血清アルブミン bovine serum albumin (BSA)
ウシノケグサ→シープフェスク
ウシハコベ water starwort, *Stellaria aquatica* (L.) Scop.
うじょうふくよう 羽状複葉 pinnate compound leaf
うずせい 渦性【オオムギ】 uzu type
うすまき 薄播き, 疎播 sparse sowing (seeding), thin sowing (seeding)
うちかえしこう 内返し耕 gathering plowing, throw-in plowing
うちぐわ 打くわ (鍬) digging hoe
うちこ 打ち粉【ソバ】 uchiko flour
うちぬきほう 打抜法 leaf-punch method
うちひきぐわ 打引ぐわ (鍬) digging

and pulling hoe
ウチワサボテン prickly pear, *Opuntia ficus-indica* (L.) Mill. 他多種
うちわら 打わら (藁) tempered straw
うてきしんしょく 雨滴侵食 raindrop erosion
ウド (独活) udo, *Aralia cordata* Thunb.
うとう 烏稲 black rice
うどんこびょう うどんこ病 powdery mildew
ウニコナゾール uniconazole
うね (畝) ridge
うねあげ うね (畝) 揚げ ridging up
うねかた うね (畝) 肩 ridge shoulder (side)
うねさく うね (畝) 作, うね (畝) 仕立て ridge planting
うねじたて うね (畝) 仕立て, うね (畝) 作 ridge planting
うねだか うね (畝) 高 ridge height
うねたて うね (畝) 立て ridging, ridge plowing
うねたてき うね (畝) 立て機 ridger, lister, ditcher
うねたてこう うね (畝) 立て耕 ridging
うねたてさいばい うね (畝) 立て栽培 ridge culture
うねどこ うね (畝) 床 ridge bed, ridge-up bed
うねどこはば うね (畝) 床幅 ridge bed width
うねはば うね (畝) 幅 ridge distance, row width, row space
うねま うね (畝) 間 furrow
うねまかんがい うね (畝) 間灌漑, けい (畦) 間灌漑 furrow irrigation
うねまき うね (畝) 播き ridge sowing (seeding)
ウマゴヤシ→バークローバ
ウメ (梅) mume, Japanese apricot, *Prunus mume* Sieb. et Zucc.

うら [がわ] の 裏 [側] の【植物】⇒背軸 [側] の abaxial
うらきり うら (梢) 切り【イグサ】 topping
うらさく 裏作 off-season crop[ping]
ウラシル uracil (U)
ウラジロラフィア→ラフィアヤシ
ウリカワ *Sagittaria pygmaea* Miq.
ウリジンさんリンさん ウリジン三リン酸 uridine triphosphate (UTP)
ウリジンにリンさん ウリジン二リン酸 uridine diphosphate (UDP)
ウーリーベッチ→ヘアリベッチ
うりょう 雨量 amount of rainfall
うりょうけい 雨量計 rain gauge, pluviometer
ウルーコ ulluco, papa lisas, *Ullucus tuberosus* Caldas
ウルシ (漆) 1)【狭義】Japanese lacquer tree, *Rhus verniciflua* Stokes 2)【広義】lacquer tree, varnish tree, *Rhus* spp.
うるし【製品】 lacquer
うるしかき うるし掻き lacquer tapping
うるちせいの 粳性の nonglutinous
うるちまい 粳米 nonglutinous rice
ウルトラミクロトーム ultramicrotome
うろほう 雨露法 dew retting
ウーロンちゃ ウーロン (烏龍) 茶 oolong
うわずみ [えき] 上ずみ (澄) [液] supernatant [liquid]
うわね うわ根 superficial root
うわは 上葉【タバコ】 undertips
ウンカ planthopper
ウンシュウミカン satsuma mandarin, *Citrus unshiu* Marcovitch
ウンダイ (蕓薹)→ナタネ

[え]

エアロゾル aerosol

えい　穎　glume
えいか　穎花　glumaceous flower
えいか　穎果　caryopsis
えいきゅうしおれ　永久しお(萎)れ　permanent wilting
えいきゅうそうち　永久草地　permanent grassland
えいきゅうそしき　永久組織　permanent tissue
えいきゅうとうけつきこう　永久凍結気候　ice climate
えいきゅうプレパラート　永久プレパラート　permanent preparation
えいきゅうほうぼくち　永久放牧地　permanent pasture
エイジ, 齢　age
エイジング, 加齢　aging, ageing
えいぞくせい　永続性【牧草】　persistency
えいぞくへんい　永続変異　permanent modification
えいねん[せい]さくもつ　永年[生]作物⇒多年生作物　perennial crop
えいねんせい[の]　永年生[の]⇒多年生[の]　perennial
えいねんそうち　永年草地　permanent grassland, permanent pasture
えいのう　営農　farming, farm management
エーイーほう　AE法　acoustic emissin method
えいよう　栄養　nutrition
えいようか[ち]　栄養価[値]　nutritive value
えいようかく　栄養核　vegetative nucleus
えいようきかん　栄養器官　vegetative organ
えいようけい　栄養系, クローン　clonal strain, clone
えいようけいせんばつ　栄養系選抜　clonal selection
えいようけいぶんり　栄養系分離　clonal separation
えいようさいぼう　栄養細胞　vegetative cell
えいようざっしゅ　栄養雑種　vegetative hybrid
えいようしょうがい　栄養障害　nutrient disorder
えいようしんだん　栄養診断　nutritional diagnosis
えいようせいちょう　栄養成長　vegetative growth
えいようせいちょうき　栄養成長期　vegetative (growth) stage
えいようせいり　栄養生理　nutriophysiology
えいようそ　栄養素, 養分　nutrient
えいようそう　栄養相　vegetative phase
えいようはんしょく　栄養繁殖　vegetative propagation, cloning
えいようはんしょくしょくぶつ　栄養繁殖植物　vegetatively propagated plant
えきか　液果, 多肉果　berry, sap fruit, succulent fruit
えきが　腋芽　axillary bud
えきざい　液剤　liquid formulation
えきしん　液浸　immersion
エキスパートシステム　expert system
えきそう　液相　liquid phase
エキソン　exon
えきたいクロマトグラフィー　液体クロマトグラフィー　liquid chromatography
えきたいシンチレーションカウンター　液体シンチレーションカウンター　liquid scintillation counter
えきたいちっそ　液体窒素　liquid nitrogen
えきたいばいよう　液体培養　liquid culture
えきたいひりょう　液体肥料, 液肥　liquid fertilizer
えきちく　役畜　draft animal

えきひ　液肥, 液体肥料　liquid fertilizer
えきびょう　疫病　1) late blight【ナス科】2) blight【ソバ】
えきほう　液胞　vacuole
えきほうまく　液胞膜, トノプラスト　tonoplast
エゴマ（荏胡麻, 荏）　perilla, *Perilla frutescens* (L.) Britton var. *japonica* (Hassk.) H. Hara (= *P. ocymoides* L.)
エゴマゆ　エゴマ油　perilla oil
えし　え（壊）死, ネクロシス　necrosis
エジプシャンクローバ　Egyptian clover, Berseem clover, *Trifolium alexandrinum* L.
エジプトメン（エジプト棉）【カイトウメンの一系統】Egyptian cotton, *Gossypium barbadense* L.
エス-アデノシルメチオニン　S-アデノシルメチオニン　S-adenosylmethionine (SAM)
エスエヌひ　SN比　signal-to-noise ratio (SN ratio)
エスき　S期　S phase, 合成期　synthetic phase
エスじがたきょくせん　S字形曲線, S字状曲線　sigmoid curve
エスじじょうきょくせん　S字状曲線, S字形曲線　sigmoid curve
エステラーゼ　esterase
エストラゴン→タラゴン
エスワンマッピング　S1マッピング　S1 mapping
エゾノギシギシ　broadleaf dock, *Rumex obtusifolius* L.
エゾノキツネアザミ　creeping thistle, *Breea setosa* (Bieb.) Kitam.
エゾノサヤヌカグサ　rice cutgrass, *Leersia oryzoides* (L.) Sw. ssp. *oryzoides*
えだ　枝　branch
えだがわり　枝変り, 芽条[突然]変異　bud mutation, bud variation, bud sport

えだざし　枝挿し, 茎挿し　stem cutting
えだつぎ　枝接ぎ　scion grafting
エタノール　ethanol ⇒エチルアルコール　ethyl alcohol
えだばり　枝張り　branch spread
エダホロジー　edaphology
えだマメ　枝豆　green soybean
えだわかれぶんるい　枝分かれ分類　nested classification, 階層分類　hierarchical classification
エチオプラスト　etioplast
エチルアルコール　ethyl alcohol
エチレン　ethylene
エックスせんかいせつ　X線回折　X-ray diffraction
えっとう　越冬　overwintering
えっとういちねんせいしょくぶつ　越冬一年生植物, 越年生植物　winter annual
えっとうが　越冬芽　overwintering bud, winter bud
えっとうせい　越冬性　overwintering ability
えっとうよう　越冬葉【チャ】overwintering leaf
えつねんせいさくもつ　越年生作物　winter annual crop
えつねんせいざっそう　越年生雑草　winter annual weed
えつねんせいしょくぶつ　越年生植物, 越冬一年生植物　winter annual
えつねんそう　越年草, 冬一年草　biennial[s], winter annual[s]
エーディーへんかん　A-D変換　analog-digital (A-D) conversion
エーディーへんかんき　A-D変換器　analog-to-digital converter
エテフォン　ethephon, 2-クロロエチルホスホン酸　2-chloroethyl phosphonic acid (CEPA)
エーテルちゅうしゅつぶつ　エーテル抽出物　ether extract
エネルギーこうりつ　エネルギー効率

energy efficiency
エノキアオイ　*Malvastrum coromandelianum* (L.) Garcke
エノキグサ　threeseeded copperleaf, *Acalypha australis* L.
エノキタケ　enokitake fungus, *Flammulina velutipes* (Fr.) Karst.
エノコログサ　green foxtail, *Setaria viridis* (L.) P. Beauv.
えのゆ　え(荏)油⇒エゴマ油　perilla oil
エビスグサ　oriental senna, *Senna obtusifolia* (L.) H.S. Irwin et Barneby (= *Cassia obtusifolia* L.)
エピブラスト　epiblast
エフけんてい　F検定　F-test
エフち　F値　F-value
エフとうけいりょう　F統計量　F-statistic
エフひょう　F表　F-table
エフぶんぷ　F分布　F-distribution
エフワンざっしゅ　F_1雑種　F_1 hybrid
エマーソンこうか　エマーソン効果　Emerson effect
エムき　M期　M phase, 分裂期 mitotic phase
エルカさん　エルカ酸, エルシン酸　erucic acid
エルシンさん　エルシン酸, エルカ酸　erucic acid
エレクトロポレーション, 電気穿孔法　electroporation
エレファントグラス→ネピアグラス
えんあん　塩安(塩化アンモニウム)　ammonium chloride
えんえきてきすいろん　演繹的推論　deductive inference
えんか[かり]　塩加(塩化加里)⇒塩化カリウム　potassium chloride
えんかアンモニウム　塩化アンモニウム(塩安)　ammonium chloride
えんがい　塩害　salt damage (injury)

えんかカリ[ウム]　塩化カリ[ウム](塩加)　potassium chloride
えんかくそくてい　遠隔測定, 隔測, リモートセンシング　remote sensing
えんかナトリウム　塩化ナトリウム　sodium chloride
えんき　塩基　base
えんきせいしきそ　塩基性色素　basic dye
えんきせいひりょう　塩基性肥料　basic fertilizer
えんきちかんようりょう　塩基置換容量　base exchange capacity ⇒陽イオン交換容量　cation exchange capacity (CEC)
えんきはいれつ　塩基配列【遺伝子】　base sequence
えんきほうわど　塩基飽和度　base saturation
えんげい　園芸　horticulture
えんげいがく　園芸学　horticultural science
えんげいさくもつ　園芸作物　horticultural crop, garden crop
えんげいしゅ　園芸種　garden species, horticultural species
えんげいひんしゅ　園芸品種　garden variety
えんこうこうどけい　炎光光度計　flame photometer
エンサイ→ヨウサイ
えんさん　塩酸　hydrochloric acid
エンシレージ　ensilage
えんしんき　遠心機　centrifuge
えんしんてき　遠心的　centrifugal
えんしんぶんり　遠心分離　centrifugation
えんすいせん　塩水選　seed selection with salt solution
えんすいとう　塩水稲　salt-tolerant rice
えんせいしょくぶつ　塩生植物　halophyte
えんせき　塩析　salting out

えんそ　塩素　chlorine (Cl)
えんぞう　塩蔵　salting
えんちんかん　遠沈管　centrifuge tube
エンドウ (豌豆)　pea, garden pea, *Pisum sativum* L.
エントロピー　entropy
エンバク (燕麦)【≠カラスムギ】　oats【通常 *pl.*】, *Avena sativa* L.
エンハンサー　enhancer
えんばんプラウ　円盤プラウ⇒ディスクプラウ　disk plow
エンピツビャクシン　eastern red cedar, pencil cedar, *Sabina virginiana* (L.) Antoine (= *Juniperus virginiana* L.)
えんぷうがい　塩風害⇒潮風害　salty wind damage (injury)
えんぶんストレス　塩分ストレス, 塩類ストレス　salt stress
えんぶんのうど　塩分濃度【土壌】salinity
エンマーコムギ　Emmer, *Triticum dicoccum* Schubl.
エンメイソウ (延命草)→ヒキオコシ
えんりゅうしゅ　円粒種【イネ】round-grain rice
えんるいしゅうせき　塩類集積　salt accumulation
えんるいしょうがい　塩類障害⇒塩害　salt damage (injury)
えんるいストレス　塩類ストレス, 塩分ストレス　salt stress
えんるいどじょう　塩類土壌　saline soil

[お]

おいこみば　追込み場　paddock, yard
オイチシカ　oiticica, *Licania rigida* Benth.
おいまき　追播き　oversowing, overseeding
おうか　黄化　etiolation, yellowing
おうかいしゅくびょう　黄化萎縮病【イネ】downy mildew
おうかびょう　黄化病, 萎黄[病] yellows
オウギヤシ (扇椰子)　palmyra palm, *Borassus flabellifer* L.
おうじゅく　黄熟　yellow ripe
おうじゅくき　黄熟期　yellow ripe stage
おうしょくかんそう　黄色乾燥【タバコ】flue-curing
おうしょくしゅ　黄色種【タバコ】flue-cured tobacco
おうしょくど　黄色土　Yellow soil
おうだんせっぺん　横断切片　cross section
オウトウ (桜桃)　cherry, *Prunus avium* L. 他数種
おうとう　応答, 反応　response, reaction
おうとうきょくせん　応答曲線, 反応曲線　response curve
おうふくこう　往復耕　return plowing
おうぶんれつ　横分裂　transverse division
おうへんき　黄変期【タバコ】yellowing stage
おうへんまい　黄変米　yellowed rice
オウレン (黄連)　*Coptis japonica* (Thunb.) Makino
オオアカウキクサ　*Azolla japonica* Franch. et Savat.
オオアキノキリンソウ→オオアワダチソウ
オオアブノメ　*Gratiola japonica* Miq.
オオアレチノギク　tall fleabane, *Conyza sumatrensis* (Retz.) Walker (= *Erigeron sumatrensis* Retz.)
オオアワ→アワ
オオアワガエリ→チモシー
オオアワダチソウ　late goldenrod, *Solidago virgaurea* L. ssp. *gigantea* (Nakai) Kitam. (= *S. gigantea* Ait. var. *leiophylla* Fern.)

おおいしたちゃ　おおい(覆)下茶　shaded tea
オオイヌタデ　*Persicaria lapathifolia* (L.) S. F. Gray (= *Polygonum lapathifolium* L. ssp. *nodosum* Kitam.)
オオイヌノフグリ　Persian speedwell, *Veronica persica* Poir.
オオウシノケグサ→レッドフェスク
オオカナダモ　Brazilian elodea, *Egeria densa* Planch.
オオカニツリ→トールオートグラス
おおがま　大がま(鎌)　scythe
オオカラスノエンドウ→コモンベッチ
オオクサキビ　fall panicum, *Panicum dichotomiflorum* Michx.
オオクログワイ　Chinese water chestnut, *Eleocharis dulcis* (Burm. f.) Trin. ex Hensch. var. *tuberosa* (Roxb.) T. Koyama
オオジシバリ　*Ixeris debilis* (Thunb. ex Murray) A. Gray (= *I. japonica* (Burm.) Nakai
オオスズメノカタビラ→ラフブルーグラス
オオスズメノテッポウ→メドーフォックステール
オオチドメ　*Hydrocotyle ramiflora* Maxim.
オオバコ　Asiatic plantain, *Plantago asiatica* L. (= *P. major* L. var. *asiatica* (L.) Dec.)
オオバナソケイ　Italian jasmine, Spanish jasmine, royal jasmine, Catalonian jasmine, *Jasminum officinale* L.f. *grandiflorum* (L.) Kobuski (= *J. grandiflorum* L.)
オオバボンテンカ　aramina, *Urena lobata* L. var. *lobata*
オオフサモ　parrot's-feather, *Myriophyllum aquaticum* (Vell.) Verdc. (= *M. brasiliense* Camb.)
オオブタクサ, クワモドキ　giant ragweed, *Ambrosia trifida* L.
オオフトモモ→レンブ
オオマツヨイグサ　evening primrose, *Oenothera erythrosepala* Borbás
オオミズオオバコ→ミズオオバコ
オオミノトケイソウ　giant granadilla, square-stalked passion flower, *Passiflora quadrangularis* L.
オオムギ(大麦)　barley, *Hordeum vulgare* L.
オカ　oca, *Oxalis tuberosa* Mol.
おかなわしろ(りくなわしろ)　陸苗代⇒畑苗代　upland rice-nursery
おかぶ　雄株　male plant
おかほ(りくとう)　陸稲　upland rice
オキサロさくさん　オキサロ酢酸　oxaloacetic acid
オキシダーゼ, 酸化酵素　oxidase
オーキシン　auxin
オーキシンけつごうタンパクしつ　オーキシン結合タンパク質　auxin-binding protein
おきべり　置減り　storage loss
おくて(ばんせい)ひんしゅ　晩生品種　late [maturing] variety
オクラ　okra, lady's fingers, *Abelmoschus esculentus* (L.) Moench (= *Hibiscus esculentus* L.)
オグルマ　elecampane, *Inula helenium* L.
おくれぼ　遅れ穂　late emerging head
オケラ　*Atractylodes lancea* (Thunb.) DC.
おしつぶしほう　押しつぶし法　squash method
おしばひょうほん　おし(押)葉標本　herbarium specimen
おしべ　雄しべ(蕊), 雄ずい(蕊)　stamen
おしむぎ　押麦　pressed barley, rolled barley
おす[の]　雄[の]　male
オスミウム　osmium (Os)

オスミウムさん　オスミウム酸　osmic acid
おせん　汚染　pollution, contamination
おせんげん　汚染源　pollution source, source of contamination
おせんほじょう　汚染圃場　contaminated field, polluted field
おせんまい　汚染米　1) polluted rice 2) stained rice【外観】
おそがり　晩刈り　late harvesting, late cutting
おそじも　遅霜, 晩霜 (ばんそう)　late frost
おそまき　晩播き　late sowing (seeding)
オタネニンジン (御種人参)→ヤクヨウニンジン
オーチャードグラス　orchardgrass, cocksfoot, *Dactylis glomerata* L.
おでい　汚泥　sludge
オート→エンバク
オートクレーブ, 高圧滅菌器　autoclave
オトコヨモギ　western mugwort, *Artemisia japonica* Thunb.
オートミール　oatmeal
オートラジオグラフィー　autoradiography
オドリコソウ　*Lamium album* L. var. *barbatum* (Sieb. et Zucc.) Franch. et Savat. (= *L. barbatum* Sieb. et Zucc.)
オナモミ　common cocklebur, *Xanthium strumarium* L.
オニウシノケグサ→トールフェスク
オニタビラコ　Asiatic hawk's-beard, *Youngia japonica* (L.) DC.
オニノゲシ　spiny sowthistle, *Sonchus asper* (L.) Hill
おのみ　お(苧)の実　hemp seed
おばな (ゆうか)　雄花　male flower, staminate flower
オヒシバ　goosegrass, wiregrass, *Eleusine indica* (L.) Gaertn.
おびじょうかんさく　帯状間作, 帯状栽培　strip cropping, lane cropping
おびじょうさいばい　帯状栽培, 帯状間作　strip cropping, lane cropping
オペレーティングシステム　operating system (OS)
オペロン　operon
オモダカ　arrowhead, *Sagittaria trifolia* L.
おもて[がわ]の　表[側]の【植物】⇒向軸[側]の　adaxial
おもてさく　表作　main season crop[ping]
おもてどし　表年⇒成り年　on-year
おもみつきへいきん　重み付き平均, 加重平均　weighted mean
おやいも　親いも　mother tuber, mother corm
おやかぶ　親株　mother plant
おやけいとう　親系統　parental line (strain)
おやこそうかん　親子相関　parent-offspring correlation
おやづる　親づる (蔓)　main vine
おやどこ　親床【タバコ】　primary bed
おやね　親根　parent root
おやほ　親穂　main stalk ear
オランダガラシ→クレソン
オランダゼリ→パセリ
オランダハッカ→スペアミント
オランダミミナグサ　sticky chickweed, *Cerastium glomeratum* Thuill.
オリエントコムギ　*Triticum turanicum* Jakubz.
オリエントしゅ　オリエント種【タバコ】　Oriental tobacco
オリゴとう　オリゴ糖, 少糖　oligosaccharide
オリーブ　olive, *Olea europaea* L.
オリーブゆ　オリーブ油　olive oil
オルガネラ, 細胞[小]器官　organelle
オールスパイス　allspice, pimento, *Pimenta dioica* (L.) Merr. (= *P. officinalis* Lindl.)
オールドファスチク　old fustic,

Chlorophora tinctoria Gaud.
おれいごえ　お礼肥　topdressing after harvest
オレインさん　オレイン酸　oleic acid
オレガノ→ハナハッカ
おんしつ　温室　greenhouse
おんしつこうか　温室効果　greenhouse effect
おんしゅうせい　温周性，温度周期性　thermoperiodism
おんしょう　温床　frame, hotbed
おんしょうなわしろ　温床苗代　hot rice-nursery bed, warm nursery bed
おんすいち　温水池　water warming pond
おんすいでん　温水田　water warming paddy field
おんすいろ　温水路　water warming canal
おんたい　温帯　temperate zone
おんたいきこう　温帯気候　temperate climate
おんたいしつじゅんきこう　温帯湿潤気候　temperate humid climate
おんたい[せい]さくもつ　温帯[性]作物　temperate crop
おんたい[せい]しょくぶつ　温帯[性]植物　temperate plant
おんとうじょゆう[ほう]　温湯除雄[法]　hot water emasculation [method]
おんとうしんぽう　温湯浸法　hot water disinfection method (treatment)
おんどけい　温度計　thermometer
おんどけいすう　温度係数　temperature coefficient
おんどしゅうきせい　温度周期性，温周性　thermoperiodism
おんりょうしすう　温量指数，暖かさの指数　warmth index

[か]

か　科　family
がい　外衣　tunica
がいいないたいせつ　外衣内体説　tunica-corpus theory
がいいんせいないぶんぴつかくらんぶっしつ　外因性内分泌かく乱物質　endocrine disrupting chemicals (EDC), 環境ホルモン　environmental hormones
がいえい　外穎　lemma
カイエンボアドローズ　Cayenne bois de rose →パウローサ
かいか　開花　flowering, anthesis
かいかき　開花期　flowering time, flowering stage, blooming season
かいかしゅうせい　開花習性　flowering habit
かいかじゅんい　開花順位，開花順序　flowering order
かいかじゅんじょ　開花順序，開花順位　flowering order
かいかちょうせつ　開花調節　regulation of flowering, flowering regulation
がいかひ　外果皮　epicarp
かいかほう　灰化法　ashing method
かいかホルモン　開花ホルモン⇒花成ホルモン　flowering hormone
ガイガーミュラーカウンター　Geiger-Müller counter
がいかん　外観　appearance
かいき　回帰　regression
かいききょくせん　回帰曲線　regression curve
かいきけいすう　回帰係数　regression coefficient
かいきしき　回帰式　regression equation
かいきちょくせん　回帰直線　regression line

かいきパラメータ 回帰パラメータ regression parameter
かいきぶんせき 回帰分析 regression analysis
かいきもけい 回帰模型 regression model
かいきゅう 階級【統計】 1)【区間】⇒級 class 2)【順位】⇒順位 rank
がいきんこん 外菌根⇒外生菌根 ectomycorrhiza, ectotrophic mycorrhiza
かいけい 塊茎 tuber
かいけいけいせい 塊茎形成 tuberization
かいけいひだい 塊茎肥大 [tuber] bulking
がいげんけい 外原型 exarch
カイコ(蚕), 家蚕 silkworm, *Bombyx mori* L.
かいこうすう 開口数【顕微鏡】 numetrical aperture
かいこん 塊根 tuberous root
かいこん 開墾 land reclamation, clearing
かいざいせいちょう 介在成長, 部間成長 intercalary growth
かいざいはいれつ 介在配列 intervening sequence, イントロン intron
かいざいぶんれつそしき 介在分裂組織, 部間分裂組織 intercalary meristem
カイじじょう カイ自乗⇒カイ二乗 Chi-square (χ^2)
がいじつリズム 概日リズム circadian rhythm
かいしほう 下位子房 inferior ovary
がいしゅひ 外種皮 outer seed coat
かいじょ 開絮 opening of boll, blowing of boll
かいじょう 階乗 factorial
かいしょく 改植, 植え直し replanting
がいしょくたい 外植体, 外植片 explant
がいしょくへん 外植片, 外植体 explant
かいしんがた 開心型 open-center type
かいせい 回青【カンキツ類】 regreening
がいせいの 外生の, 外与の exogenous
がいせいきんこん 外生菌根 ectomycorrhiza, ectotrophic mycorrhiza
かいせつ 回折 diffraction
かいせつこうし 回折格子 diffraction grating
かいせんうんどう 回旋運動 circumnutation
かいぞう 灰像 spodogram
がいそう 害草 harmful weed, hazardous weed, noxious weed
かいそうこうぞう 階層構造 stratification
かいそうど 解像度 resolution
かいそうぶんるい 階層分類 hierarchical classification, 枝分かれ分類 nested classification
かいそうほう 灰像法 spodography
がいそう[ほう] 外挿[法] extrapolation
かいぞうりょく 解像力 resolving power
かいたくち 開拓地 reclaimed land
かいだんこうさく 階段耕作, テラス栽培 terrace culture, terrace farming
がいちゅう 害虫 insect pest
がいちゅうぼうじょ 害虫防除 insect pest control
かいちょうがた 開張型【チャ】 spread type
かいちょうせいの 開張性の spreading
かいでん 開田 paddy field reclamation

かいてんじょそうき　回転除草機　rotary weeder
かいてんだっこくき　回転脱穀機　rotary thresher
かいど　開度　1) divergence【葉序】2) aperture【気孔】
かいとう　解糖　glycolysis
カイトウメン（海島棉）　sea-island cotton, *Gossypium barbadense* L.
カイにじょう　カイ二乗　chi-square (χ^2)
カイにじょうけんてい　カイ二乗 (χ^2) 検定　chi-square (χ^2) test
カイにじょうとうけいりょう　カイ二乗 (χ^2) 統計量　chi-square (χ^2) statistic
カイにじょうぶんぷ　カイ二乗 (χ^2) 分布　chi-square (χ^2) distribution
カイネチン　kinetin
がい[はい]にゅう　外[胚]乳⇒周乳　perisperm
かいはた　開畑　field reclamation
かいひ　回避　escape
がいひ　外皮　exodermis
かいひさいばい　回避栽培　evasion culture
かいぶん　灰分　ash
がいほうえい　外苞穎　outer glume
かいぼうがく　解剖学　anatomy
かいほうけい　開放系　open system
かいぼうけんびきょう　解剖顕微鏡　dissecting microscope
かいぼうざら　解剖皿　dissecting pan
かいぼうばさみ　解剖ばさみ　dissecting scissors
かいめんかっせいざい　界面活性剤　surface-active agent, surfactant
かいめんじょうそしき　海綿状組織　spongy tissue
かいやく　開葯　anther dehiscence
がいよの　外与の, 外生の　exogenous
かいよう　下位葉　lower leaf
かいようき　開葉期【チャ】　opening time of first leaf
がいらいざっそう　外来雑草　exotic weed
がいらいしゅ　外来種　exotic species
がいらいしょくぶつ　外来植物　exotic plant
かいり　解離　1) maceration【組織学】2) dissociation【化学】
かいりょうそうち　改良草地　improved grassland (pasture), modified grassland (pasture)
かいりょうひんしゅ　改良品種　improved variety
カウピー→ササゲ
カウレン　kaurene
カエデとう　カエデ糖　maple sugar
かえりざき　返り咲き　reflorescence
かえんじょそう　火炎除草　flame weeding
カオス　chaos
かおり　香り, 香気　aroma, scent, fragrance
カオリナイト　kaolinite
かおりまい　香り米　aromatic rice, scented rice
かが（はなめ）　花芽　flower bud
カカオ　cacao, cocoa, *Theobroma cacao* L.
カカオし　カカオ脂, カカオバター　cacao butter
カカオまめ　カカオ豆　cacao bean
かがくくっせい　化学屈性　chemotropism
かがくしき　化学式　chemical formula
かがくしんか　化学進化　chemical evolution
かがくてきさんそようきゅうりょう　化学的酸素要求量　chemical oxygen demand (COD)
かがくてきせいぎょ　化学的制御, 化学[的]防除　chemical control
かがく[てき]ぼうじょ　化学[的]防除, 化学的制御　chemical control

かがくひりょう 化学肥料 chemical fertilizer
かが(はなめ)ぶんか 花芽分化 flower bud initiation (differentiation), floral differentiation
かかん 架乾 rack drying
かかん 火乾 fire curing
かかん 花冠 corolla
かき 花き(卉) flowers and ornamental plants
かき 花器 floral organ
カキ(柿) kaki, Japanese persimon, *Diospyros kaki* Thunb.
かきえんげい 花き(卉)園芸 floriculture, flower gardening, ornamental horticulture
かききゅうかん 夏期休閑 summer fallow
カキドオシ *Glechoma hederacea* L. ssp. *grandis* (A.Gray) Hara
かきなえ 掻き苗【サツマイモ】 slip
かぎゃくはんのう 可逆反応 reversible reaction
かきゅうたいようぶん 可給態養分 available nutrient
かく 核 nucleus (*pl.* nuclei)
がく(萼) calyx
かくいしょく[ほう] 核移植[法] nuclear transplantation, nucleus transplantation
かくか 核果 stone fruit, drupe
かくか 殻果, 堅果 nut
かくがく 核学 karyology, caryology
かくがた 核型 karyotype
かくけいさいばい 隔畦栽培 skipped-row culture
かくこう 核孔, 核膜孔 nuclear pore
かくこくのうぎょうけんきゅうこくさいサービス 各国農業研究国際サービス International Service for National Agricultural Research (ISNAR)
かくさん 拡散 diffusion
かくさん 核酸 nucleic acid

かくさんがたオーキシン 拡散型オーキシン diffusible auxin
かくさんがたホルモン 拡散型ホルモン diffusible hormone
かくさんかてい 拡散過程 diffusion process
かくさんていこう 拡散抵抗 diffusion resistance
かくさんでんどうど 拡散伝導度 diffusion conductance
かくじききょうめい 核磁気共鳴 nuclear magnetic resonance (NMR)
かくしょうたい 核小体, 仁 nucleolus (*pl.* nucleoli)
かくそうこうたい 核相交代 alternation of nuclear phases
かくそく 隔測, 遠隔測定, リモートセンシング remote sensing
かくタンパクしつ 核タンパク質 nucleoprotein
かくちかん[ほう] 核置換[法] nuclear substitution, nucleus substitution
かくづけ 格付 grading
かくねんけっか 隔年結果 alternate [year] bearing, biennial bearing
かくひ 殻皮, クラスト crust
かくぶん 画分, フラクション, 分画 fraction
かくぶんれつ 核分裂 nuclear division
かくへき 隔壁, 隔膜【節】 nodal diaphragm
がくへん がく(萼)片 sepal
かくへんかん 角変換⇒逆正弦変換 arcsine transformation
かくまく 隔膜, 隔壁【節】 nodal diaphragm
かくまく 核膜 nuclear membrane, nuclear envelope
かくまくこう 核膜孔, 核孔 nuclear pore
がくめい 学名 scientific name
かくようたい 核様体 nucleoid,

karyoid
かくらん　かく(攪)乱　disturbance
かくり　隔離　isolation
かくりさいしゅ　隔離採種　isolated seed production
かくりつ　確率　probability
かくりつかてい　確率過程　stochastic process
かくりつてきどくりつせい　確率的独立性　stochastic independence
かくりつぶんぷ　確率分布　probability distribution
かくりつへんすう　確率変数　random variable, 変量　variate
かくりつみつど　確率密度　probability density
かくりつみつどかんすう　確率密度関数　probability density function
かくりほじょう　隔離圃場　isolation field (plot)
かけい　花茎　flower stalk, scape
かけい　果形　fruit shape
かけい　家系　family, kindred
かけいせんばつ　家系選抜　family selection
かけいぶんせき　家系分析　family analysis
かけながしかんがい　掛流し灌漑　flow irrigation
かこう　加工　processing
かこう　花梗, 花柄　peduncle
かこうてきせい　加工適性　processing suitability
かこうとう　加工糖　reprocessed sugar
かこうようひんしゅ　加工用品種　variety for processing
かこく[るい]　禾穀[類]　cereals, cereal crops
かさいるい　花菜類　flower vegetables
かさいるい　果菜類　fruit vegetables
かさみつど　かさ密度, 容積重　bulk density
かさん　家蚕, カイコ(蚕)　silkworm, *Bombyx mori* L.
かさんかこうそ　過酸化酵素, ペルオキシダーゼ　peroxidase
かさんかすいそ　過酸化水素　hydrogen peroxide
かざんばいどじょう　火山灰土壌　volcanic ash soil
かし　花糸　filament
カシア　cassia, Chinese cinnamon, *Cinnamomum cassia* J. Presl (= *C. cassia* Blume)
かしきず　花式図　floral diagram
かじく　花軸　floral axis, rachis
かじくぶんし　仮軸分枝　sympodial branching
カシグルミ　common walnut, *Juglans regia* L. var. *orientis* (Dode) Kitam.
かしけいしつ　可視形質　visual trait
かしこう[せん]　可視光[線]　visible radiation, visible light
かじつ　果実　fruit
カジノキ(梶の木)　paper mulberry, *Broussonetia papyrifera* (L.) L'Hér. ex Vent.
かじゅ　果樹　fruit tree
カシュー　cashew, cashew-nut tree, *Anacardium occidentale* L.
かじゅう　果汁　fruit juice
カシュウイモ　aerial yam, *Dioscorea bulbifera* L.
カシューナッツ→カシュー
かじゅうへいきん　加重平均, 重み付き平均　weighted mean
かじゅえん　果樹園　orchard, grove
かじゅえんげいがく　果樹園芸学, 果樹学　fruit science, pomology
かじゅがく　果樹学, 果樹園芸学　fruit science, pomology
かじゅく　花熟　ripeness to flower
かじゅく　過熟　overripe
かじゅしけんじょう　果樹試験場　National Institute of Fruit Tree Science

ガジュツ (莪蒁) zedoary, *Curcuma zedoaria* (Christm.) Roscoe
かじょ　花序　inflorescence
かじょ　果序　infructescence
かしょう　か (寡) 照　poor sunshine
かしょうかエネルギー　可消化エネルギー　digestible energy (DE)
かしょうかかんぶつ　可消化乾物　digestible dry matter
かしょうかそタンパクしつ　可消化粗タンパク質　digestible crude protein (DCP)
かしょうかゆうきぶつ　可消化有機物　digestible organic matter
かしょうかようぶん　可消化養分　digestible nutrient
かしょうかようぶんそうりょう　可消化養分総量　total digestible nutrients (TDN)
かじょうきゅうしゅう　過剰吸収　excess absorption
かじょうしょうがい　過剰障害　excess damage (injury)
かじょうしょう [じょう]　過剰症 [状]　excess symptom
がじょう [とつぜん] へんい　芽条 [突然] 変異, 枝変り　bud mutation, bud variation, bud sport
かしょく　仮植　temporary planting, provisional planting
かしょくせいの　可食性の　palatable
かしん　果心　core
かすいぶんかい　加水分解　hydrolysis
かすいぶんかいこうそ　加水分解酵素　hydrolase
ガスえきたいクロマトグラフィー　ガス液体クロマトグラフィー　gas-liquid chromatography (GLC)
ガスクロマトグラフィー　gas chromatography (GC)
ガスクロマトグラフィーしつりょうぶんせきほう　ガスクロマトグラフィー質量分析法　gas chromatography-mass spectrometry (GC-MS)
ガスこうかん　ガス交換　gas exchange
ガスじょうおせんぶっしつ　ガス状汚染物質　gaseous pollutant
カズノコグサ　American sloughgrass, *Beckmannia syzigachne* (Steud.) Fernald
カスパリーせん　カスパリー線　Casparian strip
カスマグサ　fourseeded vetch, sparrow vetch, *Vicia tetrasperma* (L.) Schreb.
かせい　花成　flower formation, flowering
かせいひりょう　化成肥料　compound fertilizer
かせいホルモン　花成ホルモン　flowering hormone
かせき　過石⇒過リン (燐) 酸石灰　superphosphate
カゼクサ　*Eragrostis ferruginea* (Thunb.) P. Beauv.
かせつ　仮説　hypothesis (*pl.* hypotheses)
かせつけんてい　仮説検定　test of hypothesis
かぜよけ　風除け　windbreak, hedge
がそ　画素【画像解析】　pixel
がぞうかいせき　画像解析　image analysis
かそうげんじつ　仮想現実, バーチャルリアリティー　virtual reality
がぞうしょり　画像処理　image processing
かそうど　下層土, 心土　subsoil
かそせい　可塑性, 塑性　plasticity
かたい (禾堆), にお　stack, cock
かたがわけんてい　片側検定　one-sided test, one-tailed test
カタクリ (片栗)　Japanese dog's tooth violet, *Erythronium japonicum* Decne.
かたさ　硬さ【米飯】　hardness
かたせいじょううえ　片正条植え

semi-regular planting
かたつけうえほう　型付植法　marking method of transplanting
カタバミ　creeping woodsorrel, *Oxalis corniculata* L.
かためんきこうよう　片面気孔葉　hypostomatous leaf
かだん　花壇　flower bed
かちく　家畜　livestock, domestic animal, farm animal
かちくえいせいしけんじょう　家畜衛生試験場　National Institute of Animal Health
かちくか　家畜化　domestication
かちくせいさん　家畜生産　animal production
かちゅう　花柱　style
カッコウアザミ　tropic ageratum, white-weed, *Ageratum conyzoides* L.
カッシー→キンゴウカン
かっしょくしんりんど　褐色森林土　Brown Forest soil
かっしょくど　褐色土　Brown soil, Brown earth
かっせいアルミナ　活性アルミナ　active alumina
かっせいおでい　活性汚泥　activated sludge
かっせいか　活性化　activation
かっせいかエネルギー　活性化エネルギー　activation energy
かっせいかざい　活性化剤　activator
かっせいたん　活性炭　active charcoal (carbon)
かっちゃく　活着　rooting
かっちゃくりょく　活着力　rooting ability
かっぺん　褐変　browning
かっぺんき　褐変期【タバコ】browning stage
かつめんしょうほうたい　滑面小胞体　smooth-surfaced endoplasmic reticulum (SER)

かつもんびょう　褐紋病
 1) septoria brown spot【ダイズ】
 2) brown spot【ソバ】
かていもけい　過程模型, プロセスモデル　process-based model
かてん　果点　dot
かとう　果糖, フルクトース　fructose
かどうかん　仮導管　tracheid
カドミウム　cadmium (Cd)
かなすき　金鋤　iron spade
カナダバルサム　Canada balsam
カナダブルーグラス　Canada bluegrass, *Poa compressa* L.
カナダワイルドライ　Canada wild rye, *Elymus canadensis* L.
カナマイシン　kanamycin
カナムグラ　Japanese hop, *Humulus japonicus* Sieb. et Zucc. (= *H. scandens* (Lour.) Merrill)
カニウア　canihua, kaniwa, *Chenopodium pallidicaule* Aellen
かにく　果肉　sarcocarp, pulp
カーネルゆ　カーネル油⇒パーム核油　palm kernel oil
かのうじょうはっさんりょう　可能蒸発散量, 蒸発散位　potential evapotranspiration
カノーラ【ナタネの一系統】canola
カノコソウ　*Valeriana fauriei* Briq.
カバーグラス　cover glass
かはいじく　下胚軸　hypocotyl
かはんがたてきさいき　可搬型摘採機【チャ】power tea plucker
かはんそう　下繁草, 下草 (したくさ)　bottom grass, undergrowth
かはんも　過繁茂　overluxuriant growth, rank growth
かひ　下皮　hypodermis
かひ　花被　perianth
かひ　果皮　pericarp
かび (黴)　mold
かぶ　株　hill, stock
カブ (蕪, 蕪菁)　turnip, *Brassica rapa*

かめら　(27)

かぶあげ　株揚げ　uprooting hills from field, pulling-out hills from field
カフェイン　caffeine
かぶがり　株刈り　ground level harvesting
かぶぎり　株切り　stubble breaking
かぶじたて　株仕立て　bush training
かぶだし［さいばい］　株出し［栽培］　ratooning, ratoon cropping
かぶだしなえ　株出し苗【サトウキビ】　ratoon
かぶだち　株立ち　stand
かぶぬき　株抜き【混種, 罹病株などの】　roguing
かぶばり　株張り　hill spread
かぶほぞん　株保存　preservation of stock, clonal preservation
かぶま　株間　hill distance, interhill space
かぶまき　株播き　sowing (seeding) in hill
かぶわけ　株分け　division, suckering
かふん　花粉　pollen
かふん　果粉　[waxy] bloom
かふんおや　花粉親　pollen parent
かふんがく　花粉学　palynology
かふんかん　花粉管　pollen tube
かふんかんかく　花粉管核　pollen tube nucleus
かふんさんぷき　花粉散布器, 花粉銃, 花粉放射器　pollen gun
かふんしぶんし　花粉四分子　pollen tetrad
かふんじゅう　花粉銃, 花粉散布器, 花粉放射器　pollen gun
かふんのう　花粉嚢　pollen sac
かふんばいかいしゃ　花粉媒介者　pollinator
かふんばいよう　花粉培養　pollen culture
かふんぶんせき　花粉分析　pollen analysis
かふんほうしゃき　花粉放射器, 花粉散布器, 花粉銃　pollen gun
かふんぼさいぼう　花粉母細胞　pollen mother cell (PMC)
かふんりゅう　花粉粒　pollen grain
かへい　花柄, 花梗　peduncle
かへい　果柄　peduncle, gynophore
カーペットグラス　carpetgrass, *Axonopus compressus* (Sw.) P. Beauv.
かべん　花弁　petal
かへんせいちょう　下偏成長　hyponasty
かぼう　花房　flower cluster
かぼう　果房　fruit cluster, bunch
かほうぼく　過放牧　overgrazing, overstocking
カボチャ（南瓜）　*Cucurbita* spp.
カポック　kapok, *Ceiba pentandra* (L.) Gaertn.
カポックゆ　カポック油　kapok seed oil
カーボニックアンヒドラーゼ, 炭酸脱水酵素　carbonic anhydrase
かほんか　禾本科⇒イネ科　Gramineae, Poaceae
かま　鎌　sickle
ガマ（蒲）　cattail, *Typha latifolia* L.
ガマグラス　gama grass, *Trispsacum dactyloides* L.
かます（叺）　straw bag
カミガヤツリ　papyrus, *Cyperus papyrus* L.
かみつ　花蜜　nectar
カミツレ　German camomile (chamomile), *Matricaria chamomilla* L.
カミツレモドキ　mayweed chamomile, *Anthemis cotula* L.
カムしょくぶつ　CAM植物　CAM (crassulacean acid metabolism) plant
カメムシ　plant bugs
カメラルシダ　camera lucida, drawing

prism
カモガヤ→オーチャードグラス
カモジグサ　*Agropyron tsukushiense* (Honda) Ohwi var. *transiens* (Hack.) Ohwi
カモミール→1) ローマカミツレ　2) カミツレ
カヤ (榧)　Japanese torreya, *Torreya nucifera* Sieb. et Zucc.
カヤツリグサ　*Cyperus microiria* Steud.
かやつりぐさか　カヤツリグサ科　Cyperaceae
カヤツリグサかざっそう　カヤツリグサ科雑草　cyperaceous weed, sedge weed
かゆ (粥)　rice porridge
かようせいたんすいかぶつ　可溶性炭水化物　soluble carbohydrate
かようむちっそぶつ　可溶無窒素物　nitrogen-free extract (NFE)
から　殻　chaff, hull, husk, shell
カラーごうせい　カラー合成【画像処理】color composite
カラードギニアグラス　colored Guinea grass, Klein grass, *Panicum coloratum* L.
カラクサナズナ　swine cress, *Coronopus didymus* (L.) J.E. Smith
カラザ, 合点　chalaza
カラシナ　leaf mustard, *Brassica juncea* (L.) Czern. et Coss.
カラシゆ　カラシ油　mustard oil
ガラスか　ガラス化【培養】vitrification
ガラスしつ　ガラス室　glasshouse
ガラスしつせい　ガラス質性, 硝子質性【コムギ】glassiness
ガラスしつの　ガラス質の, 硝子質の　flinty, glassy
ガラスしょうじ　ガラス障子　glass sash
ガラスじょうだんめん　ガラス状断面 vitreous break
カラスノエンドウ　narrowleaf vetch, *Vicia angustifolia* L.
カラスビシャク　*Pinellia ternata* (Thunb.) Breit.
ガラスマメ　grass pea, chickling vetch, *Lathyrus sativus* L.
カラスムギ【≠エンバク】　wild oat, *Avena fatua* L.
ガラスりゅう　ガラス粒【コムギ】glassy kernel
ガラナ　guarana, *Paullinia cupana* Humb. et Kunth
カラムクロマトグラフィー　column chromatography
カラムシ→チョマ (苧麻)
カリ (加里) ⇒カリウム　potassium (K)
かりあとほうぼく　刈跡放牧　stubble grazing
カリウム　potassium (K)
かりかぶ　刈り株【イネ科】stubble
かりかぶなえ　刈り株苗　ratoon
かりこみ　刈込み　trimming, clipping
ガリしんしょく　ガリ侵食　gully erosion
かりた　刈田　stubble field
かりちょぞう　仮貯蔵　tentative storage, temporary storage
かりとり　刈取り　1) cutting, clipping, mowing　2) reaping【禾穀類】
かりとりかいすう　刈取り回数, 刈取り頻度　cutting frequency
かりとりかんかく　刈取り間隔　cutting interval, clipping interval, mowing interval
かりとりき　刈取り機, リーパ　reaper
かりとりけっそくき　刈取り結束機⇒バインダ　binder, reaper and binder
かりとりたかさ　刈取り高さ　cutting height, clipping height, mowing height
かりとりひんど　刈取り頻度, 刈取り回数　cutting frequency

かりはば　刈幅　cutting width, swath
カリひりょう　カリ肥料　potash fertilizer
かりょくかんそうき　火力乾燥機　heated-air dryer, fired dryer
がりん　芽鱗　bud-scale
かりんさんせっかい　過リン(燐)酸石灰　superphosphate
カルシウム　calcium (Ca)
カルシウムしょくぶつ　カルシウム植物, 好石灰植物　calciphilous plant, calcicole plant
カルス　callus (*pl.* calli)
カルダモン, ショウズク (小豆蔻)　cardamon, *Elettaria cardamomum* (L.) Maton
カルチパッカ, 鎮圧機　culti-packer
カルチベータ　cultivator
カルナウバヤシ (カルナウバ椰子)　carnauba wax palm, Brazilian wax palm, *Copernicia prunifera* H.E. Moore (= *C. cerifera* Mart.)
カルビンかいろ　カルビン回路　Calvin cycle, カルビン-ベンソン回路　Calvin-Benson cycle, 還元的ペントースリン酸回路　reductive pentose phosphate cycle
カルボキシラーゼ　carboxylase
カルボンさん　カルボン酸　carboxylic acid
カルモジュリン　calmodulin
かれあがり　枯上がり　1) dying-off 【イネ】2) burning 【タバコ】
かれい　加齢, エイジング　aging, ageing
かれいしゅし　加齢種子　aged seed
かれうれ　枯熟れ　abnormal early ripening
かれはざい　枯葉剤, 落葉剤　defoliant, defoliator
カロース　callose
カロテノイド　carotenoid
カロテン　carotene

カロリー　calorie
かわせい　皮性【オオムギ】covered grain type
カワムギ (皮麦)　hulled barley, covered barley, *Hordeum vulgare* L.
カワラスガナ　*Pycreus sanguinolentus* (Vahl) Nees (= *Cyperus sanguinolentus* Vahl)
かん　稈　culm
かんいせいちまき　簡易整地播き　minimum tillage seeding
カンエンガヤツリ　*Cyperus exaltatus* Retz. ssp. *iwasakii* (Makino)
かんおんせい　感温性　thermosensitivity, sensitivity to temperature
かんおんせいひんしゅ　感温性品種　thermosensitive variety
かんおんそう　感温相　thermophase
かんか　乾果　dry fruit
かんがい　寒害　cold damage (injury)
かんがい　干害(旱害)　drought damage (injury)
かんがい　灌漑　irrigation
かんがいすい　灌漑水　irrigation water
かんがいそうち　灌漑草地　irrigated grassland
かんき　乾季　dry season
かんき　換気　ventilation
かんき　間期, 中間期【細胞分裂】interphase, interkinesis
かんきさく　乾季作　dry-season cropping
かんきさく[もつ]　乾季作[物]　dry-season crop
かんきゅうおんど　乾球温度　dry-bulb temperature
かんきょう　環境　environment
かんきょうアセスメント　環境アセスメント, 環境影響評価　environmental impact assessment
かんきょうえいきょうひょうか　環境影響評価, 環境アセスメント

environmental impact assessment
かんきょうおせんぶっしつ　環境汚染物質　environmental pollutant
かんきょうかがく　環境科学　environmental science
かんきょうきじゅん　環境基準　environmental quality standard
かんきょうきほんほう　環境基本法　The Environment Basic Law
かんきょうきょういく　環境教育　environmental education
かんきょうしひょう　環境指標　environmental indicator
かんきょうしゅうようりょく　環境収容力, 環境容量　environmental capacity, carrying capacity
かんきょうストレス　環境ストレス　environmental stress
かんきょうせいぶつがく　環境生物学　environmental biology
かんきょうそうかん　環境相関　environmental correlation
かんきょう[とつぜん]へんいげん　環境[突然]変異源　environmental mutagen
かんきょうはくしょ　環境白書　Quality of the Environment in Japan
かんきょうぶんさん　環境分散　environmental variance
かんきょうへんい　環境変異　environmental variation
かんきょうほぜん　環境保全　environmental conservation
かんきょうホルモン　環境ホルモン　environmental hormones, 外因性内分泌かく乱物質　endocrine disrupting chemicals (EDC)
かんきょうよういん　環境要因　environmental factor
かんきょうようりょう　環境容量, 環境収容力　environmental capacity, carrying capacity
かんきょうりょくち　環境緑地　environmental green space
かんきんさくもつ　換金作物　cash crop
かんげん　還元　reduction
がんけんせい　頑健性【統計】　robustness
かんげんそう　還元層　reduction zone, reduced layer
かんげんてきペントースリンさんかいろ　還元的ペントースリン酸回路　reductive pentose phosphate cycle, カルビン-ベンソン回路　Calvin-Benson cycle, カルビン回路　Calvin cycle
かんげんとう　還元糖　reducing sugar
かんげんぶんれつ　還元分裂⇒減数分裂　meiosis reduction division
かんこうの　慣行の, 従来の　conventional
かんこうせい　感光性　photosensitivity, photoperiodic sensitivity
かんこうせいひりょう　緩効性肥料　controlled release fertilizer, slow-release fertilizer
かんこうせいひんしゅ　感光性品種　photosensitive variety
かんこうそう　感光相　photophase, photoperiod sensitive phase
かんごえ　寒肥　winter dressing
かんこん　冠根　crown root, coronal root
カンコン→ヨウサイ
かんさく　間作　intercropping, catch cropping
かんさくもつ　間作物　intercrop, catch crop
かんしつけい　乾湿計, サイクロメータ　psychrometer
カンシャ(甘蔗)→サトウキビ
かんじゅく　完熟　full ripe, full maturity
かんじゅくき　完熟期　full-ripe stage
かんじゅくたいひ　完熟堆肥　fully

fermented compost
かんじゅせい　感受性　susceptibility, sensitivity
カンショ（甘藷）→サツマイモ
カンショ（甘蔗）→サトウキビ
かんしょう　干渉　interference
かんしょうえき　緩衝液　buffer solution
かんしょうけんびきょう　干渉顕微鏡　interference microscope
かんしょうしょくぶつ　観賞植物　ornamental plant
かんじょうディーエヌエー　環状DNA　circular DNA
かんじょうはくひ　環状剥皮　girdling, ringing
かんじょうようそ　管状要素　tracheary element
かんしょく　間植　interplanting
カンショしぼりかす　甘蔗搾粕，バガス　bagasse
カンショとう　甘蔗糖　cane sugar
かんすい　冠水　flooding, submergence
かんすい　灌水　watering, irrigation
かんすいがい　冠水害　inundation damage
かんすいしん　寒水浸【イネ種子】winter soaking
かんすいたいせい　冠水耐性　flood tolerance
かんすいていこうせい　冠水抵抗性　resistance to submergence
がんすいひ　含水比　moisture weight percentage
がんすいりょう　含水量, 水分含量　water content, moisture content
かんすう　関数　function
かんせい　間性　intersex
かんせいかせい　幹成花性　cauliflory
かんせいけいたい　乾性形態　xeromorphism, xeromorphy
かんせいざっそう　乾生雑草　xerophytic weed
かんせいしょくぶつ　乾生植物　xerophyte
かんせいゆ　乾性油　drying oil
かんせつかくぶんれつ　間接核分裂　indirect nuclear division
かんせん　感染　infection
かんぜんか　完全花　perfect flower, complete flower
かんせんせい　感染性　infectivity
かんぜんまい　完全米　perfect rice grain, whole rice grain
かんぜんむさくいせっけい　完全無作為設計　completely randomized design
かんぜんゆうせい　完全優性　complete dominance
かんぜんよう　完全葉　complete leaf
かんぜんりゅう　完全粒　perfect kernel
かんそう　乾燥　1) drying, desiccation 2) curing【タバコ】
かんそう　乾草　hay
カンゾウ（甘草）　licorice, *Glycyrrhiza* spp.【*G. uralensis* Fisch. et DC., *G. glabra* L. など】
カンゾウ（萱草）→カンエンガヤツリ
かんそうか　乾草架　hay rack
かんそうかいひせい　乾燥回避性, かんばつ（旱魃）回避性　drought avoidance
かんそうき　乾燥機　drying machine, dryer
かんそうきこう　乾燥気候　arid climate, dry climate
かんそうざい　乾燥剤　desiccant, desiccating agent
かんそうしせつ　乾燥施設　drying plant
かんそうしつ　乾燥室【タバコ】curing barn
かんそうたいせい　乾燥耐性, かんばつ（旱魃）耐性　drought tolerance
かんそうち　乾燥地　arid land
かんそうちたい　乾燥地帯　arid region, arid zone
かんそうちょうせい　乾草調製　hay

making
かんそうちょぞう　乾燥貯蔵　dry storage
かんそうとうひせい　乾燥逃避性, かんばつ(旱魃)逃避性　drought escape
かんそくち　観測値【統計】　observation
かんたいの　寒帯の　arctic
かんたく　干拓　reclamation in water area
かんたくち　干拓地　polder, polder land
かんたくでん　干拓田　reclaimed paddy field
カンタラアサ(カンタラ麻)　cantala, *Agave cantala* Roxb.
かんだんかんがい　間断灌漑　intermittent irrigation
かんちがたぼくそう　寒地型牧草　temperate forages, temperate grass and/or legume
かんちのうぎょう　乾地農業, 乾地農法　dry farming
かんちのうほう　乾地農法, 乾地農業　dry farming
かんちょう　稈長　culm length
がんてつしざい　含鉄資材　iron containing material
かんでん　乾田　well-drained paddy field
かんでんちょくはん(ちょくは)　乾田直播　lowland rice culture direct sown in drained field
かんでんなわしろ　乾田苗代　drainable rice nursery
かんてんばいち　寒天培地　agar medium
かんど　乾土　oven-dry soil, dried soil
かんど　感度【計測】　sensitivity
かんど　間土【播種】　soil insulation
カントウタンポポ　*Taraxacum platycarpum* Dahlst. ssp. *platycarpum*
かんどこうか　乾土効果　air-drying effect on ammonification
かんどぶんせき　感度分析　sensitivity analysis
カントリーエレベータ　country elevator
カントンアブラギリ(広東油桐)　tung, *Aleurites montana* (Lour.) E.H. Wilson
かんにゅうていこう　貫入抵抗【土壌】　penetration resistance
かんのうしけん　官能試験　sensory test
かんぱ　乾葉【タバコ】　cured leaf
かんばつ(旱魃)　drought
かんばつかいひせい　かんばつ(旱魃)回避性, 乾燥回避性　drought avoidance
かんばつたいせい　かんばつ(旱魃)耐性, 乾燥耐性　drought tolerance
かんばつとうひせい　かんばつ(旱魃)逃避性, 乾燥逃避性　drought escape
ガンビール　gambir, *Uncaria gambir* Roxb.
かんぶ　冠部　crown
かんぷうがい　乾風害　dry wind damage (injury)
かんぷうがい　寒風害　cold wind damage (injury), chilly wind damage (injury)
かんぶつ　乾物　dry matter
かんぶつじゅう　乾物重　dry weight
かんぶつしゅうりょう　乾物収量　dry-matter yield
かんぶつせいさん　乾物生産, 物質生産　dry-matter production
かんぶつぶんぱいりつ　乾物分配率　dry-matter partitioning ratio
かんぶつりつ　乾物率　percentage dry-matter
かんぷびょう　乾腐病【コンニャク】　dry rot
かんぼく　かん(灌)木⇒低木　shrub, bush
かんぼくかんさく　灌木間作, アレイ作付け　alley cropping
かんみ　甘味　1) sweetness　2) sweet

taste【タバコ】
がんみつとう　含蜜糖　non-centrifuged sugar
かんみひ　甘味比　sugar-acid ratio
がんもんびょう　眼紋病【コムギ】　eye spot
がんゆうりょう　含有量　content
がんゆりつ　含油率　oil percentage
がんゆりょう　含油量　oil content
かんり　管理　1) management 2) tending【動植物】
かんれいしゃ　寒冷しゃ(紗)　cheese cloth
かんれいち　寒冷地　cold climate area (region)

[き]

キアイ(木藍)　indigo tree, common indigo, *Indigofera tinctoria* L.
キアズマ　chiasma (*pl.* chiasmata)
きあつ　気圧　atmospheric pressure
ぎいでんし　偽遺伝子　pseudogene
きおん　気温　air temperature
きが　飢餓　starvation
ぎか　偽果　false fruit, pseudocarp
きかいか　機械化　mechanization
きかいじょそう　機械除草　mechanical weeding
きかいづみ　機械摘み　1) mechanical picking【ワタなど】　2) mechanical plucking【チャなど】
きかいてきかくり　機械的隔離　mechanical isolation
きかいてき[ざっそう]ぼうじょ　機械的[雑草]防除　mechanical [weed] control
きかいてきふどう　機会的浮動　random drift
きかいまき　機械播き　automatic seeding, sowing by seeder
ぎかこくるい　偽(擬)禾穀類　pseudocereals

きかざっそう　帰化雑草　naturalized weed
キカシグサ　Indian toothcup, *Rotala indica* (Willd.) Koehne var. *uliginosa* (Miq.) Koehne
きかしょくぶつ　帰化植物　naturalized plant
きかねつ　気化熱　heat of vaporization
きかへいきん　幾何平均　geometric mean
きかん　器官　organ
きかんがく　器官学　organography
きかんけいせい　器官形成　organogenesis
きかんだつり　器官脱離　abscission
きかんばいよう　器官培養　organ culture
ききとりちょうさ　聞取り調査　interview survey
ききとりちょうさひょう　聞取り調査表, 質問票　questionnaire
ききゃく　棄却【仮説検定】　rejection
ききゃくいき　棄却域　critical region
ききゃくげんかい　棄却限界　critical value
キキョウ(桔梗)　Japanese bellflower, Chinese bellflower, balloonflower, *Platycodon grandiflorum* (Jacq.) A. DC.
きぎょうのうぎょう　企業農業　commercial agriculture, 農園農業　estate agriculture, プランテーション　plantation
ききん　飢饉　famine
キクイモ(菊芋)　Jerusalem artichoke, *Helianthus tuberosus* L.
キクモ　limnophila, *Limnophila sessiliflora* Blume
キクユグラス　kikuyu grass, *Pennisetum clandestinum* Hochst. ex Chiov.
きけい　奇形　malformation, deformity, deformation, terata

きけいりゅう　奇形粒　deformed grain
きげんちゅうしん　起源中心　center of origin
きけんりつ　危険率【仮説検定】⇒有意水準　significance level, level of significance
きこう　気孔　stoma (*pl.* stomata)
きこう　気候　climate
きこういんし　気候因子　climatic factor
きこうかいど　気孔開度　stomatal aperture
きこうくぶん　気候区分　climatic division
きこうコンダクタンス　気孔コンダクタンス, 気孔伝導度　stomatal conductance
きこうしすう　気候指数　climatic index
きこうじょうさん　気孔蒸散　stomatal transpiration
きこうせいさんりょくしすう　気候生産力示数　climatic productivity index
きこうたい　気候帯　climatic zone
きこうていこう　気孔抵抗　stomatal resistance
きこうてきさいばいげんかい　気候的栽培限界　climatic cultivation limit
きこうてきモデル　機構的モデル　mechanistic model
きこうでんどうど　気孔伝導度, 気孔コンダクタンス　stomatal conductance
きこうないこう　気孔内腔　substomatal cavity
きこうへんか　気候変化　climate change
きこうへんどう　気候変動　climatic fluctuation, climatic variation
きこうようそ　気候要素　climatic element
きこん　気根　aerial root
ぎざっしゅ　偽雑種　false hybrid
きざとう　生砂糖　raw sugar

キサントフィル　xanthophyll
ぎじカラー　擬似カラー　pseudo color
ギシギシ　Japanese dock, *Rumex japonicus* Houtt. (= *R. cripus* L. ssp. *japonicus* (Houtt.) Kitam.)
きしつ　基質　1) substrate　2) matrix【細胞等】
きしつしんわせい　基質親和性　affinity of substrate
きしゃく　希釈　dilution
きしゃくざい　希釈剤　diluent
きしゅ　寄主⇒宿主　host
キシュウスズメノヒエ　knotgrass, *Paspalum distichum* L.
きしゅしょくぶつ　寄主植物　host plant
ぎじゅせい　偽受精　pseudogamy, false fertilization
ぎじゅつたいけい　技術体系　systematized techniques, series of techniques, system of techniques
きじゅんひんしゅ　基準品種【食味試験】　standard variety
きしょうかんのうしけん　気象感応試験　crop response test to weather
きしょうさいがい　気象災害　meteorological disaster
きしょうじょうほうシステム　気象情報システム　weather information system
キシレン　xylene
キシログルカン　xyloglucan
きすい　汽水　brackish water
きずくっせい　傷屈性　taumatotropism
きずもみ　傷籾　wounded rough rice
きせい　寄生　parasitism
きせいき　起生期【ムギ類】　regrowing stage
きせいざっそう　寄生雑草　parasitic weed
ぎせいでん　犠牲田　sacrificed paddy field to raise water temperature
きせつ [てき] せいさんせい　季節 [的]

生産性　seasonal productivity
きせつふう　季節風　monsoon
キセニア　xenia
きそう　気相　gas phase
きたいち　期待値　expectation, expected value
きたいひんど　期待頻度　expected frequency
キダチハッカ (木立薄荷) →サマーサボリー
キダチルリソウ (木立瑠璃草) →ヘリオトロープ
きついんりょうさくもつ　喫飲料作物　beverage crop
きつえんりょうさくもつ　喫煙料作物　smoke crop
きっこうさよう　拮抗作用　antagonism
きっこうそがい　拮抗阻害, 競合阻害　competitive inhibition
キツネアザミ　*Hemistepta lyrata* Bunge
キツネノボタン　buttercup, *Ranunculus silerifolius* Lév. (= *R. quelpaertensis* (Lév.) Nakai)
キツネノマゴ　*Justicia procumbens* L.
きていひど　基底被度　basal coverage (cover)
きどうさいぼう　機動細胞　motor cell, bulliform cell
きどりわた　木採棉　stalk-cut cotton
キナ　cinchona, quinine, *Cinchona* spp. 【*C. pubescens* Vahl など】
きなこ　黄粉　soybean flour
ギニアグラス　guineagrass, *Panicum maximum* Jacq.
ギニアヤム　1) white guinea yam【白肉】, *Dioscorea rotundata* Poir. 2) yellow guinea yam【黄肉】, *Dioscorea cayenensis* Lam.
ぎねんせい　偽稔性　pseudofertility
キノア　quinoa, *Chenopodium quinoa* Willd.
きのう　機能　function
きのうてきすいろん　帰納的推論　inductive inference
キハダ　Amur cork-tree, *Phellodendron amurense* Rupr.
きはつせいの　揮発性の　volatile
キバナオモダカ　yellow velvetleaf, *Limnocharis flava* (L.) Buchenau
キバナスィートクローバ (黄花スィートクローバ)　yellow sweetclover, *Melilotus officinalis* (L.) Lam.
キバナルーピン (黄花ルーピン)　yellow lupine, *Lupinus luteus* L.
きひ (もとごえ)　基肥, 元肥　basal dressing, basal application
キビ (黍)　[common] millet, proso millet, hog millet, *Panicum miliaceum* L.
きひざい　忌避剤　repellent
きぼかくだい　規模拡大　scale expansion, farm-size expansion
きほんえいようせいちょうせい　基本栄養成長性　basic vegetative growth
きほんかくがた　基本核型　basikaryotype
きほんきかん　基本器官　fundamental organ
きほんしゅ　基本種　elementary species
きほんすう　基本数【染色体】　basic number
きほんそしき　基本組織　fundamental tissue, ground tissue
きほんばいち　基本培地　basal medium
きほんぶんれつそしき　基本分裂組　fundamental meristem, ground meristem
キマメ　pigeon pea, cajan pea, *Cajanus cajan* (L.) Millsp.
きむかせつ　帰無仮説　null hypothesis
キメラ　chimera
ぎゃくい　逆位【染色体】　inversion
ぎゃくぎょうれつ　逆行列　inverse matrix

ぎゃくこうざつ 逆交雑 reciprocal cross[ing]
ぎゃくせいげんへんかん 逆正弦変換 arcsine transformation
ぎゃくせんばつ 逆選抜 reverse selection, adverse selection
ぎゃくてんしゃこうそ 逆転写酵素 reverse transcriptase
ぎゃくてんそう 逆転層【気象】 inversion layer
ぎゃくど 客土 soil dressing
ぎゃくとうた 逆淘汰⇒逆選抜 reverse selection, adverse selection
ぎゃくとつぜんへんい 逆突然変異, 復帰突然変異 reverse mutation, back mutation
ぎゃくにこうぶんぷ 逆二項分布 inverse binomial distribution, 負の二項分布 negative binomial distribution
キャタピラートラクタ crawler tractor
キャッサバ cassava, manioc, tapioca plant, *Manihot esculenta* Crantz (= *M. utilissima* Pohl)
キャッサバデンプン cassava starch
キャベツ cabbage, *Brassica oleracea* L. var. *capitata* L.
キャラウェー, ヒメウイキョウ caraway, *Carum carvi* L.
キャリヤー, 担体 carrier
キャリブレーション, 較正 calibration
きゅう 級 class
きゅうか 球果, 毬果【ホップ】 cone
きゅうかん 休閑 fallow
きゅうかんこう 休閑耕 fallowing
きゅうかんさくもつ 休閑作物 fallow crop
きゅうかんち 休閑地 fallow, fallowed field
きゅうきてき 求基的 basipetal
きゅうくかん 級区間 class interval
きゅうけい 球茎 corm, solid bulb
きゅうこう 休耕 non-cultivation, non-cropping
ぎゅうこう 牛耕 plowing by cow
きゅうこうけいすう 吸光係数 extinction coefficient
きゅうこうさくもつ 救荒作物, 備荒作物 emergency crop
きゅうこうでん 休耕田 uncultivated paddy field
きゅうこうど 吸光度 absorbance
きゅうこうぶんこうぶんせき 吸光分光分析 spectrophotometric analysis
きゅうこん 球根 bulb
きゅうし 吸枝, ひこばえ【園芸】 sucker
きゅうじ 給餌 feeding
きゅうしが 休止芽⇒休眠芽 dormant bud, resting bud
きゅうしつすい 吸湿水 hygroscopic water
きゅうしゅう 吸収 absorption, uptake
きゅうしゅうこん 吸収根 sucking root, absorbing root, feeder root
きゅうしゅうスペクトル 吸収スペクトル absorption spectrum
きゅうしゅうにっしゃりょう 吸収日射量 absorbed radiation
きゅうしゅうのうぎょうしけんじょう 九州農業試験場 Kyushu National Agricultural Experiment Station
きゅうしゅうりつ 吸収率
1) absorptivity, absorptance【物理】
2) absorption percentage
きゅうしんてき 求心的 centripetal
きゅうすい 吸水 water absorption
きゅうすいあつ 吸水圧 suction pressure
きゅうすいけい 吸水計, ポトメーター potometer
きゅうすいりょく 吸水力 suction force
ぎゅうせい 偽優性 pseudodominance
きゅうちゃく 吸着 adsorption
きゅうちゃくクロマトグラフィー 吸着

クロマトグラフィー　adsorption chromatography
きゅうちゃくすい　吸着水　absorbed water
きゅうひ　きゅう(厩)肥　farmyard manure, barnyard manure, stable manure
きゅうひりょく　吸肥力　nutrient absorption ability
きゅうひんど　級頻度　class frequency
ぎゅうふんきゅうひ　牛糞きゅう(厩)肥　cattle manure
きゅうみん　休眠　dormancy, rest
きゅうみんが　休眠芽　dormant bud, resting bud
きゅうみんかくせい　休眠覚醒　dormancy awakening
きゅうみんかくせいざい　休眠覚醒剤　dormancy breaker
きゅうみんき[かん]　休眠期[間]　dormant period (season)
きゅうみんしゅし　休眠種子　dormant seed
きゅうみんだは　休眠打破　dormancy breaking
キュウリ　cucumber, *Cucumis sativus* L.
キュウリグサ　*Trigonotis peduncularis* (Trevir.) Benth. ex Hemsl.
きゅちょうてき　求頂的　acropetal
きょうか　莢果⇒豆果　legume
きょうかいじょうけん　境界条件　boundary condition
きょうかいそう　境界層　boundary layer
きょうかいそうていこう　境界層抵抗　boundary layer resistance
きょうかまい　強化米　enriched rice
きょうかんせい　強稈性　straw stiffness, culm stiffness
きょうかんひんしゅ　強稈品種　stiff-strawed variety, strong-strawed variety

ギョウギシバ→バーミューダグラス
きょうきゃく　鏡脚【顕微鏡】　base, foot
きょうごう　競合, 競争　competition
きょうこうそがい　強光阻害, 光阻害, 光障害　photoinhibition
きょうごうそがい　競合阻害, 拮抗阻害　competitive inhibition
きょうさく　凶作　poor crop, poor harvest
きょうざつぶつ　夾雑物, 異物　foreign matter, impurity
ぎょうしゅうりょくせつ　凝集力説　cohesion theory
きょうしょうてんそうさがたけんびきょう　共焦点走査型顕微鏡　confocal scanning microscope
きょうしょうてんレーザーけんびきょう　共焦点レーザー顕微鏡　confocal laser microscope (CLM)
きょうしんか　共進化　coevolution
きょうせい　共生　symbiosis
きょうせいえいか　強勢頴花　superior spikelet
きょうせんてい　強せん(剪)定　heavy pruning
きょうそう　競争, 競合　competition
きょうそうしゃ　競争者　competitor
きょうそうりょく　競争力　competitive ability
きょうだいけんてい　きょうだい検定　sib test
きょうだいこうはい　きょうだい交配　sib cross
きょうちゅう　鏡柱【顕微鏡】　arm
きょうとう　鏡筒【顕微鏡】　tube
きょうどうぼうじょ　共同防除　cooperative control
きょうぶんさん　共分散　covariance
きょうぶんさんぎょうれつ　共分散行列　covariance matrix
きょうぶんさんせいぶん　共分散成分　covariance component, component of

covariance
きょうぶんさんぶんせき　共分散分析　analysis of covariance
きょうよおや　供与親　doner parent
きょうよくせい　共抑制　cosuppression
きょうよたい　供与体　donor
きょうりゅう　狭粒　slender grain
きょうりきこ　強力粉【コムギ】 strong flour
きょく　極　pole
きょくかく　極核　polar nucleus
きょくすう　極数　pole number
きょくせい　極性　polarity
きょくせいいどう　極性移動　polar transport
きょくせんかいき　曲線回帰　curvilinear regression
きょくせんのあてはめ　曲線の当てはめ　curve fitting
きょくそう　極相　climax
きょくち　極値　extreme value
きょし　鋸歯　serration
ぎょどくせい　魚毒性　fish toxicity
きょようげんかい　許容限界　tolerance limit
きょようど　許容度　allowance
きよりつ　寄与率　ratio of contribution, contribution ratio
キリアサ（桐麻）→ボウマ
きりかえせんてい　切返しせん（剪）定　heading-back pruning, cutting-back pruning
きりかえばた　切替畑　shifting field
きりかぶ　切り株　stump
きりつぎ　切接ぎ　veneer grafting
きりつけ　切付け，タッピング【ゴム・ウルシなど】 tapping
キレート　chelate
きれつ　亀裂　crack
キレハイヌガラシ　yellow fieldcress, *Rorippa sylvestris* (L.) Bess.
きろくけい　記録計　recorder

きろくし　記録紙　recording chart
ぎわごうせい　偽和合性　pseudocompatibility
きわた　生綿，原綿（げんめん）　raw cotton
きんいつせい　均一性　homogeneity
きんいつせいしけん　均一性試験，一様性試験【圃場等】　homogeneity trial, uniformity trial
きんいつせいのけんてい　均一性の検定【分散等】　test of homogeneity, homogeneity test
キンエノコロ　yellow foxtail, *Setaria glauca* (L.) Beauv.
きんえんけいすう　近縁係数　coefficient of parentage
きんえんしゅ　近縁種　allied species, related species
きんえんやせいしゅ　近縁野生種　wild relatives
きんかく　菌核　sclerotium (*pl.* sclerotia)
キンカン（金柑）　kumquats, *Fortunella* spp.
キンギョモ→マツモ
きんこう　近交，近親交配　inbreeding
きんこうえんげい　近郊園芸　suburban gardening
キンゴウカン（金合歓）　cassie, sweet acacia, *Acacia farnesiana* (L.) Willd.
ギンゴウカン（銀合歓）　white popinac, ipil-ipil, *Leucaena leucocephala* (Lam.) De Wit
きんこうけい　近交系　inbred line, inbred strain
きんこうけいすう　近交係数　inbreeding coefficient, coefficient of inbreeding
きんこうじゃくせい　近交弱勢　inbreeding depression
きんこん　菌根　mycorrhiza
きんこんきん　菌根菌　mycorrhizal fungus

・きんし 菌糸 hypha (*pl.* hyphae)
きんしんこうはい 近親交配, 近交 inbreeding
ぎんせんしょくほう 銀染色法 silver staining
きんぞくイオン 金属イオン metal ion
きん(じゅん)どうしついでんしけいとう 近(準)同質遺伝子系統 near-isogenic line
ギンネム→ギンゴウカン
きんぴ 金肥, 販売肥料, 購入肥料 commercial fertilizer
きんぺい 均平 levelling
きんぼくく 禁牧区 exclosure, nongrazing area (plot)
きんるい 菌類 fungi (*sing.* fungus)
キンレンカ(金蓮花) common nasturtium, garden nasturtium, *Tropaeolum majus* L.

[く]

グアニン guanine (G)
グァユール guayule, Mexican rubber, *Parthenium argentatum* A. Gray
グアル→クラスタマメ
くうえい 空穎 empty glume, sterile glume
くうきかんそう 空気乾燥【タバコ】 air-curing
くうきかんそうしゅ 空気乾燥種【タバコ】 air-cured tobacco
ぐうぜんごさ 偶然誤差 random error
くうちゅうさんぷ 空中散布 aerial application
くうちゅうしゃしん 空中写真 aerial photographs
くうどうげんしょう 空洞現象 cavitation
ぐうはつとつぜんへんい 偶発突然変異 spontaneous mutation, natural mutation
クエンさん クエン酸 citric acid
クエンさんかいろ クエン酸回路 citric acid cycle, クレブス回路 Krebs cycle, トリカルボン酸(TCA)回路 tricarboxylic acid (TCA) cycle
くかくせいり 区画整理 land consolidation
くかんすいてい 区間推定 interval estimation
くき 茎 stem
くきざし 茎挿し, 枝挿し stem cutting
くきだちき 茎立期【ムギ類】 jointing stage
ククイノキ candlenut, *Aleurites moluccana* (L.) Willd.
クグガヤツリ annual sedge, *Cyperus compressus* L.
くさ 草 1) herb【一般】 2) grass【イネ科】
クサイ slender rush, *Juncus tenuis* Willd.
くさがた 草型 plant type
くさかり 草刈り mowing
くさかりき 草刈り機, モーア mower
くさけずり 草削り weed scraping
くさたけ 草丈 plant length【≠草高(そうこう) plant height】
クサネム(草合歓) Indian jointvetch, *Aeschynomene indica* L.
クサビコムギ *Aegilops speltoides* Tausch.
クサフジ cow vetch, *Vicia cracca* L.
クサヨシ→リードカナリーグラス
クジャクヤシ(孔雀椰子) toddy palm, wine palm, *Caryota urens* Jacq.
くじょざい 駆除剤 expellent, repellent
クス(樟) camphor tree, *Cinnamomum camphora* (L.) Presl
クズ kudzu, kudzu-vine, *Pueraria lobata* (Willd.) Ohwi (= *P. thunbergiana* (Sieb. et Zucc.) Benth.)
クズイモ yam bean, *Pachyrhizus*

erosus (L.) Urban
クズウコン (葛鬱金) →アロールート
クズデンプン　pueraria starch
くずまい　屑米　rice screenings
くずもみ　屑籾　rough rice screenings
くずりゅう　屑粒　screenings
くだけまい　砕け米, 砕米 (さいまい)　broken rice
クダモノトケイソウ→パッションフルーツ
クチクラ　cuticle
クチクラじょうさん　クチクラ蒸散　cuticular transpiration
クチクラそう　クチクラ層　cuticular layer
クチクラていこう　クチクラ抵抗　cuticular resistance
クチナシ　cape jasmine, *Gardenia jasminoides* Ellis
クチン　cutin
クチンか　クチン化　cutinization
くっかせい　屈化性⇒化学屈性　chemotropism
くっきょく　屈曲　curvature, bending
くっこうせい　屈光性⇒光屈性　phototropism
くつじつせい　屈日性⇒日光屈性　heliotropism
くっしょうせい　屈傷性⇒傷屈性　traumatotropism
くっすいせい　屈水性⇒水分屈性　hydrotropism
くっせい　屈性　tropism
くっせつけいしど　屈折計示度　refractometer index, 糖度【糖液】Brix
くっせつりつ　屈析率　refraction index
グッタペルカ　guttapercha, *Palaquium gutta* (Hook. f.) Baill.
くっちせい　屈地性　geotropism⇒重力屈性　gravitropism
くどひりょう　苦土肥料　magnesium fertilizer

くどリンあん　苦土リン(燐)安　magnesium ammonium phosphate
クノップえき　クノップ液　Knop's solution
くびまがり　首曲り【カーネーション, チューリップ等】crooked neck
くぶんキメラ　区分キメラ　sectorial chimera
クマツヅラ　vervain, *Verbena officinalis* L.
くみあわせ　組合せ　combination
くみあわせいくしゅほう　組合せ育種法　combination breeding
くみあわせのうりょく　組合せ能力　combining ability
くみかえ　組換え　recombination
くみかえか　組換え価　recombination value
くみかえがたじしょくけいとう　組換え型自殖系統　recombinant inbred line
くみかえディーエヌエー　組換え DNA　recombinant DNA
クミン　cumin, *Cuminum cyminum* L.
グライそう　グライ層　gley horizon
グライど　グライ土　Gley soil
クラウンゴール　crown gall
クラスター　cluster
クラスターぶんせき　クラスター分析　cluster analysis
クラスタマメ　cluster bean, guar, *Cyamopsis tetragonoloba* (L.) Taub.
クラスト, 殻皮　crust
グラスピー→ガラスマメ
くらつき (鞍築)　mound, hill
くらつき　鞍接ぎ　saddle grafting
グラナ　grana (*sing.* granum)
クラブコムギ　club wheat, *Triticum compactum* Host
グラベリマイネ　African rice, *Oryza glaberrima* Steud.
クラムヨモギ　kurram santonica, *Artemisia kurramensis* Quazilbash
クランツこうぞう　クランツ構造

Kranz anatomy
グランドトルース【リモートセンシング】 ground truth
クランプ, 締め具 clamp
クリ(栗) Japanese chestnut, *Castanea crenata* Sieb. et Zucc.
くりいろど 栗色土 Chestnut soil
グリオキシソーム glyoxysome, ペルオキシソーム peroxisome, ミクロボディ microbody
クリカボチャ→セイヨウカボチャ
グリコーゲン glycogen
グリコールさんかいろ(けいろ) グリコール酸回路(経路) glycolate cycle (pathway)
グリシン glycine (Gly)
クリスタルバイオレット crystal violet
クリステ cristae (*sing.* crista)
グリセリン glycerin
クリノスタット klinostat
クリーピングフェスク→レッドフェスク
クリーピングベントグラス creeping bentgrass, *Agrostis stolonifera* L.
クリムソンクローバ crimson clover, *Trifolium incarnatum* L.
クリヤシ→モモヤシ
くりわた 繰り綿⇒繰綿(そうめん) ginned cotton
くりわたぶあい 繰り綿歩合⇒繰綿歩合(そうめんぶあい) ginning percentage
グリーンけいこうタンパクしつ グリーン蛍光タンパク質 green fluorescent protein (GFP)
グリーンパニック green panic, *Panicum maximum* Jacq. var. *trichoglume* Eyles
グルコース, ブドウ糖 glucose
グルジアコムギ *Triticum paleocolchicum* Men.
グルタールアルデヒド glutaraldehyde
グルタミン glutamine (Gln)
グルタミンさん グルタミン酸 glutamic acid (Glu)
グルテリン glutelin
グルテン gluten
クルマバザクロソウ common carpetweed, *Mollugo verticillata* L.
クルミ(胡桃) walnut, *Juglans* spp.
クレイングラス→カラードギニアグラス
グレープフルーツ grapefruit, *Citrus paradisi* Macf.
グレーンドリル grain drill
グレコラテンほうかく グレコラテン方格 Greco-Latin square
クレステッド・ホィートグラス crested wheatgrass, *Agropyron cristatum* (L.) Gaertn.
クレソン(和蘭芥) water-cress, cresson, *Nasturtium officinale* R. Br. (= *Roripa nasturtium-aquaticum* (L.) Hayek.)
クレブスかいろ クレブス回路 Krebs cycle, クエン酸回路 citric acid cycle, トリカルボン酸(TCA)回路 tricarboxylic acid (TCA) cycle
くろ(畔) levee
クロガラシ(黒芥子) black mustard, *Brassica nigra* (L.) Koch
クログワイ *Eleocharis kuroguwai* Ohwi
くろざとう 黒砂糖 brown sugar
グロースキャビネット growth cabinet, グロースチャンバー growth chamber
くろだねウンダイ 黒種ウンダイ(蕓薹)→洋種ナタネ
クローニング, クローン化 cloning
くろぬり くろ(畔)塗り, あぜ(畦)塗り levee coating
くろぼくど 黒ぼく土 Andosol
くろほびょう 黒穂病【コムギ, トウモロコシ】 smut
くろまい 黒米 black rice
クロマチン, 染色質 chromatin
クロマトグラフィー chromatography

クロマトフォア　chromatophore
クロモ　hydrilla, *Hydrilla verticillata* (L.f.) Casp.
クロレラ　chlorella, *Chlorella* spp.
クロロシス　chlorosis
クロロフィル, 葉緑素　chlorophyll
クロロプラスト, 葉緑体　chloroplast
クロロホルム　chloroform
クローン, 栄養系　clone, clonal strain
クローンか　クローン化, クローニング　cloning
クワ (桑)　mulberry, *Morus bombycis* Koidz. (= *M. alba* L.)
クワイ　arrowhead, *Sagittaria trifolia* L. var. *edulis* (Sieb.) Ohwi
クワクサ　*Fatoua villosa* (Thunb.) Nakai
クワズイモ　*Alocasia odora* (Lodd.) Spach
グワバ　guava, *Psidium guajava* L.
クワモドキ, オオブタクサ　giant ragweed, *Ambrosia trifida* L.
ぐんかんぶんさん　群間分散　between-group variance
ぐんしゅう　群集　community, association
ぐんしゅうせいたいがく　群集生態学, 群落生態学　community ecology, synecology
くんじょう　くん (燻) 蒸　fumigation
くんじょうざい　くん (燻) 蒸剤　fumigant
ぐんたい　群体　colony
ぐんないぶんさん　群内分散　within-group variance
ぐんらく　群落　community, stand
ぐんらくきゅうこうけいすう　群落吸光係数, 個体群吸光係数　canopy extinction coefficient
ぐんらくこうごうせい　群落光合成, 個体群光合成　canopy photosynthesis
ぐんらくせいたいがく　群落生態学, 群集生態学　synecology, community ecology
ぐんらくていこう　群落抵抗　canopy resistance
ぐんらくてきごうど　群落適合度, 適合度【生態】　fidelity
ケアリタソウ　Mexican tea, *Chenopodium ambrosioides* L.

[け]

けい　系, システム　system
けいかさいぼう　珪化細胞　silicified cell
ケイカル　珪カル⇒ケイ酸カルシウム　calcium silicate
けいかん　景観　landscape
けいかんかんがい　けい (畦) 間灌漑, うね (畝) 間灌漑　furrow irrigation
けいかんきょうそう　けい (畦) 間競争　row competition
けいけんてきモデル　経験的モデル　empirical model
けいこう　蛍光　fluorescence
けいこうけんびきょう　蛍光顕微鏡　fluorescence microscope
けいこうこうたいほう　蛍光抗体法　fluorescent antibody technique, 免疫蛍光法　immunofluorescence technique
けいこうしょうこう　蛍光消光　fluorescence quenching
けいこうどくせい　経口毒性　oral toxicity
けいざいてきしゅうりょう　経済的収量　economic yield
けいさいるい　茎菜類　stem vegetables
ケイさん　ケイ (珪) 酸　silicic acid
ケイさんカルシウム　ケイ酸カルシウム (珪カル)　calcium silicate
ケイさんしつひりょう　ケイ (珪) 酸質肥料　silicate fertilizer
けいさんたい　珪酸体　silica body
けいしつ　形質　character, trait,

characteristic
けいしつてんかん 形質転換, トランスフォーメーション transformation
けいしつてんかんしょくぶつ 形質転換植物 transgenic plant
けいしつどうにゅう 形質導入 transduction
けいしゃちのうぎょう 傾斜地農業 hillside farming
けいしゃばた 傾斜畑 hillside farm, sloping field
けいしょうど 軽しょう(鬆)土 light soil
けいず 系図 pedigree [chart]
けいせいそう 形成層 cambium (*pl.* cambia)
ケイそ ケイ(珪)素 silicon (Si)
ケイそうど ケイ(珪)藻土 diatomaceous earth
けいたいがく 形態学 morphology
けいたいけいしつ 形態形質 morphological character
けいたいけいせい 形態形成 morphogenesis
けいだいばいよう 継代培養 subculture
けいちょう 茎長 stem length
けいちょう 茎頂 shoot apex
けいちょうばいよう 茎頂培養 shoot apex culture
けいとう 系統 line, strain, pedigree, stock
けいとういくしゅほう 系統育種法 pedigree breeding method
けいとういくせい 系統育成 line breeding
けいとうがく 系統学 phylogeny
けいとうかんこうざつ 系統間交雑 blending
けいとうかんこうはい 系統間交配 line cross
けいとうけんてい 系統検定 line test
けいとうごさ 系統誤差 systematic error
けいとうじゅ 系統樹 genealogical tree, phylogenetic tree
けいとうしゅうだんせんばつ 系統集団選抜 pedigree mass selection
けいとうせんばつ 系統選抜 pedigree selection, line selection
けいとうてきおうせいけんていしけん 系統適応性検定試験 local adaptability test, test for regional adaptability
けいとうてき[ひょうほん]ちゅうしゅつ 系統的[標本]抽出 systematic sampling
けいとうはっせい 系統発生 phylogeny
けいとうぶんり 系統分離 line separation
けいとうほぞん 系統保存 preservation of line
けいないこんさく 畦内混作 intra-row intercropping
ケイヌビエ【イヌビエの変種】 cockspur grass, barnyard grass, *Echinochloa crus-galli* (L.) Beauv.
けいはん 畦畔 levee
けいひどくせい 経皮毒性 dermal toxicity
けいふ 系譜, 血統 pedigree
けいふん 鶏糞 chicken dropping, poultry (chicken) manure
けいふんきゅうひ 鶏糞きゅう(厩)肥 poultry manure
けいめんひふく 畦面被覆 mulching
けいようしりょう 茎葉飼料【畜産】 forage, fodder
けいよう[ぶ] 茎葉[部] shoot, foliage
けいりょうけいしつ 計量形質, 量的形質 quantitative character (trait)
けいりょうせいぶつがく 計量生物学, 生物測定学 biometrics, biometry
けいれつそうかん 系列相関 serial

correlation
けいろ　経路　pathway
けいろけいすう　経路係数　path coefficient
けいろず　経路図　path diagram
けいろぶんせき　経路分析, パス解析　path analysis
げきぶつ　劇物　deleterious substance
ケシ(芥子)　opium poppy, garden poppy, *Papaver somniferum* L.
ケシゆ　ケシ油　poppy seed oil
げじゅん　下旬　the last ten-days [of a month], the third ten-days [of a month]
けずりまき　削り播き　scrape sowing (seeding)
ケチョウセンアサガオ→アメリカチョウセンアサガオ
けっか　結果【果実】　fruiting, bearing
けっかし　結果枝　fruit-bearing branch (shoot), bearing branch (shoot)
けっかしゅうせい　結果習性　fruiting habit, bearing habit
けっかぶ　欠株　vacant hill, missing plant
けっきゅう　結球　1) bulb formation, bulbing【タマネギ等】2) head formation【葉菜類】
けっきょう　結莢　podding
ゲッケイジュ(月桂樹)　laurel, bay laurel, *Laurus nobilis* L.
けつごうがた　結合型　bound type
けつごうがたオーキシン　結合型オーキシン　bound auxin
けつごうすい　結合水　bound water, combined water
けつごうたんぱくしつ　結合タンパク質　binding protein
けつごうぶい　結合部位　binding site
けっさく　結さく(萠)　bolling
けっしつ　欠失【染色体】　deficiency, deletion
けつじつ　結実　seed-setting, fructification
けつじつりつ　結実率　seed-set percentage
けっしょ　結しょ(藷)　tuberization, tuber formation
けっしょう　結晶　crystal
けっしょうさいぼう　結晶細胞　crystal cell, crystalliferous cell
けっしょがた　結しょ(藷)型【サツマイモ】　root type
けっせいはんのう　血清反応　serological reaction
けっそく　結束　1) bundling, binding 2) tying【タバコ】
けっそくき　結束機, バインダ　binder
けっそくち　欠測値, 欠損値　missing value
けっそんち　欠損値, 欠測値　missing value
けっていけいすう　決定係数　coefficient of determination
けっていろんてきモデル　決定論的モデル　deterministic model
けっとう　血統, 系譜　pedigree
けつぼう　欠乏　deficiency
けつぼうしょう[じょう]　欠乏症[状]　deficiency symptom, hunger sign
ケツルアズキ(毛蔓小豆), ブラックマッペ　black gram, urd, black matpe, *Vigna mungo* (L.) Hepper (= *Phaseolus mungo* L.)
げつれい　月齢　age of the moon, moon's age
ケナフ　kenaf, ambari hemp, *Hibiscus cannabinus* L.
ゲノム　genome
ゲノムこうせい　ゲノム構成　genome constitution
ゲノムサイズ　genome size
ゲノムぶんせき　ゲノム分析　genome analysis
ケーパー　caper, *Capparis spinosa* L.
けみ　検見, 立毛調査(りつもうちょう

さ） stand observation
ケーラーしょうめいほう　ケーラー照明法　Köhler's illumination
ゲル　gel
ゲルシフトぶんせき　ゲルシフト分析　gel shift assay
ケルダールほう　ケルダール法　Kjeldahl method
ゲルろか　ゲルろ過　gel filtration
げんあつしんじゅんほう　減圧浸潤法　vacuum infiltration method
けんあつほう　検圧法　manometry
げんえき　原液　1) formulated concentrate 2) stock solution【培養液等】
けんか　堅果, 殻果　nut
げんかいおんど　限界温度　critical temperature
げんがいけんびきょう　限外顕微鏡　ultramicroscope
げんかいち　限界値　critical value
げんかいにっちょう　限界日長　critical daylength
げんかいのうど　限界濃度　critical concentration
げんがいろか　限外ろ(濾)過　ultrafiltration
けんかしょくぶつ　顕花植物　phanerogams
げんき　原基, 始原体　primordium (*pl.* primordia)
けんきせいしゅし　嫌気性種子　anaerobic seed
けんきてきこきゅう　嫌気的呼吸, 無気呼吸, 無酸素呼吸　anaerobic respiration
けんきてきじょうけん　嫌気的条件　anaerobic condition, anoxia
けんぎょうのうか　兼業農家　part-time farmer
ゲンゲ→レンゲソウ
げんけい　原型　prototype
げんけいしつ　原形質　protoplasm
げんけいしつたい　原形質体, プロトプラスト　protoplast
げんけいしつぶんり　原形質分離　plasmolysis
げんけいしつまく　原形質膜⇒細胞膜　cell membrane, cytoplasmic membrane
げんけいしつりゅうどう　原形質流動　protoplasmic streaming
げんけいしつれんらく　原形質連絡　plasmodesmata (*sing.* plasmodesma)
げんげんしゅ　原原種　foundation seed, breeder's stock
げんげんしゅほ　原原種圃　foundation seed farm, breeder's stock farm
けんこうせいしゅし　嫌光性種子⇒暗発芽種子　dark germinater, negative photoblastic seed
けんさとうきゅう　検査等級　inspection grade
けんさとうきゅうひょうじゅんまい　検査等級標準米　standard specimen of rice kernel
げんさんち　原産地　provenance, place of origin
けんし　絹糸【トウモロコシ】　silk
げんしきそたい　原色素体, プロプラスチド　proplastid
げんしきゅうこうぶんせき　原子吸光分析　atomic absorption analysis
けんしちゅうしゅつき　絹糸抽出期　silking stage
げんしばんごう　原子番号　atomic number
げんしゅ　原種　stock seed, registered seed, original seed, foundation stock
げんしゅう　減収　yield decrease
けんしゅつりょく　検出力【統計的検定】　power
げんしゅほ　原種圃　original seed farm, stock seed field
げんしょくせい　原植生　original vegetation

げんしりょう 原子量 atomic weight
げんすいしん 減水深 water requirement in depth, water loss in depth
げんすうぶんれつ 減数分裂 meiosis (*pl.* meioses)
げんすうぶんれつき 減数分裂期 meiosis stage, reduction division stage
げんせいいでん 限性遺伝 sex-limited inheritance
げんせいしぶ 原生篩部 protophloem
げんせいち 原生地 original habitat, native land
げんせいちゅうしんちゅう 原生中心柱 protostele
げんせいもくぶ 原生木部 protoxylem
けんせっかいしょくぶつ 嫌石灰植物 calciphobous plant, calcifuge plant
けんぜんせい 健全性 healthiness
けんぜんな 健全な, 無傷の intact
げんぞう 現像 development
げんぞうえき 現像液 developer
けんソバ 玄ソバ buckwheat seed
げんそん (げんぞん) りょう 現存量, バイオマス, 生物 [体] 量 standing crop, biomass
けんだくばいよう 懸濁培養, 浮遊培養 suspension culture
ケンタッキーブルーグラス Kentucky bluegrass, *Poa pratensis* L.
ゲンチアナ yellow gentian, *Gentiana lutea* L.
ゲンチアナバイオレット gentian violet
げんちしけん 現地試験 regional trial
げんちゅうしんちゅう 原中心柱 plerome
けんていこうざつ 検定交雑 test cross
けんていしょくぶつ 検定植物 test plant, assay plant
けんていとうけいりょう 検定統計量 test statistic
けんていほ 検定圃 test field

げんていよういん 限定要因, 制限因子 limiting factor
ケンディル kendyr, ツルカ turka, *Apocynum sibiricum* Jacq.
ケンドールのじゅんいそうかん ケンドールの順位相関 Kendall's rank correlation
げんばく 玄麦 husked barely
けんびかいほう 顕微解剖 microdissection
けんびかいほうき 顕微解剖器 micromanipulator
けんびきょう 顕微鏡 microscope
けんびきょうしゃしん 顕微鏡写真 microphotography
けんびぶんこうほう 顕微分光法 microspectrophotometry
げん (ぜん) ひょうひ 原 (前) 表皮 protoderm
けんびょう 健苗 good seedling
げんまい 玄米 brown rice, hulled rice, husked rice
けんままい 研磨米 polished rice
げんめん 原綿, 生綿 (きわた) raw cotton

[こ]

コアカザ figleaved goosefoot, *Chenopodium ficifolium* Smith (= *C. serotinum* L.)
コアゼガヤツリ *Cyperus haspan* L.
コアマチャ (小甘茶) *Hydrangea serrata* (Thunb. ex Murray) Ser. var. *thunbergii* (Sieb.) H.Ohba (= *H. macrophylla* Ser. var. *thunbergii* Makino)
コアワ small foxtail millet, *Setaria italica* (L.) Beauv. var. *germanicum* Trin.
コイチゴツナギ→カナダブルーグラス
こいも 子いも 1) daughter tuberous root【根】 2) daughter tuber【地下茎】

コイル【ココヤシ】 coir
こう 綱【分類】 class
こうあつめっきんき 高圧滅菌器, オートクレーブ autoclave
こういきてきおうせい 広域適応性 wide [regional] adaptability
こういしゅうかくでん 高位収穫田 high-yielding paddy field
こういせいさんち 高位生産地 high-productivity land (area)
こういぶんげつ 高位分げつ upper nodal tiller, tillers at high nodal position
こううにっすう 降雨日数 number of rainy days
こううりょう 降雨量⇒雨量 amount of rainfall
こううん 耕うん(耘) tillage, tilling
こううんき 耕うん(耘)機 power tiller
こうオーキシン 抗オーキシン, アンチオーキシン antiauxin
こうオスミウムかりゅう 好オスミウム顆粒 osmiophilic globule
こうおんき 恒温器, 定温器 incubator
こうおんしょうがい 高温障害 heat damage
こうか 硬化, 順化, ハードニング【培養】 hardening
こうがい 公害 public hazards, environmental pollution
こうがい 鉱害 mining (mine) pollution
コウガイゼキショウ *Juncus prismatocarpus* R. Br. (= *J. leschenaultii* Gay)
こうかがくオキシダント 光化学オキシダント photochemical oxidant
こうかがくけい 光化学系 photosystem, photochemical system
こうかがくスモッグ 光化学スモッグ photochemical smog
こうかがくはんのう 光化学反応 photochemical reaction
こうがくけんびきょう 光学顕微鏡 light microscope
こうかくさいぼう 厚角細胞 collenchyma cell, collenchymatous cell
こうがくセンサー 光学センサー optical sensor
こうかくそしき 厚角組織 collenchyma
こうかとくせい 硬化特性 hardening property
こうかんさんど 交換酸度 exchange acidity
こうかんせいえんき 交換性塩基 exchangeable base
こうかんせいようイオン 交換性陽イオン exchangeable cation
こうかんぶんごう 交換分合 land consolidation
こうかんようりょう 交換容量 exchange capacity
こうき 後期【細胞分裂】 anaphase
こうき 耕起 plowing
こうき 香気, 香り aroma, scent, fragrance
こうきさいど 耕起砕土, 耕は(耙) plowing and harrowing
コウキシタン（紅木紫檀） red sandalwood, red sanders, *Pterocarpus santalinus* L. f.
こうきせだい 後期世代 advanced generation
こうきち 耕起地 tilth
こうきついひ 後期追肥 topdressing at a later growth stage
こうきてき 好気的 aerobic
こうきてき 向基的⇒求基的 basipetal
こうきてきこきゅう 好気的呼吸, 酸素呼吸, 有気呼吸 aerobic respiration
こうきてきじょうけん 好気的条件 aerobic condition
コウキヤガラ *Scirpus planiculmis* Fr.

こうげいさくもつ 工芸作物 industrial crop
こうけいしつ 後形質 metaplasm
こうげき 孔げき(隙) pore space, void
こうげきりつ 孔げき(隙)率 porosity
こうけっせい 抗血清 antiserum
こうげん 抗原 antigen
こうごうせい 光合成 photosynthesis
こうごうせいかっせい 光合成活性 photosynthetic activity
こうごうせいきかん 光合成器官 photosynthetic organ
こうごうせいさんぶつ 光合成産物 photosynthate, photosynthetic product
こうごうせいしきそ 光合成色素 photosynthetic pigment
こうごうせいしゅし 好光性種子⇒光発芽種子 light germinater, photoblastic seed
こうごうせいそくど 光合成速度 photosynthetic rate
こうごうせいそしき 光合成組織 photosynthetic tissue
こうごうせいてきリンさんか 光合成的リン酸化 photosynthetic phosphorylation, 光リン酸化 photophosphorylation
こうごうせいのうりょく 光合成能力 photosynthetic capacity, photosynthetic ability
こうごうせいゆうこうこうりょうしそくみつど 光合成有効光量子束密度 photosynthetic photon flux density (PPFD)
こうごうせいゆうこうほうしゃ 光合成有効放射 photosynthetically active radiation (PAR)
こうごうせいゆうこうほうしゃきゅうしゅうりつ 光合成有効放射吸収率 fraction of photosynthetically active radiation absorbed by a canopy (fAPAR)
こう(ひかり)こきゅう 光呼吸 photorespiration
こうごさく 交互作 alternating cropping
こうごさよう 交互作用【統計】 interaction
こうこてい 後固定 post-fixation
こうこん 梗根, ごぼう根【サツマイモ】pencil-like root, cylindrical root
こうさ 交さ(又), 乗換え crossing-over
こうさい 鉱さい(滓), スラグ slag
こうさか 交さ(又)価, 乗換え価 crossing-over value
こうさく 耕作 cultivation
こうざつ 交雑 hybridization, cross, crossing
こうざついくしゅ 交雑育種 hybridization breeding, cross breeding
こうざつくみあわせ 交雑組合せ, 交配組合せ cross combination
こうざつのうりょく 交雑能力 crossability
こうざつふねんぐん 交雑不稔群, 交配不稔群 cross sterile group
こうざつふわごうせい 交雑不和合性, 交配不和合性 cross incompatibility
こうざつりつ 交雑率 crossing rate
こうざつわごうせい 交雑和合性, 交配和合性 cross compatibility
こうさよう 後作用 aftereffect
こうしがたせっけい 格子型設計 lattice design
ごうしき 合糸期, ザイゴテン期【減数分裂】zygotene stage
こうじく[がわ]の 向軸[側]の【植物】adaxial
こうじつ 硬実 hard seed
こうしつあきまきあかコムギ 硬質秋播赤コムギ hard red winter wheat
こうしつこむぎ 硬質コムギ hard wheat

こうしつせんい　硬質繊維　hard fiber
こうしつデンプンさいぼう　硬質デンプン細胞【コムギ】　hard starch cell
こうしつふん　硬質粉　hard flour
こうしつまい　硬質米　hard-textured rice
こうしつりゅう　硬質粒　vitreous grains
こうじぶんげつ　高次分げつ　high order tiller
こうしゅ　耕種　crop cultivation
こうしゅうき　光周期　photoperiod
こうしゅうせい　光周性　photoperiodism
こうしゅうせいひんしゅ　高収性品種, 多収性品種　high-yielding variety
こうしゅうりつ　光周律⇒光周性　photoperiodism
こうしゅがいよう　耕種概要　cropping directory (manual)
こうじゅく　後熟　afterripening
こうしゅこうがい　耕種梗概⇒耕種概要　cropping directory (manual)
こうしゅてきざっそうぼうじょ　耕種的雑草防除　cultural weed control
こうしゅほう　耕種法　cultivation method
こうじょ　耕鋤⇒耕うん（耘）　tillage, tilling
こうしん　更新　1) regeneration, renewal【チャ, クワ】2) renovation【草地など】
こうしんりょう　香辛料　spice, condiment
こうしんりょうさくもつ　香辛料作物　spice crop
こうすいりょう　降水量　[amount of] precipitation
ごうせいオーキシン　合成オーキシン　synthetic auxin
ごうせいかいこうレーダ　合成開口レーダ　synthetic aperture radar (SAR), side-looking radar

ごうせいき　合成期　synthetic phase, S期　S phase
こうせいこきゅう　構成呼吸, 成長呼吸　growth respiration
こうせいしぶ　後生篩部　metaphloem
ごうせいしゅ　合成種　synthetic species, synthetic breed, synthetic strain
こう（ひかり）せいちょうはんのう　光成長反応　light growth reaction
こうせいのうえきたいクロマトグラフィー　高性能液体クロマトグラフィー　high performance liquid chromatography (HPLC), 高速液体クロマトグラフィー　high speed liquid chromatography
ごうせいひんしゅ　合成品種　synthetic variety
こうせいぶっしつ　抗生物質　antibiotics
こうせいもくぶ　後生木部　metaxylem
こうせきそう　洪積層　diluvium
こうせきち　洪積地　diluvial land
こうせきど　洪積土　diluvial soil
こうせっかいしょくぶつ　好石灰植物, カルシウム植物　calciphilous plant, calcicole plant
こうそ　酵素　enzyme
コウゾ（楮）　paper mulberry, *Broussonetia kazinoki* Sieb.
こうぞういでんし　構造遺伝子　structural gene
こうぞうしき　構造式　structural formula
こうぞうせいたんすいかぶつ　構造性炭水化物　structural carbohydrate
こうぞうタンパクしつ　構造タンパク質　structural protein
こうぞうてきざっしゅ　構造的雑種　structural hybrid
こうそかっせい　酵素活性　enzyme activity
こうそくえきたいクロマトグラフィー

高速液体クロマトグラフィー　high speed liquid chromatography, 高性能液体クロマトグラフィー　high performance liquid chromatography (HPLC)
こうそめんえきそくていほう　酵素免疫測定法　enzyme immunoassay (EIA)
こうたい　抗体　antibody
こうだい　後代　progeny, offspring, descendant
こうだいけんてい　後代検定　progeny test
こうタンパクしつまい　高タンパク質米　high protein rice
こうち　耕地　cultivated land, arable land
こうちざっそう　耕地雑草　arable land weed
こうちどじょう　耕地土壌, 耕土　cultivated soil, arable soil, topsoil
こうちのうぎょう　高地農業　highland agriculture
こうちはくとう　耕地白糖　plantation white sugar
こうちゃ　紅茶　black tea
こう (ひかり) ちゅうだん　光中断　light interruption, light break
こうちょうてき　向頂的⇒求頂的 acropetal
こうちりつ　耕地率　arable land rate
こうちりようりつ　耕地利用率　utilization rate of arable land
ごうてん　合点, カラザ　chalaza
こうど　硬度【土壌・水】　hardness
こうど　黄土, レス　loess
こうど　耕土, 耕土土壌　cultivated soil, arable soil, topsoil
こうど　高度　altitude, elevation
こうどかせいひりょう　高度化成肥料　high-analysis mixed fertilizer
こうどく　鉱毒　mine pollutant, mineral pollutant
こうどばいよう　耕土培養　improvement of soil fertility
こうにゅうしりょう　購入飼料　purchased feed
こうにゅうひりょう　購入肥料, 金肥, 販売肥料　commercial fertilizer
こうは　耕は (耙), 耕起砕土　plowing and harrowing
こうはい　交配　mating, cross, crossing
こうはいくみあわせ　交配組合せ, 交雑組合せ　cross combination
こうはいち　荒廃地, 荒れ地　wasteland, barrens
こうはいふねんぐん　交配不稔群, 交雑不稔群　cross sterile group
こうはいふわごうせい　交配不和合性, 交雑不和合性　cross incompatibility
こうはいぼん　交配母本　parent
こうはいわごうせい　交配和合性, 交雑和合性　cross compatibility
こう (ひかり) はつが　光発芽　light germination
こう (ひかり) はつがしゅし　光発芽種子　light germinater, photoblastic seed
こうばん　硬盤　hardpan
こうばん　耕盤　plow sole, plow pan, furrow pan
こうはんいさんざいはんぷくはいれつ　広範囲散在反復配列　long interspersed repetitive sequence (LINE)
こうはんいていこうせい　広範囲抵抗性　broad-spectrum resistance
こう (ひかり) ぶんかい　光分解　photodecomposition, photolysis
こうへきさいぼう　厚壁細胞　sclerenchyma cell, sclerenchymatous cell
こうへきそしき　厚壁組織　sclerenchyma
こうへんさいぼう　孔辺細胞　guard cell
こうぼ　酵母　yeast

こう(ひかり)ほうわ　光飽和　light saturation
こうぼく　高木　tree, arbor tree
こうぼじんこうせんしょくたいベクター　酵母人工染色体ベクター　yeast artificial chromosome (YAC) vector
コウマ(黄麻)→ジュート
こうまくさいぼう　厚膜細胞⇒厚壁細胞　sclerenchyma cell, sclerenchymatous cell
こうまくそしき　厚膜組織⇒厚壁組織　sclerenchyma
こうまふ　黄麻布, ヘシアンクロース　Hessian cloth
ごうもう　剛毛　bristle
こうよう　紅葉　red coloring of leaves
こうようざっそう　広葉雑草　broadleaved weed, broadleaf weed
こうようじゅ　硬葉樹　sclerophyllous tree
こうようじゅ　広葉樹　broad-leaved tree, broadleaf tree
こうらく　交絡【統計】　confounding
こうりゅう　硬粒　hard grain
こうりゅうぶんぱい　交流分配　counter-current distribution
こうりょうさくもつ　香料作物　aromatic crop
こうりょうし　光量子　photon, quantum
こうりょうしけい　光量子計　quantum meter
こうりょうしそくみつど　光量子束密度　photon flux density (PFD)
こうりょうゼラニウム　香料ゼラニウム→ゼラニウム
こうリンさんか　光リン酸化　photophosphorylation, 光合成的リン酸化　photosynthetic phosphorylation
こうれいち　高冷地　cool-climate highland
こえ　肥　manure
ごえい　護穎　glume

こえおけ　肥桶　nightson pail, night-soil pail
こえぎれ　肥切れ　nutrient deficiency (hunger)
こえだめ　肥溜　manure pool, night-soil reservoir
こえやけ　肥焼け　fertilizer injury, burning
コエンドロ　coriander, *Coriandrum sativum* L.
コオニタビラコ　*Lapsana apogonoides* Maxim.
こか　糊化　gelatinization
こかおんど　糊化温度　gelatinization temperature
こかとくせい　糊化特性　gelatinization property
コカナダモ　western elodea, *Elodea nuttallii* (Planch.) St. John
コガネバナ　Baical skullcap, *Scutellaria baicalensis* Georgi
コガマ　*Typha orientalis* Presl
こきばし　こ(扱)き箸　threshing sticks
こきゅう　呼吸　respiration
こきゅうかっせい　呼吸活性　respiratory activity
こきゅうこん　呼吸根　respiratory root
こきゅうしょう　呼吸商, 呼吸率　respiratory quotient (RQ)
こきゅうそくど　呼吸速度　respiratory rate, respiration rate
こきゅうりつ　呼吸率, 呼吸商　respiratory quotient (RQ)
ごくおくて(ごくばんせい)ひんしゅ　極晩生品種　extremely late [maturing] variety
こくさいアグロフォレストリーけんきゅうセンター　国際アグロフォレストリー研究センター　International Council for Research in Agroforestry (ICRAF)
こくさいいねけんきゅうしょ　国際稲研

究所　International Rice Research Institute (IRRI)
こくさいかんそうちのうぎょうけんきゅうセンター　国際乾燥地農業研究センター　International Center for Agricultural Research in the Dry Areas (ICARDA)
こくさいきょうりょくじぎょうだん　国際協力機構(旧国際協力事業団)　Japan International Cooperative Agency (JICA)
こくさいこんちゅうせいりせいたいセンター　国際昆虫生理生態センター　International Center of Insect Physiology and Ecology (ICIPE)
こくさいしょくぶついでんしげんけんきゅうしょ　国際植物遺伝資源研究所　International Plant Genetic Resources Institute (IPGRI)
こくさいしょくりょうせいさくけんきゅうしょ　国際食糧政策研究所　International Food Policy Research Institute (IFPRI)
こくさいすいさんしげんかんりセンター　国際水産資源管理センター　International Center for Living Aquatic Resources Management (ICLAM)
こくさいちくさんけんきゅうしょ　国際畜産研究所　International Livestock Research Institute (ILRI)
こくさいトウモロコシ・コムギかいりょうセンター　国際トウモロコシ・コムギ改良センター　International Maize and Wheat Improvement Center, Centro Internacional de Mejoramiento de Maiz y Trigo (CIMMYT)
こくさいどじょうけんきゅうかんりひょうぎかい　国際土壌研究・管理評議会　International Board for Soil Research and Management (IBSRAM)
こくさいねったいのうぎょうけんきゅうしょ　国際熱帯農業研究所　International Institute of Tropical Agriculture (IITA)
こくさいねったいのうぎょうけんきゅうセンター　国際熱帯農業研究センター　International Center for Tropical Agriculture, Centro Internacional de Agricultura Tropical (CIAT)
こくさいのうぎょうけんきゅうきょうぎグループ　国際農業研究協議グループ　Consultative Group on International Agricultural Research (CGIAR)
こくさいのうりんすいさんぎょうけんきゅうセンター　国際農林水産業研究センター　Japan International Research Center for Agricultural Sciences (JIRCAS)
こくさいバレイショセンター　国際バレイショセンター　International Potato Center, Centro Internacional de Papa (CIP)
こくさいはんかんそうねったいさくもつけんきゅうしょ　国際半乾燥熱帯作物研究所　International Crop Research Institute for the Semi-Arid Tropics (ICRISAT)
こくさいひりょうかいはつセンター　国際肥料開発センター　International Fertilizer Development Center (IFDC)
こくさいみずかんりけんきゅうしょ　国際水管理研究所　International Water Management Institute (IWMI)
こくさいりんぎょうけんきゅうセンター　国際林業研究センター　Center for International Forestry Research (CIFOR)
こくさいれんごうしょくりょうのうぎょうきかん　国際連合食糧農業機関　Food and Agriculture Organization of the United Nations (FAO)
こくさく　穀作　cereal cropping, cereal (grain) production

コクサギス→ゴムタンポポ
こくじつ　穀実　grain, kernel
こくじつしゅうりょう　穀実収量, 子実収量　grain yield
こくしょくきゅうかん　黒色休閑⇒裸地休閑　bare fallow
こくしょくど　黒色土　Black soil
こくそう　穀倉　granary
こくそうしきのうほう　穀草式農法　convertible husbandry, alternate husbandry, ley farming
こくでいど　黒泥土　Muck soil
こくとう　黒稲　black rice
こくどすうちじょうほう　国土数値情報　digital national land information
こくないさんまい　国内産米　domestic rice
こくないしょうひ　国内消費　domestic consumption
こくないせいさん　国内生産　domestic production
ごくばんせい (ごくおくて) ひんしゅ　極晩生品種　extremely late [maturing] variety
こくはんびょう　黒斑病【サツマイモ】　black rot
こくひ　穀皮　hull, husk
こくもつ　穀物　grain crops, cereals
こくもつせいさん　穀物生産　cereal production, cereal cropping
こくもつねんど　穀物年度　crop year
こくりゅう　穀粒　kernel, grain
こくりゅうけいすうき　穀粒計数機　grain counter
こくりゅうせつだんき　穀粒切断機　grain cutter
こくりゅうせんべつき　穀粒選別機　grain separator
こくりゅうひんしつ　穀粒品質　grain quality
こくるい　穀類　cereals, cereal crops
ごくわせひんしゅ　極早生品種　extremely early [maturing] variety

こけいしりょう　固形飼料, ペレット　pellet
こけいばいち　固形培地　solid medium (*pl.* media)
こけいばいちこう　固形培地耕　substrate culture, solid medium culture, aggregate culture
こけいひりょう　固形肥料　solid fertilizer
ココア→カカオ
コゴメガヤツリ　rice flatsedge, *Cyperus iria* L.
ココヤシ (ココ椰子)　coconut palm, *Cocos nucifera* L.
ココヤシゆ　ココヤシ油, ヤシ油　coconut oil, copra oil
ごさ　誤差　error
こさく　小作　tenant farming
こさくにん　小作人　tenant
こさくのう　小作農　tenant farmer
ごさのへいきんへいほう　誤差の平均平方⇒残差平均平方　residual mean-square
ごさのへいほうわ　誤差の平方和⇒残差平方和　residual sum of squares
ごさぶんさん　誤差分散　error variance
こし　枯死　death, dying, plant death
こじゅく　糊熟　dough ripe
こじゅく　枯熟　dead ripe
こじゅくき　糊熟期　dough[-ripe] stage
こじゅくき　枯熟期　dead-ripe stage
こしゅし　古種子　aged seed
ゴシュユ (呉茱萸)　*Euodia ruticarpa* (Juss.) Benth.
コショウ (胡椒)　pepper, *Piper nigrum* L.
こじょうしたて　弧状仕立て【チャ】　arc-shaped bush formation
コスズメノチャヒキ→スムーズブロムグラス
コストかんすう　コスト関数　cost function
コスミドベクター　cosmid vector

こせいたいがく 個生態学 idioecology, 種生態学 autecology, species ecology
ごせいの 互生の alternate
こそう 固相 solid phase
こそうこうそめんえきけんていほう 固相酵素免疫検定法 enzyme-linked immunosorbent assay (ELISA)
こたい 個体 individual
こたいぐん 個体群, 集団 population
こたいぐんきゅうこうけいすう 個体群吸光係数, 群落吸光係数 canopy extinction coefficient
こたいぐんこうごうせい 個体群光合成, 群落光合成 canopy photosynthesis
こたいぐんこうぞう 個体群構造, 草冠構造 canopy structure, canopy architecture
こたいぐんせいたいがく 個体群生態学 population ecology
こたいぐんせいちょうそくど 個体群成長速度 crop growth rate (CGR)
こたいぐんみつど 個体群密度, 集団密度 population density
こたいけんてい 個体検定 individual test
こたいせんばつ 個体選抜 individual selection
こたいはっせい 個体発生 ontogenesis, ontogeny
こたいへんい 個体変異 individual variation
ごたんとう 五炭糖, ペントース pentose
コックこうか コック効果 Kok effect
こっぷん 骨粉【肥料】 bone meal
こてい 固定 fixation
こていえき 固定液 fixative, fixative solution
こていけいとう 固定系統 fixed line
こてい[こうか]もけい 固定[効果]模型, 定数[効果]模型 fixed [effects] model
コド→コドラ
こどこ 子床【タバコ】 secondary bed
コドラ Kodo millet, *Paspalum scrobiculatum* L.
コドラート, 方形区, わく(枠) quadrat
コードりょういき コード領域 coding region
コドン codon
コナギ monochoria, *Monochoria vaginalis* (Burm.f.) C. Presl
コナスビ *Lysimachia japonica* Thunb.
コニシキソウ prostrate spurge, *Euphorbia supina* Raf.
コヌカグサ→レッドトップ
ごばいたい 五倍体 pentaploid
コハクさん コハク酸 succinic acid
コハコベ→ハコベ
コーパス⇒内体 corpus
ごはんべつ 誤判別 misclassification
コーヒー coffee, *Coffea* spp.
コーヒーまめ コーヒー豆 coffee bean
コヒメビエ→ワセビエ
コヒルガオ Japanese bindweed, *Calystegia hederacea* Wall.
ごぶづき[せい]まい 五分づき[精]米 half-polished rice
コブナグサ jointhead arthraxon, *Arthraxon hispidus* (Thunb.) Makino
コプラ copra
こふんそう 糊粉層 aleurone layer
こふんりゅう 糊粉粒 aleurone grain
ゴボウ(牛蒡) edible burdock, *Arctium lappa* L.
ごぼうね ごぼう根, 梗根(こうこん)【サツマイモ】 pencil-like root, cylindrical root
ゴマ(胡麻) sesame, gingelly, *Sesamum indicum* L.
こまい 古米 old [crop] rice
ゴマかす ゴマ粕 sesame meal
こまごめピペット 駒込ピペット dropping pipet, Pasteur pipet

ごまはがれびょう　ごま(胡麻)葉枯病　1)helminthosporium leaf spot, brown spot【イネ】 2) leaf spot【トウモロコシ, アワ】
ゴマゆ　ゴマ油　sesame oil
コミカンソウ　*Phyllanthus urinaria* L.
こみにしたすいていりょう　こみにした推定量　pooled estimator
ゴム　gum, rubber
コムギ(小麦)　wheat, *Triticum* spp.
コムギこ　小麦粉　wheat flour
コムギそさいふん　コムギ粗砕粉　wheat meals
コムギデンプン　小麦デンプン　wheat starch
ゴムタンポポ　Russian dandelion, kok-saghyz, *Taraxacum kok-saghyz* L.E. Rodin
ゴムりょうさくもつ　ゴム料作物　rubber crop
こめ　米　rice
こめだわら　米俵　1) straw rice bag【表装のみ】 2) straw rice bale【内容を含む】
コメツブウマゴヤシ→ブラックメディック
こめデンプン　米デンプン　rice starch
こめぬか　米ぬか(糠)　rice bran
こめぬかゆ　米ぬか(糠)油　rice bran oil
コモチマンネングサ　*Sedum bulbiferum* Makino
コモンベッチ　common vetch, *Vicia sativa* L.
こゆうち　固有値　eigenvalue
こゆうベクトル　固有ベクトル　eigenvector
こよう　個葉　single leaf
ゴヨウドコロ　five-leaved yam, *Dioscorea pentaphylla* L.
コーラ　cola, kola, *Cola nitida* (Vent.) Schott et Endl.
コリアンダー→コエンドロ
コリヤナギ(杞柳)　osier, *Salix koriyanagi* Kimura (= *S. purpurea* L. var. *multinervis* Fr. et Sav.)
こりょうさくもつ　糊料作物　paste crop
コルク　cork
コルクか　コルク化, スベリン化　suberization
コルクけいせいそう　コルク形成層　cork cambium, phellogen
コルクそしき　コルク組織　cork tissue, phellem
ゴルジたい　ゴルジ体　Golgi body, Golgi apparatus
コルチカム→イヌサフラン
コルヒチン　colchicine
コルメラ　columella
コルモゴロフ-スミルノフけんてい　コルモゴロフ-スミルノフ検定　Kolmogorov-Smirnov test
ゴレンシ(五斂子)　carambola, star fruit, *Averrhoa carambola* L.
コロニアルベントグラス　colonial bentgrass, *Agrostis capillaris* L. (= *A. tenuis* Sibth.)
コロニー　colony
ころびなえ　転び苗【イネ】　turned down seedling
こわは　こわ葉【チャ】　over-matured leaf for plucking
こんあつ　根圧　root pressure
こんいき　根域　rooting zone
こんかん　根冠　root cap
こんぐん　根群⇒根系　root system
こんけい　根系　root system
こんけい　根茎　rhizome
こんけい　混系　mixed line
こんけん　根圏　rhizosphere
こんごういくしゅほう　混合育種法　bulk method
こんごうじゅふん　混合受(授)粉　mass pollination, mixed pollination
こんごうもけい　混合模型【統計】

mixed model
コンゴーコーヒー→ロブスタコーヒー
こんさい　根菜　root crop, root vegetable
こんさいるい　根菜類　root vegetables
こんさく　混作　mixed cropping
こんじく　根軸　root axis
コンシステンシー【土壌】　consistency
こんしゅ　混種　seed contamination
こんしゅつよう　根出葉,根生葉　radical leaf
こんしょう　根鞘　coleorhiza
こんしょく　混植　companion planting, mixed planting
こんすうせい　混数性,混倍数性　mixoploidy
こんせいざっそう　根生雑草　root weed
こんせいよう　根生葉,根出葉　radical leaf
こんぜつ　根絶　eradication
こんそうこう　混層耕　layer-mixing tillage
コンダクタンス,伝導度　conductance
こんたん　根端　root apex, root tip
こんたんぶんれつそしき　根端分裂組織　root apical meristem
こんちゅうがく　昆虫学　entomology
こんちゅうでんぱん　昆虫伝搬　insect transmission
こんちゅうでんぱんウイルス　昆虫伝搬ウイルス　insect-borne virus, insect-transmitted virus
こんちょうこんじゅうひ　根長/根重比　specific root length
こんちょうみつど　根長密度　root length density, rooting density
コンデンサー,集光器【顕微鏡】　condenser
コンニャク(蒟蒻)　konjak, elephant foot, *Amorphophallus konjac* K. Koch
コンニャクだま　コンニャク玉　corm of konjak

コンニャクマンナン　konjak mannan
こんぱ(こんぱん)　混播　mix-sowing (-seeding), mixed sowing (seeding)
こんばいすうせい　混倍数性,混数性　mixoploidy
コンバイン　combine harvester
こんぱ[ん]そうち　混播草地　mixed sward, mixed pasture
コンピュータ,電子計算機　computer
コンピュータグラフィクス　computer graphics
コンピュータネットワーク　computer network
コンピュータビジョン　computer vision
コーンフレーク　corn flakes
こんぼくりん　混牧林,林内草地　woodland pasture, wood-pasture, grazable forestland
こんめん　根面　rhizoplane, root-soil interface
こんもう　根毛　root hair
コーンゆ　コーン油⇒トウモロコシ油　corn oil
こんりゅう　根粒　root nodule
こんりゅうきん　根粒菌　root-nodule bacteria, Rhizobia
こんりゅうひちゃくせいけいとう　根粒非着生系統【ダイズ】　non[-]nodulating line
こんりょう　根量　root mass

[さ]

ざい　材　wood
さいか　催花　flower induction
さいが　催芽　hastening of germination, forcing of sprouting
さいがしゅし　催芽種子　pregerminated seed
さいきん　細菌　bacteria (*sing.* bacterium)
さいきんりんこうか　最近隣効果　nearest neighborhood effect

サイクリン　cyclin
サイクロメータ, 乾湿計　psychrometer
さいげんきかん　再現期間　return period
さいこうおんど　最高温度　maximum temperature
さいこうさいていおんどけい　最高最低温度計　maximun and minimum thermometer
さいこうぶんげつき　最高分げつ期　maximum tiller number stage
サイコセル　cycocel (chlorocholine chloride (CCC), 2-chloroethyltrimethyl-ammonium chloride)
ザイゴテンき　ザイゴテン期, 合糸期【減数分裂】　zygotene stage
さいこん　細根　1) rootlet, fine root　2) fibrous root, feeder root【チャ】
さいさ　細砂　fine sand
サイザル　sisal, *Agave sisalana* Perr. ex Engelm.
さいしゅ　採種　seed production
さいしゅうさんぶつ　最終産物　end product
さいしゅうさんぶつそがい　最終産物阻害　end product inhibition
さいしゅさいばい　採種栽培　seed production culture, seed growing
さいしゅほ　採種圃　seed [production] farm, propagation farm
さいしょうにじょうすいていりょう　最小二乗推定量　least-squares estimator
さいしょうにじょうほう　最小二乗法　least-squares method
さいしょうゆういさ　最小有意差　least significant difference (LSD)
さいしょうりつ　最少律【植物栄養】　law of minimum
さいしょく　採食　1) browsing【木本の小枝, 葉, 芽など】　2) grazing【草本の茎葉など】
さいしょく　栽植　planting
さいしょくきょり　栽植距離　plant spacing, planting distance
さいしょくじき　栽植時期　planting time
さいしょく[そう]りょう　採食[草]量, 食草量　herbage intake, forage intake, grazing intake
さいしょくみつど　栽植密度　planting density, planting rate
さいしょくようしき　栽植様式　planting pattern, spacial arrangement [of plants]
さいせい　再生　1) regeneration　2) regrowth【草地】
さいせいけいとう　再生系統　recovered line
さいせいこたい　再生個体　regenerated plant
さいせいさん　再生産　reproduction
さいせいそう　再生草【牧草】　aftermath
さいせいちょう　再成長　regrowth
さいせいりょく　再生力　regrowth vigor
さいそうち　採草地　meadow
さいだいげんかいしゅうりょう　最大限界収量　maximum potential yield
さいだいようすいりょう　最大容水量, 飽和水分量　maximum water holding capacity
さいていおんど　最低温度　minimum temperature
さいてきおんど　最適温度, 適温　optimum temperature
さいてきち　最適値　optimum value
さいてきのうど　最適濃度　optimum concentration
さいてきようめんせきしすう　最適葉面積指数　optimum leaf area index
さいど　砕土　harrowing, pulverization
さいど　細土　fine earth, fine soil

サイトウ (菜豆) →インゲンマメ
サイドオートグラマ　side-oats grama, *Bouteloua curtipendula* (Michx.) Torr.
サイトカイニン　cytokinin
さいどき　砕土機⇒ハロー　harrow, clod crusher, pulverizer
さいは　再播　reseeding, resowing
さいばい　栽培　plant husbandry, cultivation
さいばいいちりゅうけいコムギ　栽培一粒系コムギ　cultivated einkorn wheat, *Triticum monococcum* L.
さいばいか　栽培化　domestication
さいばいがく　栽培学　crop husbandry
さいばいがた　栽培型　cultivation type
さいばいぎじゅつ　栽培技術　cultivation technique
さいばいげんかい　栽培限界　cultivation limit
さいばいしけん　栽培試験　cultivation experiment, cultivation test
さいばいしゅ　栽培種　cultivated species
さいばいたいけい　栽培体系　cultivation system
さいばいチモフェービけいコムギ　栽培チモフェービ系コムギ→チモフェービコムギ
さいばいひんしゅ　栽培品種　cultivar (cv.), cultivated variety
さいばいめんせき　栽培面積　growing area
さいびょう　採苗　pulling of seedling
さいひんち　最頻値, モード　mode
さいぶつだい　載物台⇒ステージ【顕微鏡】　stage
サイブリッド　cybrid, 細胞質雑種　cytoplasmic hybrid
さいぶんか　再分化　redifferentiation
さいぼう　細胞　cell
さいぼういでんがく　細胞遺伝学　cytogenetics
さいぼうえき　細胞液　cell sap
さいぼうかがく　細胞化学　cytochemistry
さいぼうがく　細胞学　cytology
さいぼうかくだい　細胞拡大　cell expansion
さいぼうかんげき　細胞間隙　intercellular space
さいぼうかんげきにさんかたんそのうど　細胞間隙二酸化炭素濃度　intercellular carbon dioxide concentration (Ci)
さいぼうけいふ　細胞系譜　cell lineage
さいぼうこうがく　細胞工学　cell engineering, cell technology
さいぼうこっかく　細胞骨格　cytoskeleton
さいぼうしつ　細胞質　cytoplasm
さいぼうしついでん　細胞質遺伝　cytoplasmic inheritance
さいぼうしついでんし　細胞質遺伝子　cytoplasmic gene, cytogene, plasmagene
さいぼうしつざっしゅ　細胞質雑種　cytoplasmic hybrid, サイブリッド　cybrid
さいぼうしつとつぜんへんい　細胞質突然変異　cytoplasmic mutation
さいぼうしつぶんれつ　細胞質分裂　cytokinesis
さいぼうしつゆうせいふねん　細胞質雄性不稔　cytoplasmic male sterility
さいぼうしゅうき　細胞周期　cell cycle
さいぼう[しょう]きかん　細胞[小]器官, オルガネラ　organelle
さいぼうしんちょう　細胞伸長　cell elongation
さいぼうせいぶつがく　細胞生物学　cell biology
さいぼうないきょうせい　細胞内共生　endosymbiosis
さいぼうないようぶつ　細胞内容物　cell contents, cellular contents

さいぼうばいよう　細胞培養　cell culture
さいぼうばん　細胞板　cell plate
さいぼうぶんか　細胞分化　cell differentiation
さいぼうぶんかくほう　細胞分画法　cell fractionation
さいぼうぶんれつ　細胞分裂　cell division
さいぼうへき　細胞壁　cell wall
さいぼうへき [こうせい] ぶっしつ　細胞壁 [構成] 物質, 細胞壁成分　cell wall constituents
さいぼうへきしんてんせい　細胞壁伸展性　cell wall extensibility
さいぼうへきせいぶん　細胞壁成分, 細胞壁 [構成] 物質　cell wall constituents
さいぼうまく　細胞膜　cell membrane, cytoplasmic membrane
さいぼうゆうごう　細胞融合　cell fusion
さいまい　砕米, 砕き米　broken rice
さいみゃく　細脈　small vein, veinlet
さいゆうすいていりょう　最尤推定量　maximum-likelihood estimator
さいゆうほう　最尤法　maximum-likelihood method
ざいらいしゅ　在来種　1)native species 2)【タバコ】domestic tobacco
ざいらいしゅナタネ　在来種ナタネ　rape, *Brassica campestris* L. (= *B. rapa* L. var. *campestris* (L.) Clapham)
ざいらいひんしゅ　在来品種　native variety, indigenous variety, local variety
サイラトロ　siratro, *Macroptilium atropurpureum* (DC.) Urb.
さいりゅう　砕粒　broken kernel
サイレージ　silage
サイレージちょうせい　サイレージ調製　silage making, ensiling
サイレンサー　silencer
サイロ　silo
さかまい　酒米　rice for sake brewery, brewers' rice
さきがけばな　さきがけ花【タバコ】first flower
さきがり　先刈り【イグサ】top clipping
サギゴケ (鷺苔)　*Mazus miquelii* Makino
さきゅう　砂丘　sand dune
さぎょうたいけい　作業体系　1) operation sequence【狭義】2) work system【広義】
さく (蒴)　boll, capsule
さくか　さく (蒴) 果　capsule
さくがた　作型　cropping type
さくがら　作柄, 作況　crop situation
さくき　作期, 作季　cropping season
さくこん　索根【サトウキビ】rope root
さくさん　酢酸　acetic acid
さくじゅう　搾汁　pressed juice
さくじょう　作条　1) seed furrow 2) seed furrow making【作業】
さくじょうかんかく　作条間隔⇒条間　row distance, interrow space
さくじょうそしき　柵状組織　palisade tissue
さくつけ　作付け　cropping
さくつけじゅんじょ　作付順序　cropping sequence
さくつけたいけい　作付体系　cropping system
さくつけめんせき　作付面積　planted area
さくつけようしき　作付様式　cropping pattern
さくど　作土　top soil, plow layer
さくもく　作目　kind of crop
さくもつ　作物　crop [plant]
さくもついくしゅ　作物育種　crop breeding
さくもついたい　作物遺体　crop residue, crop remain

さくもつがく　作物学　crop science
さくもつさいばいがく　作物栽培学　crop husbandry, agronomy
さくもつしんだん　作物診断　crop diagnosis
さくもつせいちょうモデル　作物成長モデル　crop growth model
さくもつせいりがく　作物生理学　crop physiology
さくゆ　搾油　oil extraction
さくようひょうほん　さく(腊)葉標本　⇒おし(押)葉標本　herbarium specimen
サクランボ→オウトウ
ザクロ(石榴)　pomegranate, *Punica granatum* L.
ザクロソウ　*Mollugo pentaphylla* L.
さこう　砂耕　sand culture, sandponics
サゴヤシ(サゴ椰子)　sago palm, non-spiny sago palm, *Metroxylon sagu* Rottb.
ササゲ(豇豆, 大角豆)　cowpea, southern pea, *Vigna unguiculata* (L.) Walp. (= *V. sinensis Endl.*)
サザンブロットほう　サザンブロット法　Southern blot technique, Southern blotting
ざし　座止　remaining in rosette state
サジオモダカ　oriental water plantain, *Alisma plantago-aquatica* L. var. *orientale* Sam.
さしき　挿し木　cutting
さしきどこ　挿木床　propagation bed, propagation bench
さしきなえ　挿木苗　rooted cutting
さしきはんしょく　挿木繁殖　cuttage
さしつどじょう　砂質土壌　sandy soil
さしほ　挿し穂　scion, cutting
さじょうど　砂壌土　sandy loam
ざせつていこう　挫折抵抗　breaking resistance, resistance to breaking
サッカー【サゴヤシ苗】　sucker
さっきょう　作況, 作柄　crop situation

さっきょうしすう　作況指数　crop situation index, crop index
さっきん　殺菌　sterilization
ざっきんこんにゅう　雑菌混入　contamination
さっきんざい　殺菌剤　1) fungicide【真菌】　2) bactericide【細菌】
さっきんざいたいせい　殺菌剤耐性　fungicide resistance
ざっこく　雑穀　miscellaneous cereals, millets
ざっしゅ　雑種　hybrid
ざっしゅきょうせい　雑種強勢　hybrid vigor, ヘテロシス　heterosis
ざっしゅクローン　雑種クローン　hybrid clone
ざっしゅけいせい　雑種形成, ハイブリダイゼーション　hybridization
ざっしゅじゃくせい　雑種弱勢　hybrid weakness, pauperization
ざっしゅしゅうだん　雑種集団　hybrid population
ざっしゅだいいちだい　雑種第一代　first filial generation (F_1)
さつせんちゅうざい　殺線虫剤　nematocide, nematicide
ざっそう　雑草　weed
ざっそうがい　雑草害　weed loss, loss from weed
ざっそうがく　雑草学　weed science
ざっそうかんり　雑草管理　weed management
さっそうスペクトル　殺草スペクトル　weed control spectrum, weeding spectrum, herbicidal spectrum
ざっそうぼうじょ　雑草防除　weed control
ざっそうぼうじょたいけい　雑草防除体系　weed control program
さっそざい　殺鼠剤　rodenticide
さつダニざい　殺ダニ剤　acaricide
さっちゅうざい　殺虫剤　insecticide
サツマイモ(薩摩芋)　sweet potato,

Ipomoea batatas (L.) Lam.
サツマイモデンプン　sweet potato starch
さど　砂土　sand, sandy soil
サトイモ(里芋)　eddoe, *Colocasia esculenta* (L.) Schott var. *antiquorum* Hubbard & Rehder
サトウカエデ(砂糖楓)　sugar maple, *Acer saccharum* Marsh.
さどうがた　差働型【ガス分析】　differential type
サトウキビ(砂糖黍)　sugarcane, *Saccharum officinarum* L.
さどうきょり　作動距離【顕微鏡】　working distance
サトウダイコン(砂糖大根)→テンサイ
サトウナツメヤシ　sugar date palm, *Phoenix sylvestris* Roxb.
サトウモロコシ(砂糖蜀黍)　sweet sorghum, sugar sorghum, sorgo, *Sorghum bicolor* (L.) Moench var. *saccharatum* (L.) Mohlenbr.
サトウヤシ(砂糖椰子)　1) gomuti palm, *Arenga pinnata* (Kuntze) Merr. 2)【サトウヤシ(gomuti palm), オオギヤシなど, 砂糖を採るヤシ類】sugar palm
サナエタデ　*Persicaria scabra* Mold. (= *Polygonum scabrum* Moench)
さなぎ(蛹)　pupa (*pl.* pupae)
さなご, 末粉(すえこ)【ソバ】　sanago, dark flour
さばくか　砂漠化　desertification
さばくしょくぶつ　砂漠植物　desert plant, eremophyte
サバンナ　savanna
さびびょう　さび(銹)病　rust
さびまい　さび米⇒茶米　rusty rice [kernel]
ざひょう　座標　co-ordinates
ざひょうじく　座標軸　axis [of co-ordinates]
サブクローバ　subclover, サブタレニアンクローバ　subterranean clover, *Trifolium subterraneum* L.
サブソイラ, 心土プラウ　subsoiler
サブタレニアンクローバ　subterranean clover, サブクローバ　subclover, *Trifolium subterraneum* L.
サブユニット　subunit
サフラニン　safranin
サフラワー→ベニバナ
サフラン(泪夫藍)　saffron, *Crocus sativus* L.
サブルーチン　subroutine
サプレッサー　suppressor
サポジラ　sapodilla, naseberry, *Manilkara zapota* (L.) P. Royen (= *Achras zapota* L.)
サマーサボリー　summer savory, *Satureja hortensis* L.
サーミスタおんどけい　サーミスタ温度計　thermistor thermometer
サーモカップルサイクロメータ　thermocouple psychrometer
さや　莢　1) pod【インゲン, エンドウ, ナタネなど】2) shell【ラッカセイ】
さやせんじゅく　莢先熟　delayed stem senescence
サヤヌカグサ　*Leersia oryzoides* (L.) Sw. spp. *japonica* (Hack.) T. Koyama (= *L. oryzoides* Sw. var. *sayanuka* Ohwi)
サヤバナ　hausa potato, *Coleus parviforus* Benth.
さようきこう　作用機構　mechanism of action
さようスペクトル　作用スペクトル　action spectrum
サラカヤシ　salak, *Salacca edulis* Reinw.
さらしなこ　さらしな粉【ソバ】　sarashina flour
サラダゆ　サラダ油　salad oil
ざるた　ざる田⇒漏水田　water-leaking paddy field, high permeable paddy

field
サルノパンノキ→バオバブ
さん　酸　acid
さんか　酸化　oxidation
さんか　酸価【油脂】　acid value
さんかかんげんでんい　酸化還元電位　oxidation-reduction potential (Eh), redox potential
サンカクイ→タイコウイ
さんかくうえ　三角植え　triangular planting
さんかくす　三角洲, デルタ　delta
さんかくフラスコ　三角フラスコ　Erlenmeyer flask
さんかこうそ　酸化酵素, オキシダーゼ　oxidase
さんかせんしょくたい　三価染色体　trivalent [chromosome]
さんかそう　酸化層　oxidized layer
さんかてきリンさんか　酸化的リン酸化　oxidative phosphorylation
ざんかん　残稈【タバコ】　residual stalks
さんきしゅ　三基種　trigenomic species
さんきろくばいたい　三基六倍体　trigenomic hexaploid
さんけいかじょ　散形花序　umbel
さんけいこうはい　三系交配, 三元交配　three-way cross, triple cross
さんけいざっしゅ　三系雑種　triple hybrid
さんげんこうはい　三元交配, 三系交配　three-way cross, triple cross
さんこう　散光⇒散乱光　diffused light, scattered light
ざんこう　残効　residual effect, residual activity
ざんごうほう　ざんごう (塹壕) 法　trench method
ざんさ　残差　residual
ざんさぶんさん　残差分散　residual variance
ざんさへいきんへいほう　残差平均平方　residual mean-square
ざんさへいほうわ　残差平方和　residual sum of squares
さんさぼう　三叉芒, 僧帽芒　hooded awn
さんし・こんちゅうのうぎょうぎじゅつけんきゅうしょ　蚕糸・昆虫農業技術研究所　National Institute of Sericultural and Entomological Sciences
サンジャクバナナ (三尺バナナ)　dwarf banana, *Musa cavendishii* Lamb.
さんじゅうすいそ　三重水素, トリチウム　tritium
さんじゅうせんしょく　三重染色　triple staining
さんしゅつふくよう　三出複葉　ternately compound leaf
さんじゅつへいきん　算術平均　average, arithmetic mean
サンショウ (山椒)　Japanese pepper, Japanese prickly ash, *Zanthoxylum piperitum* (L.) DC.
サンショウモ　*Salvinia natans* (L.) All.
さんすい　散穂　spreading panicle
さんすいかんがい　散水灌漑　spray irrigation
ざんすいほう　残穂法　remnant method
さんせいアミノさん　酸性アミノ酸　acidic amino acid
さんせいう　酸性雨　acid rain
さんせいざっしゅ　三性雑種　trihybrid
さんせいちょう　酸成長　acid growth
さんせいデタージェントせんい　酸性デタージェント繊維　acid detergent fiber (ADF)
さんせいどじょう　酸性土壌　acid soil
さんせいひりょう　酸性肥料　acid fertilizer
サンセベリア　sansevieria, bowstring

hemp, *Sansevieria nilotica* Baker 他数種
さんせんしょくたいしょくぶつ　三染色体植物　trisomic plant
さんそ　酸素　oxygen (O)
ざんそう　残草　residual herbage
さんそこきゅう　酸素呼吸, 好気的呼吸, 有気呼吸　aerobic respiration
さんそでんきょく　酸素電極　oxygen electrode
さんそほうしゅつ　酸素放出　oxygen evolution
ざんぞんしゅ　残存種　relic species, relict
さんたんとう　三炭糖, トリオース　triose
さんちらくのう　山地酪農　mountain dairy
さんど　酸度　acidity
さんとうりょう　酸当量　acid equivalent
さんにさんかぶつ　三二酸化物　sesquioxide
さんにゅう［りょう］　産乳［量］　milk production
さんねんりんさく　三年輪作　three-year rotation
さんぱ　散播　broadcast sowing (seeding), broadcast
さんばいせい　三倍性　triploidy
さんばいたい［しょくぶつ］　三倍体［植物］　triploid [plant]
さんぱき　散播機　broadcaster
さんばんこ　三番粉【ソバ】　No.3 flour
さんぷ　散布　1) application, spray 2) dissemination【種子】
さんぷず　散布図　scatter diagram
さんぷん　散粉　dusting
さんぷんき　散粉機　duster
サンヘンプ　sun (sunn) hemp, *Crotalaria juncea* L.
さんぽうコック　三方コック　three-way stopcock
さんぽしきのうほう　三圃式農法　three-course rotation, three-field system
さんほんぐわ　三本ぐわ(鍬)　three prong digging hook
さんもうさく　三毛作　triple cropping, three-crop system
さんようそ　三要素⇒肥料三要素　three major nutrients, NPK elements
さんらんこう　散乱光　diffused light, scattered light
さんらんたいようほうしゃ　散乱太陽放射, 散乱日射　diffused solar radiation
さんらんにっしゃ　散乱日射, 散乱太陽放射　diffused solar radiation
さんりゅうき　散粒機　granule applicator
ざんりゅうどくせい　残留毒性　residual toxicity
ざんりゅうぶつ　残留物　residue

[し]

シアし　シア脂　shea butter
ジアゾメタン　diazomethane (CH_2N_2)
シアナミドたいちっそ　シアナミド態窒素　cyanamide nitrogen
シアノバクテリア　cyanobacteria, 藍藻類　blue green algae
シアバター⇒シア脂　shea butter
シアバターノキ(シアバターの木)　shea butter tree, shea tree, *Vitellaria paradoxa* (A. DC.) C. F. Gaertn. (= *Butyrospermum parkii* Don Kotschy)
シアンかぶつ　シアン化物　cyanide
しいき　篩域　sieve area
シイタケ　shiitake fungus, *Lentinus edodes* (Berk.) Sing.
ジーいちき　G_1期【細胞周期】　G_1 phase
しいな (粃, 秕)　empty grain, abortive grain

シェッフェのけんてい　シェッフェの検定　Scheffe's test
シーエヌひ　C-N比　C-N ratio, carbon-nitrogen ratio
シェパードのほせい　シェパードの補正　Sheppard's correction
ジェフレーえき　ジェフレー液　Jeffrey's solution
ジオウ(地黄)→アカヤジオウ
しおれ　しお(萎)れ　wilting
しおれけいすう　しお(萎)れ係数　wilting coefficient
しおれてん　しお(萎)れ点　wilting point
しか(めばな)　雌花　female flower, pistillate flower
しがいほうしゃ　紫外放射　ultraviolet radiation (UV)
しかくうえ　四角植え⇒正方形植え　square planting
シカクマメ　winged bean, four-angled bean, goa bean, asparagus pea, *Psophocarpus tetragonolobus* (L.) DC.
じかさいしゅ　自家採種　home seed-raising
じかじゅせい　自家受精　self-fertilization, selfing
じかじゅせいしょくぶつ　自家受精植物, 自殖性植物　autogamous plant, self-fertilizing plant
じかじゅふん　自家受粉　self-pollination
じかせいしょく　自家生殖　autogamy
じかふけっかせい　自家不結果性　self-unfruitfulness
じかふねんせい　自家不稔性　self-sterility
じかふわごうせい　自家不和合性　self-incompatibility
じかまき(じきまき)　直播き, 直播(ちょくは, ちょくはん)　direct sowing (seeding)

じかわごうせい　自家和合性　self-compatibility
しかん　篩管　sieve tube
じきおんしつどけい　自記温湿度計　hygrothermograph
しきくさ　敷草　1) grass mulch 2) mulching【作業】
しきざきの　四季咲きの　ever-blooming, ever-flowering, perpetual flowering
しきそ　色素　1) pigment　2) dye【染料】
しきそたい　色素体, プラスチド　plastid
しきそたいとつぜんへんい　色素体突然変異　plastid mutation
しきたく　色沢　color and gloss
ジギタリス　digitalis, foxglove, *Digitalis purpurea* L.
[じき]ディスク　[磁気]ディスク　magnetic disk
しきべつしゅ　識別種, 判別種　differential species
しきべつひんしゅ　識別品種, 判別品種　differential variety
じきまき(じかまき)　直播き, 直播(ちょくは, ちょくはん)　direct sowing (seeding)
しきゅう　子球　1) bulblet【ユリ等】2) cormel【グラジオラス等】3) dry set【タマネギ等】
じきゅう　自給　self-supply
じきゅうさくもつ　自給作物　home-consuming crop, crop for home consumption
じきゅうじそく　自給自足　self-sufficiency
じきゅうしりょう　自給飼料　self-supplied feed
じきゅうひりょう　自給肥料　self-supplied manure
じきゅうりつ　自給率　self-sufficiency rate

しきりょう 敷料【畜産】 bedding, litter
しきわら 敷わら(藁) straw mulch
シグナルでんたつ シグナル伝達 signal transduction
じけいれつ 時系列 time series
じけいれつかいせき 時系列解析, 時系列分析 time series analysis
じけいれつぶんせき 時系列分析, 時系列解析 time series analysis
しげき 刺激 stimulus (*pl.* stimuli)
しけんかん 試験管 test tube
しけんく 試験区 [experimental] plot
しげんけい 始原型⇒原型 prototype
しげんさいぼう 始原細胞 initial cell
しげんたい 始原体, 原基 primordium (*pl.* primordia)
しけんほ[じょう] 試験圃[場] experimental field
しこう 枝梗 rachis branch
しこう 試行【確率論】 trial
しこう 篩孔 sieve pore
しこうせい 嗜好性 palatability, acceptability
しこうりょうさくもつ 嗜好料作物 recreation crop
じこかいきかてい 自己回帰過程 autoregressive process
しこくのうぎょうしけんじょう 四国農業試験場 Shikoku National Agricultural Experiment Station
シコクビエ(龍爪稷) finger millet, African millet, *Eleusine coracana* (L.) Gaertn.
じこそうかん 自己相関 autocorrelation
しこん 支根, 支持根, 支柱根 prop root, brace root
じさく 自作 owner-operated farming
じさくのう 自作農 owner farmer
シーさんしょくぶつ C_3植物 C_3 plant
しじこん 支持根, 支根, 支柱根 prop root, brace root

ししつ 脂質 lipid
しじつ 子実 1) grain【穀類】 2) bean【マメ類】
しじつさくもつ 子実作物 grain crop
しじつしゅうりょう 子実収量, 穀実収量 grain yield
しじつせいさん 子実生産 grain production
しじやく 指示薬 indicator
じしょう 事象 event
じしょく 自殖 selfing, self-fertilization, self-propagation, self-reproduction
じしょくけいとう 自殖系統 selfed line
じしょくせいしょくぶつ 自殖性植物, 自家受精植物 autogamous plant, self-fertilizing plant
しすい 雌穂【トウモロコシ】 ear
しずい 雌ずい(蕊) pistil
しずいせんじゅく 雌ずい(蕊)先熟 protogyny, proterogyny
しすう[がた]せいちょう 指数[型]成長 exponential growth
しすう[がた]せいちょうきょくせん 指数[型]成長曲線 exponential growth curve
しすうせいちょうき 指数成長期 exponential growth phase
しすうぶんぷ 指数分布 exponential distribution
シスエレメント cis-element
シスチン cystine $((Cys)_2)$
システイン cysteine (Cys)
システム, 系 system
シストセンチュウ cyst nematode
シストせんちゅうびょう シスト線虫病 cyst nematode disease
じすべりしんしょく 地滑り侵食 land slide, slip erosion
じせいの 自生の indigenous, native, spontaneous
せいふねん 雌性不稔 female

sterility
じせき　耳石⇒平衡石　statolith
しせついくびょう　施設育苗　raising of seedling under structure
しせつえんげい　施設園芸　protected horticulture, horticulture under structure
しせつのうぎょう　施設農業　greenhouse agriculture, protected cultivation, farming under structure
ジーゼロき　G_0期【細胞周期】　G_0 phase
しぜんかしゅ　自然下種　natural seeding, self seeding
しぜんかんそう　自然乾燥　natural drying
しぜんきゅうみん　自然休眠　natural dormancy
しぜんこうざつ　自然交雑　natural crossing, natural hybridization
しぜんさいがい　自然災害　natural disaster
しぜんしたてちゃえん　自然仕立て茶園　natural-shaped tea bush
しぜんしゅうだん　自然集団　natural population
しぜんじゅふん　自然受(授)粉　open pollination
しぜんせんたく　自然選択　natural selection
しぜんそうち　自然草地　natural grassland, natural pasture
しぜんとうた　自然淘汰⇒自然選択　natural selection
しぜんとつぜんへんい　自然突然変異　natural mutation, spontaneous mutation
シソ(紫蘇)　perilla, *Perilla frutescens* (L.) Britton var. *crispa* (Thunb. ex Murray) W. Decne
じぞくがたのうぎょう　持続型農業　sustainable agriculture
シダーゆ　シダー油　cedar oil

じだい　次代　progeny, offspring
じだいけんてい　次代検定　progeny test, offspring test
したくさ　下草, 下繁草　bottom grass, undergrowth
じだつがたコンバイン　自脱型コンバイン　head feeding combine
したて　仕立て　training
したは　下葉【タバコ】　primings, flyings
したば　下葉⇒下位葉　lower leaf
シタン(紫檀)　rosewood, *Dalbergia cochinchinensis* Pierre ex Laness. など数種
シチトウイ(七島藺)　Chinese mat grass, three-cornered grass, *Cyperus malaccensis* Lam. ssp. *brevifolius* (Boeck.) T. Koyama
しちぶづきまい　七分づ(搗)き米　70%-polished rice
しちゅうこん　支柱根, 支根, 支持根　prop root, brace root
しつがい　湿害　excess-moisture injury, wet injury
しっき(しっけ)　湿気　moisture
しっきゅうおんど　湿球温度　wet-bulb temperature
しっけ(しっき)　湿気　moisture
じっけんけいかく　実験計画　experimental design, design of experiment
じっけんごさ　実験誤差　experimental error
じっしゅう[りょう]　実収[量]　actual yield
しつじゅんきこう　湿潤気候　humid climate
しつじゅんねったいちいきそりゅうこくもつ・まめるい・ちかさくもつけんきゅうかいはつちいきちょうせいセンター　湿潤熱帯地域粗粒穀物・豆類・地下作物研究開発地域調整センター　Center for Coarse Grains, Pulses,

Roots, and Tuber Crops (CGPRT Center)
しっせいざっそう　湿生雑草　hygrophytic weed
しっせいしょくぶつ　湿生植物　hygrophyte, hygrophytic plant
じったい[かいぼう]けんびきょう　実体[解剖]顕微鏡　stereoscopic microscope
しつてきけいしつ　質的形質　qualitative character (trait)
しつでん　湿田　ill-drained paddy field
しつど　湿度　humidity
しつどけい　湿度計　hygrometer
しつないいくびょう　室内育苗　indoor rasing of seedling
じつめん　実棉　seed cotton
しつもんひょう　質問票, 聞取り調査表　questionnaire
しつゆうしつむけいしつ　しつゆうしつむ(悉有悉無)形質　all-or-none trait
じつようけいしつ　実用形質　economic character (trait)
じつようひんしゅ　実用品種　commercial variety
しつりょうぶんせき　質量分析　mass spectrometry (MS)
しつりょうぶんせきけい　質量分析計　mass spectrometer
じていすう　時定数　time costant
じどうかんすい　自動灌水　automatic watering, automatic irrigation
じどうだっこくき　自動脱穀機　automatic thresher, mechanical feeding thresher, self-feeding thresher
シトクロム, チトクロム　cytochrome
じどこ　地床【タバコ】　ground seedbed
シードテープ　seed tape
シトロネラソウ　citronella grass, 1) *Cymbopogon nardus* (L.) Rendle (セイロンシトロネラソウ) 2) *C. winterianus* Jowitt (ジャワシトロネラソウ)
シナアブラギリ(支那油桐)　tung tree, *Aleurites fordii* Hemsl.
シナダレスズメガヤ→ウィーピングラブグラス
シナモン　cinnamon, Ceylon cinnamon, *Cinnamomum verum* J. Presl (= *C. zeylanicum* (Garc.) Bl.)
じならしき　地ならし機　land leveler
ジーにき　G₂期【細胞周期】　G_2 phase
しにまい　死米　opaque rice-kernel
ジネンジョ(自然薯)→ヤマノイモ
シバ(芝)　Japanese lawngrass, *Zoysia japonica* Steud.
しばち　芝地　sod, sward, turf
しばふ　芝生　lawn, turf
シバムギ　quackgrass, *Agropyron repens* (L.) P. Beauv.
しばん　篩板　sieve plate
しはんしゅし　市販種子　commercial seed
しはんびょう　紫斑病【ダイズ】　purple stain of seed
しひょうしょくぶつ　指標植物　indicator plant, plant indicator
しぶ　篩部　phloem
シープフェスク　sheep fescue, *Festuca ovina* L.
しぶみ　渋み　astringency
しぶんいてん　四分位点　quartile
しぶんいへんさ　四分位偏差　quartile deviation
しぶんし　四分子　tetrad, quartet
しぶんせんしょくたい　四分染色体　[chromosome] tetrad
ジベレリン　gibberellin
しぼう　子房　ovary
しぼう　脂肪　fat
しぼうさん　脂肪酸　fatty acid
しぼうばいよう　子房培養　ovary culture

じぼし　地干し　drying on ground
じぼしれつ　地干し列　windrow
しぼり　絞り　1)【光学機器】diaphragm　2)【模様】flake, variegation
しま(島)【イネ, ソバなど】stack
しまいしゅくびょう　縞萎縮病【オオムギ】yellow mosaic
シマスズメノヒエ→ダリスグラス
しまだてかんそう　島立て乾燥　stack drying
シマツナソ　nalta jute, Corchorus olitorius L.
シマツルアズキ→タケアズキ
しまはがれびょう　縞葉枯病【イネ】stripe
しみゃく　支脈⇒側脈　lateral vein
シミュレーション　simulation
しめぐ　締め具, クランプ　clamp
しめそ(乾麻)　dry stem of hemp
しめん　死綿　dead cotton
じもう　地毛, 短毛　fuzz, linters
しもごえ　下肥　night soil
しもばしら　霜柱　frost pillars
しもよけ　霜除け, 防霜　frost protection
しゃかい　舎飼　housing
ジャガイモ　potato, Irish potato, Solanum tuberosum L.
ジャガイモデンプン　potato starch
シャカトウ(釈迦頭)→バンレイシ
しやく　試薬　reagent
じゃくかんひんしゅ　弱稈品種　weak-strawed variety
じゃくせいえいか　弱勢頴花　inferior spikelet
シャクチリソバ(赤地利蕎麦)　perennial buckwheat, Fagopyrum cymosum Meisn.
しゃくど　尺度　scale, measure
しゃくびん　試薬びん　reagent bottle
シャクヤク(芍薬)　Chinese paeony, Chinese peony, Paeonia lactiflora Pall. (= P. albiflora Pall.)
しゃこう　遮光　shading
しゃこうさいばい　遮光栽培　shade culture
ジャスミン　jasmine, Jasminum spp.
ジャスモンさん　ジャスモン酸　jasmonic acid
ジャックナイフすいてい　ジャックナイフ推定　jackknife estimation
ジャックフルーツ→パラミツ
シャドウイング【電顕】shadowing
シャトルベクター　shuttle vector
ジャノヒゲ　Ophiopogon japonicus (L. f.) Ker-Gawl.
しゃぶんれつ　斜分裂　oblique division
しゃへいじゅ　遮へい(蔽)樹　shelter tree
シャーレ　culture dish, ペトリ皿　petri dish
ジャワがた　ジャワ型【イネ】javanica type
ジャワシトロネラソウ　Java citronella grass, Cymbopogon winterianus Jowitt
ジャワフトモモ→レンブ
ジャワムカゴコンニャク　Amorphophallus oncophyllus Prain ex Hook. f.
しゅ　種　species
しゆういかどうしゅ[せい]　雌雄異花同株[性]　monoecism, diclinism
しゆいさく　周囲作　fence cropping
しゆいしゅ[せい]　雌雄異株[性]　dioecism
しゆいじゅく　雌雄異熟　dichogamy
じゅうえき　汁液　sap
しゅうえんキメラ　周縁キメラ　periclinal chimera
しゅうえんくぶんキメラ　周縁区分キメラ　marginal sectorial chimera
しゅうえんこうか　周縁効果　border effect, 周辺効果　marginal effect
しゅうえんさくもつ　周縁作物　border

crop
しゅうえんせいちょう　周縁成長　marginal growth
しゅうえんぶんれつそしき　周縁分裂組織　marginal meristem
じゅうかいき　重回帰　multiple regression
じゅうかいきぶんせき　重回帰分析　multiple regression analysis
しゅうかく　収穫　harvest, harvesting
しゅうかくき　収穫期　harvest time
しゅうかくき　収穫機, ハーベスタ　harvester
しゅうかくしすう　収穫指数　harvest index
しゅうかくめんせき　収穫面積　harvested area
しゅうき　終期【細胞分裂】　telophase
しゅうきせい　周期性　periodicity
じゅうきんぞく　重金属　heavy metal
しゅうこう　秋耕　autumn plowing
しゅうこう　縦溝, 腹溝【ムギ類】　crease, furrow
しゅうごうか　集合果, 多花果　multiple fruit, aggregate fruit, syncarp
しゅうこうき　集光器, コンデンサー【顕微鏡】　condenser
しゅうこうせいしきそ　集光性色素　light harvesting pigment, アンテナ色素　antenna pigment
しゅうこうせいふくごうたい　集光性複合体【光合成】　light-harvesting complex (LHC)
じゆうさいしょく［りょう］　自由採食［量］　voluntary intake
じゅうさいぼう　柔細胞　parenchyma cell, parenchymatous cell
シュウさん　シュウ酸　oxalic acid
じゅうじたいせい　十字対生　decussate
じゆうすい　自由水【土壌】　free water
じゅうそうかん　重相関　multiple correlation
じゅうそうかんけいすう　重相関係数　multiple correlation coefficient
じゅうぞくへんすう　従属変数　dependent variable
じゅうそしき　柔組織　parenchyma
しゅうだん　集団, 個体群　population
しゅうだんいくしゅほう　集団育種法　mass method of breeding, bulk method [of breeding]
しゅうだんいでんがく　集団遺伝学　population genetics
しゅうだんかいりょう　集団改良　population improvement
しゅうだんさいしゅ　集団採種　mass production of seed, mass seed production
しゅうだんじゅふん　集団受(授)粉　mass pollination
しゅうだんじょゆう　集団除雄　mass emasculation, bulk emasculation
じゅうだんせっぺん　縦断切片　longitudinal section
しゅうだんせんばつ［ほう］　集団選抜［法］　mass selection [method of breeding]
しゅうだんみつど　集団密度, 個体群密度　population density
じゅうてんようせんい　充填用繊維　filling fiber
じゅうてんりょうさくもつ　充填料作物　wadding crop
じゆうど　自由度　degree of freedom
しゆうどうしゅ［せい］　雌雄同株［性］　monoecism
しゆうどうじゅく　雌雄同熟　homogamy
しゅうにゅう　周乳　perisperm
しゅうねんきょうきゅう　周年供給　year-round supply
しゅうねんさいばい　周年栽培　year-round cropping, year-round culture
じゅうねんど　重粘土　heavy [clay] soil

しゅうのう　収納【干草, 穀物など】 barning
しゅうひ　周皮　periderm
じゅうぶんれつ　縦分裂　longitudinal division
しゅうへんこうか　周辺効果　marginal effect, 周縁効果　border effect
しゅうへんぶんれつそしき　周辺分裂組織　peripheral meristem
しゅうみんうんどう　就眠運動, 睡眠運動　sleep movement, nyctinastic movement, nyctinasty
しゅうやくさいばい　集約栽培　intensive cultivation
じゅうらいの　従来の, 慣行の　conventional
しゅうりょう　収量　yield
しゅうりょうこうせいようそ　収量構成要素　yield component
しゅうりょうしけん　収量試験　yield trial
しゅうりょうせい　収量性　yielding ability
しゅうりょうぜんげんのほうそく　収量漸減の法則, 報酬漸減の法則　law of diminishing returns
しゅうりょうちょうさ　収量調査　yield survey
しゅうりょうポテンシャル　収量ポテンシャル　yield potential
しゅうりょうよそく　収量予測　yield prediction, yield forecast
じゅうりょくくっせい　重力屈性　gravitropism
じゅうりょくすい　重力水　gravitational water
じゅうりょくポテンシャル　重力ポテンシャル　gravitational potential
ジュウロクササゲ (十六豇豆) asparagus bean, *Vigna unguiculata* (L.) Walp. var. *sesquipedalis* (L.) H. Ohashi
しゅが　珠芽⇒むかご (零余子)　bulbil, aerial tuber
しゅかん　主稈　main culm
じゅかん　樹冠　crown
しゅかんきょうそう　種間競争　interspecific competition
しゅかんこうざつ　種間交雑　interspecific crossing, species cross
しゅかんさ　種間差　interspecific difference
しゅかんざっしゅ　種間雑種　interspecific hybrid, species hybrid
しゅかんようすう　主稈葉数　number of leaves on the main culm
しゅきゅう　種球　seed bulb, seed corm
しゅくこくるい　しゅく (菽) 穀類⇒マメ類　pulses, pulse crops, leguminous crops
しゅくしゅ　宿主　host
しゅくじゅう　縮重　degeneracy
じゅくせい　熟性　maturity
じゅくせい　熟成【タバコ】aging, ageing
じゅくでん　熟田　mature paddy field
じゅくど　熟度　grade of maturity
じゅくばた　熟畑　mature [upland] field
じゅくばたか　熟畑化　field maturing process
しゅけい　主茎　main stem
しゅげい　種芸　crop cultivation
しゅこう　珠孔　micropyle
じゅこう　受光　light interception
しゅこうか　主効果　main effect
じゅこうかくど　受光角度　leaf angle
じゅこうたいせい　受光態勢　light-intercepting characteristics, stand geometry
じゅこうりょう　受光量　intercepted radiation
ジュコブスキーけいコムギ　ジュコブスキー系コムギ　zhukovskyi wheat
ジュコブスキーコムギ　*Triticum*

zhukovskyi Men. et Er.
しゅこん　主根　main root, taproot
しゅこんがたこんけい　主根型根系　taproot system
じゅこんさくもつ　需根作物　root crop
しゅさくもつ　主作物, 主要作物　major crop, main crop
しゅし　種子　seed
じゅし　樹脂　resin
しゅしかつりょく　種子活力　seed vigor
しゅしかんべつ　種子鑑別　seed identification
しゅしきゅうみん　種子休眠　seed dormancy
しゅじく　主軸　main axis, primary axis
しゅしけんさ　種子検査　seed inspection, seed testing
しゅしけんてい　種子検定　seed certification
しゅしこうかん　種子交換　exchange of seeds
しゅしこうしん　種子更新　renewal of seeds
しゅしこん　種子根　seminal root
しゅししゅうりょう　種子収量　seed yield
しゅしじゅみょう　種子寿命　seed longevity
しゅししゅんか　種子春化　seed vernalization
しゅししょうどく　種子消毒　seed disinfection
しゅししょくぶつ　種子植物　seed plant, spermatophyte
しゅししんせき　種子浸漬 ⇒ 浸種　seed soaking
しゅしせいさん　種子生産　seed production
しゅしせいさんりょく　種子生産力　seed productivity
しゅしせいじゅく　種子成熟　seed maturity
しゅしせいせん　種子精選　seed cleaning
しゅしタンパクしつ　種子タンパク質　seed protein
しゅしちょぞう　種子貯蔵　seed storage
しゅじつ　種実⇒子実　grain
じゅじつさくもつ　需実作物　fruit and seed crop
しゅしでんせん　種子伝染　seed infection, seed transmission
しゅしバンク　種子バンク　seed bank
しゅしはんしょく　種子繁殖　seed propagation
しゅしひんしつ　種子品質　seed quality
しゅしふんい　種子粉衣　seed coating, seed dressing
しゅしほぞん　種子保存　seed preservation
じゅじょうず　樹状図, デンドログラム　dendrogram
しゅしょく　主食　staple food
じゅしりょうさくもつ　樹脂料作物　resin crop
しゅしれい　種子齢　seed age
しゅしん　珠心　nucellus
ジュズダマ　Job's-tears, *Coix lacryma-jobi* L.
じゅせい　受精, 授精　fertilization
じゅせいきょうそう　受精競争　certation
しゅせいたいがく　種生態学　autecology, species ecology, 個生態学　idioecology
しゅせいぶつがく　種生物学　species biology
しゅせいぶん　主成分【統計】　principal component
しゅせいぶんスコア　主成分スコア　principal component score
しゅせいぶんぶんせき　主成分分析

principal component analysis (PCA)
しゅちん　種枕　caruncle, strophiole
しゅつえき　出液　1) bleeding, exudation【現象】 2) bleeding sap, exudate【物質】
しゅつが　出芽　emergence, emergence of seedling
しゅつがき　出芽器, 育苗器　nursery chamber
しゅっこんさくもつ　宿根作物⇒多年生作物　perennial crop
しゅっこんせいざっそう　宿根性雑草⇒多年生雑草　perennial weed
しゅっこんせいの　宿根性の⇒多年生の perennial
シュッコンソバ (宿根蕎麦)→シャクチリソバ
しゅっすい　出穂　heading, ear emergence
しゅっすいき　出穂期　heading time, heading stage
しゅっすいしき　出穂始期　first heading time
しゅっすいせいき　出穂盛期　middle heading time
しゅっすいぜんにっすう　出穂前日数　number of days before heading
しゅっすいちえん　出穂遅延　delayed heading
しゅつよう　出葉　leaf emergence, leaf unfolding
しゅつようかんかく　出葉間隔 phyllochron
しゅつようそくど　出葉速度　leaf emergence rate
しゅつようてんかんてん　出葉転換点 turning point of leaf emergence rate
ジュート　jute, white jute, *Corchorus capsularis* L.
しゅどういでんし　主働遺伝子　major gene
じゅどうてききゅうしゅう　受動的吸収　passive absorption

じゅどうてききゅうすい　受動的吸水 passive water absorption
しゅとくいせい　種特異性　species specificity
しゅないきょうそう　種内競争 intraspecific competition
しゅひ　珠皮　integument
しゅひ　種皮　seed coat, testa
じゅひ　樹皮　bark
しゅひしょり　種皮処理　scarification
しゅびょう　種苗　seed and seedling
じゅふん　受粉, 授粉　pollination
しゅぶんか　種分化　speciation
じゅふんじゅ　授粉樹　pollinizer
しゅへい　珠柄　funicle, funiculus, ovule stalk
しゅみゃく　主脈, 中央脈　main vein
じゅみょう　寿命　longevity, life duration
しゅもくざし　しゅ(撞)木挿し mallet cutting
しゅもくどり　しゅ(撞)木取り horizontal layering, continuous layering
しゅよう　腫瘍　tumor
じゅようき　受容器, 受容体, レセプター　receptor
しゅようさくもつ　主要作物, 主作物 major crop, main crop
じゅようさくもつ　需葉作物　leaf crop
じゅようたい　受容体, 受容器, レセプター　receptor
しゅようひんしゅ　主要品種　leading variety, main variety
シュルツェえき　シュルツェ液 Schultze solution
シュロ (棕櫚)　windmill palm, *Trachycarpus fortunei* (Hook.) Wendl.
じゅん　旬　ten-days
じゅん(きん)どうしついでんしけいとう　準(近)同質遺伝子系統 near-isogenic line

じゅんい 順位 rank
じゅんいそうかん 順位相関 rank correlation
じゅんいちじせいさんりょく 純一次生産力 net primary productivity
じゅんか 順化 acclimation, acclimatization
じゅんか 順化, 硬化, ハードニング【培養】 hardening
じゅんか 馴化 ⇒順化 acclimation, acclimatization
しゅんかしょうきょ 春化消去 devernalization
しゅんか [しょり] 春化 [処理], バーナリゼーション vernalization
じゅんかんせんばつほう 循環選抜法 recurrent selection
じゅんぐりせんばつほう 順繰り選抜法 tandem selection
じゅんけい 純系 pure line
じゅんけいせつ 純系説 pure line theory
じゅんけいせんばつ 純系選抜 pure line selection
じゅんけいぶんり 純系分離 pure line separation
しゅんこう 春耕 spring plowing
じゅんこうごうせい 純光合成 net photosynthesis, みかけの光合成 apparent photosynthesis
じゅんじょとうけいりょう 順序統計量 order statistics
じゅんせいさん 純生産 net production
じゅんどうかりつ 純同化率 net assimilation rate (NAR)
じゅんほうしゃ 純放射 net radiation
じゅんれつ 順列 permutation
しよう 飼養 feeding
しよう 子葉 cotyledon
じょう 条 row
しょうあん 硝安 ⇒硝酸アンモニウム ammonium nitrate

しょういかんそく 小維管束 small vascular bundle
じょういしぼう 上位子房 superior ovary
じょういも 上いも【ジャガイモ】 marketable tuber
じょういよう 上位葉 upper leaf
しょうか しょう(漿)果 berry
しょうか 小花 floret
しょうか 消化 digestion
しょうか 硝化⇒硝酸化成 [作用] nitrification
ショウガ (生姜) ginger, *Zingiber officinale* Rosc.
しょうがいがたれいがい 障害型冷害【イネ】 floral sterility caused by low temperature, cool summer damage due to floral impotency
しょうかきん 硝化菌⇒硝酸化成菌 nitrifying bacteria, nitrifier
じょうかく 娘核, 嬢核 daughter nucleus
しょうかへい 小花柄 pedicel
しょうかりつ 消化率 digestibility
しょうかるい 小果類 small fruits
じょうかん 条間 row distance, interrow space
じょうきあつ 蒸気圧 vapor pressure
じょうぎうえ 定規植え regular planting with ruler
しょうきょくてききゅうしゅう 消極的吸収⇒受動的吸収 passive absorption
じょうけん [つき] かくりつ 条件 [付] 確率 conditional probability
じょうけん [つき] きたいち 条件 [付] 期待値 conditional expectation
じょうけんぶんぷ 条件分布 conditional distribution
じょうさいぼう 娘細胞, 嬢細胞 daughter cell
じょうさく 上作⇒豊作 good harvest, bumper crop

しょうさん　硝酸　nitric acid
じょうさん　蒸散　transpiration
しょうさんアンモニウム　硝酸アンモニウム　ammonium nitrate
しょうさんイオン　硝酸イオン　nitrate ion
しょうさんえん　硝酸塩　nitrate
しょうさん[えん]ちゅうどく　硝酸[塩]中毒　nitrate toxicity
しょうさんかせい[さよう]　硝酸化成[作用]　nitrification
しょうさんかせいきん　硝酸化成菌　nitrifying bacteria, nitrifier
しょうさんカルシウム　硝酸カルシウム, 硝酸石灰　calcium nitrate
しょうさんかんげん　硝酸還元　nitrate reduction
しょうさんかんげんこうそ　硝酸還元酵素　nitrate reductase
じょうさんけいすう　蒸散係数　transpiration coefficient
じょうさんこうりつ　蒸散効率　transpiration efficiency
しょうさんせっかい　硝酸石灰, 硝酸カルシウム　calcium nitrate
じょうさんそくど　蒸散速度　transpiration rate
しょうさんたいちっそ　硝酸態窒素　nitrate nitrogen
じょうさんひ　蒸散比　transpiration ratio
じょうさんよくせいざい　蒸散抑制剤　antitranspirant
しょうじ　小耳, 葉耳　auricle
しょうしこう　小枝梗　pedicel
しょうししつの　硝子質の, ガラス質の　flinty, glassy
しょうししつりゅう　硝子質粒【ムギ類】　glassy kernel
しょうしせい　硝子性, ガラス質性【コムギ】　glassiness
じょうじたんすい　常時湛水　continuous flooding

しょうしゃ　照射　irradiation
じょうじゅん　上旬　the first ten-days [of a month]
しようしょう　子葉鞘, 鞘葉　coleoptile
しょうじょうふくよう　掌状複葉　palmate compound leaf
しょうしりつ　硝子率　percentage of glassy kernel
しょうすい　小穂　spikelet
ショウズク (小豆蔻), カルダモン　cardamon, *Elettaria cardamomum* (L.) Maton
しょうぜつ　小舌, 葉舌　ligule
しょうせっかい　消石灰　slaked lime
じょうせんしょくたい　常染色体　autosome
じょうそう　条桑【クワ】　mulberry shoot
じょうぞう　醸造　brewing
じょうぞうようオオムギ　醸造用オオムギ　malting barley
じょうたいへんすう　状態変数　state variable
しょうてん　焦点　focus
しょうてんしんど　焦点深度　focal depth
しょうど　焼土　calcination of soil
しょうど　照度　illuminance
じょうど　壌土　loam
しょうとう　少糖, オリゴ糖　oligosaccharide
しょうどく　消毒　disinfection
しょうどけい　照度計　illuminometer, photometer
じょうねつおんしょう　醸熱温床　manure-heated seedbed
じょうは　条播　drilling, row sowing (seeding)
じょうはいじく　上胚軸　epicotyl
じょうはき　条播機, すじ播き機　drill
じょうはくまい　上白米　head pollished rice, head rice
じょうはつ　蒸発　evaporation

じょうはつけい 蒸発計 evaporimeter, evaporation pan, atmometer
じょうはっさん 蒸発散 evapotranspiration
じょうはっさんい 蒸発散位, 可能蒸発散量 potential evapotranspiration
じょうはっさんそくど 蒸発散速度 evapotranspiration rate
じょうはつよくせいざい 蒸発抑制剤 evaporation suppressor
じょうはんそう 上繁草 top grass
じょうはんびょう 条斑病【コムギ, オオムギ】 cephalosporium stripe
しょうひしゃ 消費者 consumer
しょうひしゃべいか 消費者米価 consumer rice price
しようひょうじゅん 飼養標準 feeding standard
しょうひんさくもつ 商品作物 commercial crop
じょうへんせいちょう 上偏成長 epinasty
じょうほう 情報 information
じょうほうけんさく 情報検索 information retrieval
じょうほうシステム 情報システム information system
じょうほうしょり 情報処理 information processing
しょうほうたい 小胞体 endoplasmic reticulum (ER)
しょうほう[よう] 小苞[葉] bractlet, bracteole
じょうほうりょう 情報量 amount of information
じょうまい 上米 high-grade rice
じょうみょうしゅし 常命種子 mesobiotic seed
しょうゆ（醤油） soy sauce, soybean sauce
しょうよう 小葉 leaflet
しょうよう 鞘葉, 子葉鞘 coleoptile
じょうようがたてきさいき 乗用型摘採機【チャ】 riding-type tea plucker
しょうようじゅりん 照葉樹林 laurel forest, laurilignosa
じょうようトラクタ 乗用トラクタ riding tractor
しょうようへい 小葉柄 petiolule
じょうりゅうすい 蒸留水 distilled water
しょうりゅうひんしゅ 小粒品種 small grain variety
しょうりょくか 省力化 labor-saving
しょうりょくさいばい 省力栽培 labor-saving cultivation
じょうりょくじゅ 常緑樹 evergreen tree
しょうりょくてき 省力的, 労働節約的 labor-saving
じょうりょくの 常緑の evergreen
しょうれいひんしゅ 奨励品種 recommended variety
しょうれいひんしゅけっていしけん 奨励品種決定試験 performance test for recommendable varieties
じょうろ（如雨露） watering pot
じょえん 除塩 desalinization
しょきじょうけん 初期条件 initial condition
しょきせだい 初期世代 early generation
しょきち 初期値 initial value
しょくえん 食塩⇒塩化ナトリウム sodium chloride
しょくじょうど 埴壌土 clay loam
しょくせい 植生 vegetation
しょくせいしすう 植生指数, 植生指標 vegetation index
しょくそう 食草 grazing, foraging
しょくそうりょう 食草量, 採食[草]量 herbage intake, forage intake, grazing intake
しょくど 埴土 clay soil, clay
しょくひ 植被 vegetation cover
しょくひんそうごうけんきゅうしょ

食品総合研究所　National Food Research Institute
しょくひんてんかぶつ　食品添加物　food additive
しょくぶついくしゅ　植物育種　plant breeding
しょくぶついくしゅがく　植物育種学　plant thremmatology
しょくぶついたい　植物遺体, 植物残さ (渣)　plant residue, plant remain
しょくぶつえいよう[がく][学]　植物栄養[学]　plant nutrition
しょくぶつかい　植物界　plant kingdom
しょくぶつかいぼうがく　植物解剖学　plant anatomy
しょくぶつがく　植物学　botany
しょくぶつきかんがく　植物器官学　plant organography
しょくぶつぐんらく　植物群落　plant community
しょくぶつけい　植物計　phytometer
しょくぶつけいたいがく　植物形態学　plant morphology, phytomorphology
しょくぶつけんえき　植物検疫　plant quarantine
しょくぶつざんさ　植物残さ (渣), 植物遺体　plant residue
しょくぶつし　植物脂　vegetable fat
しょくぶつしきそ　植物色素　plant pigment
しょくぶつしゃかいがく　植物社会学　plant sociology
しょくぶつしんひんしゅほごこくさいどうめい　植物新品種保護国際同盟　International Union for the Protection of New Varieties of Plants (UPOV)
しょくぶつせいたいがく　植物生態学　plant ecology
しょくぶつせいちょうちょうせつぶっしつ　植物成長調節物質　plant growth regulator
しょくぶつせいりがく　植物生理学　plant physiology
しょくぶつそう　植物相, フロラ　flora
しょくぶつそしきがく　植物組織学　plant histology
しょくぶつたい　植物体　plant body
しょくぶつたんぱくせき　植物タンパク石, プラントオパール　plant opal
しょくぶつちりがく　植物地理学　plant geography, phytogeography
しょくぶつどく　植物毒　phytotoxin, plant poison
しょくぶつどくせい　植物毒性　phytotoxicity
しょくぶつどくそ　植物毒素　phytotoxin
しょくぶつびょうりがく　植物病理学　plant pathology, phytopathology
しょくぶつぶんるいがく　植物分類学　plant taxonomy
しょくぶつほご　植物保護　plant protection
しょくぶつホルモン　植物ホルモン　plant hormone, phytohormone
しょくぶつゆ　植物油　vegetable oil
しょくぶつゆし　植物油脂　vegetable oil and fat
しょくぶん　植分　stand
しょくみ　食味　palatability, eating quality, taste
しょくみけんていしゃ　食味検定者　taste panelist
しょくみしけん　食味試験　eating quality test
しょくもつもう　食物網　food web
しょくもつれんさ　食物連鎖　food chain
ショクヨウガヤツリ　tiger nut, *Cyperus esculentus* L.
しょくようカンナ　食用カンナ　edible canna, purple arrowroot, Queensland arrowroot, achira, *Canna edulis* Ker-Gawl.
しょくようさくもつ　食用作物　food

crop
しょくようばな 食用花 edible flower
しょくりょうじきゅうりつ 食糧(食料)自給率 food self-sufficiency rate
しょくりょうじゅきゅう 食糧(食料)需給 demand and supply of food
しょくりょう・のうぎょう・のうそんきほんほう 食料・農業・農村基本法 The Basic Law on Food, Agriculture and Rural Areas
しょくりょう・ひりょうぎじゅつセンター 食糧・肥料技術センター Food and Fertilizer Technology Center (FFTC)
しょこう 藷梗, なり首【サツマイモ】 joint (upper) part of tuberous root
じょこうそ 助酵素⇒補酵素 coenzyme
じょさいぼう 助細胞, 助胎細胞 synergid
じょしつ 除湿 dehumidifying
しょじょせいしょく 処女生殖, 単為発生, 単為生殖 parthenogenesis
しょせいしぶ 初生篩部⇒一次篩部 primary phloem
しょせいへきこういき 初生壁孔域 primary pit-field
しょせいもくぶ 初生木部⇒一次木部 primary xylem
しょせいよう 初生葉 primary leaf
じょそう 除草 weeding
じょそうき 除草機 weeder
じょそうざい 除草剤 herbicide
じょそうざいたいせい 除草剤耐性 herbicide tolerance
じょそうたいけい 除草体系 weeding system
じょたいさいぼう 助胎細胞, 助胎細胞 synergid
ジョチュウギク(除虫菊) insectpowder plant, insect flower, Dalmatian chrysanthemum, *Pyrethrum cinerariifolium* Trevir.

しりょ (77)

(= *Chrysanthemum cinerariaefolium* Visiani)
ショトウ ショ(蔗)糖, スクロース sucrose
しょはつしおれ 初発しお(萎)れ incipient wilting
じょゆう 除雄 emasculation, castration
しょり 処理 treatment
ジョルダンにっしょうけい ジョルダン日照計 Jordan's sunshine recorder
シーよんけいろ C_4 経路 C_4 pathway, C_4 ジカルボン酸回路 C_4-dicarboxylic acid cycle
シーよんジカルボンさんかいろ C_4 ジカルボン酸回路 C_4-dicarboxylic acid cycle, C_4 経路 C_4 pathway
シーよんしょくぶつ C_4 植物 C_4 plant
ジョンソングラス Johnsongrass, *Sorghum halepense* (L.) Pers.
しらきぬびょう 白絹病【トウモロコシ, ダイズ, インゲンマメ】 stem rot, southern blight
シラゲガヤ(白毛茅)→ベルベットグラス
シーラゴム⇒マニホットゴム
しらしめゆ 白絞油 refined oil
しらはがれびょう 白葉枯病【イネ】 bacterial leaf blight
しらほ 白穂 white head
しりぐされ 尻腐れ blossom-end rot
しりょう 飼料 feed [stuff]
しりょうか[ち] 飼料価[値] feeding value, forage value
しりょうカブ 飼料カブ(蕪, 蕪菁) turnip, *Brassica rapa* L.
しりょうさくもつ 飼料作物 forage crop, fodder crop
しりょうせいぶん 飼料成分 feed composition
しりょうばた 飼料畑 forage crop field
しりょうぼく 飼料木 fodder tree

(shrub)
しりょうようビート　飼料用ビート　fodder beet, mangold, field beet, *Beta vulgaris* L. var. *alba* DC.
ジリンマメ　jiring, *Pithecellobium jiringa* (Jack) Prain
シルト　silt
しろかき　代か(掻)き　puddling and levelling
しろかきき　代か(掻)き機　implement for puddling
シロガラシ(白芥子)　white mustard, *Sinapis alba* L. (= *Brassica alba* (L.) Boiss.)
シロクローバ　white clover, *Trifolium repens* L.
しろこ　白子⇒アルビノ　albino
シロザ　common lamb's-quarters, *Chenopodium album* L. var. *album*
シロツメクサ→シロクローバ
シロバナスィートクローバ(白花スィートクローバ)　white sweetclover, *Melilotus albus* Medik.
シロバナヨウシュチョウセンアサガオ→ヨウシュチョウセンアサガオ
シロバナルーピン(白花ルーピン)　white lupine, *Lupinus albus* L.
じん　仁, 核小体　nucleolus (*pl.* nucleoli)
じん　仁【種子】　kernel
じんいこうはい　人為交配, 人工交配　artificial crossing
じんいせんたく　人為選択, 人為選抜　artificial selection
じんいせんばつ　人為選抜, 人為選択　artificial selection
じんいとうた　人為淘汰⇒人為選択　artificial selection
じんいとつぜんへんい　人為突然変異　artificial mutation
しんか　進化　evolution
じんか　仁果, ナシ状果　pome, pomaceous fruit
しんかくせいぶつ　真核生物　eucaryote
じんかるい　仁果類　pomaceous fruits
しんかろん　進化論　evolution theory
シンク　sink
しんぐされ　心腐れ　heart rot
しんこう　深耕　deep plowing, deep tillage
しんごう　信号　signal
じんこうかんそう　人工乾燥　artificial drying
じんこうきしょうしつ　人工気象室　climatron, artificial climate room
じんこうこう　人工光　artificial light
じんこうこうげん　人工光源　artificial light source
じんこうこうはい　人工交配, 人為交配　artificial crossing
じんこうじゅふん　人工授粉　artificial pollination, hand pollination
じんこうしょうど　人工床土　commercial bed soil, artificial bed soil
じんこうしょうめい　人工照明　artificial lighting, artificial illumination
じんこうせっしゅ　人工接種　artificial inoculation
じんこうそうち　人工草地　artificial grassland (pasture), sown grassland (pasture), tame pasture
じんこうちのう　人工知能　artificial intelligence (AI)
じんこうらくよう　人工落葉　chemical defoliation
しんこんせいさくもつ　深根性作物　deep-rooted crop
しんこんせいの　深根性の　deep-rooted
ジンジャーグラス　gingergrass, *Cymbopogon martini* Stapf var. *sofia*
しんし　浸種　seed soaking
しんしょく　侵食　erosion
しんしょくぼうし　侵食防止　erosion control

しんすいせいの　親水性の　hydrophilic
しんすいせいれん　浸水精練【繊維作物】　water retting
しんすいとう ⇒ ふかみずいね　深水稲　deepwater rice
しんせいしゅし　真正種子　true [potato] seed (TPS)
しんせいちゅうしんちゅう　真正中心柱　eustele
しんせいていこうせい　真正抵抗性　true resistance
しんせいラベンダー　真正ラベンダー → ラベンダー
しんせんじゅう　新鮮重 ⇒ 生体重　fresh weight
しんそうせひ　深層施肥　deep placement of fertilizer, deep application of fertilizer
しんそうついひ　深層追肥　supplement application to deep layer
じんちくどくせい　人畜毒性　toxicity to mammals
しんちょう　伸長　elongation
しんちょうせい　伸長性　extensibility
しんでん　新田　reclaimed paddy field
しんてんせい　伸展性　extensibility
しんど　心土, 下層土　subsoil
しんとう　浸透　1) osmosis【水分生理】2) percolation【土壌水分】
しんとう　浸透, 透入　penetration
しんとうあつ　浸透圧　osmotic pressure
しんとう[あつ]けい　浸透[圧]計　osmometer
しんとうあつストレス　浸透圧ストレス　osmotic stress
しんとうあつちょうせつ　浸透圧調節, 浸透調整　osmotic adjustment
しんとうこうざつ　浸透交雑, 移入交雑, 導入交雑　introgression, introgressive hybridization
しんとうせいじょそうざい　浸透性除草剤　systemic herbicide

しんとうちょうせい　浸透調整, 浸透圧調節　osmotic adjustment
しんとうポテンシャル　浸透ポテンシャル　osmotic potential
しんどこう　心土耕　subsoiling, subsoil plowing
しんどじめ　心土締め ⇒ 床じめ　subsoil compaction
しんどはさい　心土破砕　subsoiling
しんどはさいき　心土破砕機　subsoil breaker
しんどプラウ　心土プラウ, サブソイラ　subsoiler
しんどまり　心止り　self-topping
しんどめ　心止め, 摘心【タバコ】　topping
しんにゅう　侵入【雑草】　invasion, intrusion
しんにゅうせいちょう　侵入成長, 割込み成長　intrusive growth
しんのこうごうせい　真の光合成　true photosynthesis, 総光合成　gross photosynthesis
しんぱくまい　心白米　white core rice
しんぴ　心皮　carpel
じんぴ　じん(靭)皮　bast
じんぴせんい　じん(靭)皮繊維　bast fiber
シンプラスト　symplast
しんまい　新米　new [crop] rice
しんめ　新芽【チャ】　new shoot, sprouting shoot
しんよう　新葉　new leaf
しんらいくかん　信頼区間　confidence interval
しんらいげんかい　信頼限界　confidence limit
しんりんそうごうけんきゅうしょ　森林総合研究所　Forestry and Forest Products Research Institute
しんわせい　親和性　affinity

[す]

す (鬆)【ダイコン等】 pithy tissue
す 酢 vinegar
ずい 髄 pith
すいいかくりつ 推移確率 transition probability
すいおん 水温 water temperature
スイカ (西瓜) watermelon, *Citrullus lanatus* (Thunb.) Matsum. et Nakai (= *C. vulgaris* Schrad.)
すいがい 水害 flood damage
すいぎん 水銀 mercury (Hg)
すいぎんおんどけい 水銀温度計 mercury thermometer
すいこう 水耕 water culture
すいこう 水孔 water pore
ずいこう 髄腔 medullary cavity, pith cavity
すいこうさいばい 水耕栽培 hydroponics, water culture
すいさんかカリウム 水酸化カリウム potassium hydroxide
すいさんかカルシウム 水酸化カルシウム calcium hydroxide
すいさんかだいにてつ 水酸化第二鉄 ferric hydroxide
すいじく 1) 穂軸【イネ・ムギ類他】 rachis 2) 穂軸, 穂心【トウモロコシ】 cob
すいしつおだく 水質汚濁 water pollution
すいじゅん 水準 level
すいじょうかじょ 穂状花序 spike
すいじょうきあつさ 水蒸気圧差 vapour pressure deficit (VPD)
すいしょく 水食 water erosion
すいしん 穂心, 穂軸【トウモロコシ】 cob
すいせいざっそう 水生雑草 aquatic weed
すいせいしょくぶつ 水生植物 aquatic plant, hydrophyte
すいせん 水洗 rinsing
すいせん 水選 selection with water
すいそイオンのうど 水素イオン濃度 hydrogen ion concentration
すいそうぶんれつ 垂層分裂 anticlinal division
すいちょくていこうせい 垂直抵抗性 vertical resistance
すいちょくようがたの 垂直葉型の erectophyll
スイッチグラス switchgrass, *Panicum virgatum* L.
すいてい 推定 estimation
すいていち 推定値 estimate, estimated value
すいていりょう 推定量 estimator
すいでん 水田 paddy field
すいでんうらさく 水田裏作 winter cropping on drained paddy field
すいでんさく 水田作 paddy-field cropping
すいでんさくもつ 水田作物 paddy-field crop
すいでんてんかんさくもつ 水田転換作物 switch crop from rice culture
すいでんどじょう 水田土壌 paddy soil
すいとう 水稲 paddy rice, lowland rice, paddy
スイートオレンジ sweet orange, *Citrus sinensis* Osbeck
スイートコーン sweet corn, *Zea mays* L. var. *saccharata* Bailey
スイートバーナルグラス sweet vernalgrass, *Anthoxanthum odoratum* L.
スイバ green sorrel, *Rumex acetosa* L.
すいばいでんせん 水媒伝染 water transmission
ずいはんさくもつ 随伴作物, 同伴作物 companion crop
ずいはんざっそう 随伴雑草

companion weed
すいはんとくせい 炊飯特性 boiling characteristics of rice
すいぶん 水分 moisture
すいぶんがんりょう 水分含量, 含水量 water content, moisture content
すいぶんくっせい 水分屈性 hydrotropism
すいぶんけいざい 水分経済 water economy
すいぶんけつぼう 水分欠乏 water deficit
すいぶんせいり 水分生理, 水関係 water relations
すいへいうえ 水平植え horizontal planting
すいへいうえ 水平植え, 水平挿し【サツマイモ】 horizontal planting
すいへいざし 水平挿し, 水平植え【サツマイモ】 horizontal planting
すいへいていこうせい 水平抵抗性 horizontal resistance
すいへいぶんぷ 水平分布 horizontal distribution
すいへいようがたの 水平葉型の planophyll
すいみんうんどう 睡眠運動, 就眠運動 sleep movement, nyctinastic movement, nyctinasty
すいもんがく 水文学 hydrology
すいようざい 水溶剤 water-soluble concentrate
すいようせいたんすいかぶつ 水溶性炭水化物 water-soluble carbohydrate
すいり す(鬆)入り【ダイコン等】 pithiness
すいりくみあい 水利組合 water utilization association, irrigation association
すいりけん 水利権 water right
すいろ 水路 creek, canal
すいわざい 水和剤 water-dispersible powder, wettable agent (powder)

スウェーデンカブ→ルタバガ
すうがくモデル 数学モデル mathematical model
スーダングラス Sudan grass, *Sorghum sudanense* (Piper) Stapf
すうちせきぶん 数値積分 numerical integration
すうちちず 数値地図 digital map
すうちひょうこうちず 数値標高地図 digital elevation map
すうりょうか 数量化 quantification
すうりょうぶんるい[がく] 数量分類[学] numerical taxonomy
すえこ 末粉, さなご【ソバ】 sanago, dark flour
スオウ(蘇芳) sappanwood, *Caesalpinia sappan* L.
スカシタゴボウ marsh yellowcress, *Rorippa islandica* (Oeder) Borb.
すき(鋤) spade
すきおこし すき(犂)起こし plowing
すきこみ すき(犂)込み plowing-in
すきどこ すき(犂)床⇒耕盤 plow sole (pan), furrow pan
スギナ field horsetail, *Equisetum arvense* L.
スクロース, ショ(蔗)糖 sucrose
スクロースリンさんシンターゼ スクロースリン酸シンターゼ sucrose-phosphate synthase
スコア, 評点 score
すじはがれびょう すじ(条)葉枯病【イネ】 cerospora leaf spot
すじまき すじ播き⇒条播 drilling, row sowing (seeding)
すじまきき すじ播き機, 条播機 drill
ずじょうかんすい 頭上灌水 overhead watering, overhead irrigation
ススキ(薄) Japanese plume-grass, eulalia grass, *Miscanthus sinensis* Andersson
スズメノエンドウ tiny vetch, *Vicia*

hirsuta (L.) S. F. Gray
スズメノカタビラ　annual bluegrass, *Poa annua* L.
スズメノコビエ→コドラ
スズメノチャヒキ　Japanese brome, *Bromus japonicus* Thunb. ex. Murray
スズメノテッポウ　water foxtail, *Alopecurus aequalis* Sobol. var. *amurensis* (Komar.) Ohwi
スズメノトウガラシ　*Vandellia anagallis* (Burm.f.) Yamazaki var. *verbenaefolia* (Colsm.) Yamazaki
スズメノヒエ　*Paspalum thunbergii* Kunth ex Steud.
すすもんびょう　すす紋病【トウモロコシ】leaf blight, northern leaf blight
スターフルーツ→ゴレンシ
スタイロ　stylo, *Stylosanthes guianensis* (Aubl.) Sw. 他数種
スチューデントかされたはんい　スチューデント化された範囲　Studentized range
ステアリンさん　ステアリン酸　stearic acid
ステージ【顕微鏡】stage
ステップ　steppe
ステビア, アマハステビア　stevia, kaa he-e, *Stevia rebaudiana* (Bertoni) Hemsl.
ストレス　stress
ストレスたいせいしゅ　ストレス耐性種　stress tolerator
ストロベリクローバ　strawberry clover, *Trifolium fragiferum* L.
ストロマ　stroma (*pl.* stromata)
ストロン, ほふく (匍匐) 枝, ふく (匐) 枝　stolon, runner
スパイクラベンダー　spike lavender, broadleaved lavender, *Lavandula latifolia* Medik. (= *L. spica* DC.)
スパチュラ, へら　spatula
スピアマンのじゅんいそうかん　スピアマンの順位相関　Spearman's rank correlation
スピードスプレーヤ　speed sprayer
スブタ　*Blyxa ceratosperma* Maxim.
スプリングフラッシュ【牧草】spring flush
スプリンクラ　sprinkler
スペアミント　spearmint, *Mentha spicata* L. (= *M. viridis* L.)
スペクトル　spectrum (*pl.* spectra)
スペクトルかいせき　スペクトル解析　spectral analysis
スベリヒユ　common purslane, *Portulaca oleracea* L.
スベリン　suberin
スベリンか　スベリン化, コルク化　suberization
スペルトコムギ　spelt [wheat], *Triticum spelta* L.
スペルミジン　spermidine
スペルミン　spermine
すまき　巣播き⇒摘播 (てきは)　nest sowing (seeding)
すみくろほびょう　墨黒穂病【イネ】kernel smut
スムーズブロムグラス　smooth bromegrass, *Bromus inermis* Leyss.
すやきばち　素焼鉢　clay pot, unglazed pot
スライドグラス　slide glass, slide
スラグ, 鉱さい (滓)　slag
スラリー　slurry
スレッシャ, 脱穀機　thresher
スレンダーホィートグラス　slender wheatgrass, *Agropyron trachycautum* Link (= *A. pauciflorum* Hitchc.)
スワンプタロ　[giant] swamp taro, *Cyrtosperma chamissonis* (Schott) Merr.
スンプほう　スンプ法　SUMP method

[せ]

ゼアチン　zeatin

せいいく　生育　growth and development, growth
せいいくかてい　生育過程　growing process
せいいくしんだん　生育診断　growth diagnosis
せいいくそくしん　生育促進　growth promotion, growth stimulation
せいいくち　生育地　habitat
せいえん　成園【チャ】　mature tea field
せいかがく　生化学　biochemistry
せいかがくてきさんそようきゅうりょう　生化学的酸素要求量　biochemical oxygen demand (BOD)
せいかく　精核　sperm nucleus, 雄核 male nucleus
せいかつかん　生活環　life cycle
せいかつし　生活史　life history
せいきこうがく　生気候学　bioclimatology
せいきしょうがく　生気象学　biometeorology
せいきせいけんてい　正規性検定　test of normality
せいきぶんぷ　正規分布　normal distribution
せいきほうていしき　正規方程式　normal equation
せいきぼしゅうだん　正規母集団　normal population
せいきみつど　正規密度　normal density
せいき[みつど]きょくせん　正規[密度]曲線　normal [density] curve
せいぎゃくこうざつ　正逆交雑, 相反交雑　reciprocal crosses (crossing)
せいぎょ　制御　control
せいきらんすう　正規乱数　normal random number
せいぐんしゅうだんせんばつ　成群集団選抜　group-mass selection
せいぐんせんばつほう　成群選抜法　group selection

せいけいばいち　成形培地　nursery mat
せいげんいんし　制限因子, 限定要因　limiting factor
せいげんこうそ　制限酵素　restriction enzyme
せいげんだんぺんちょうたけい　制限断片長多型　restriction fragment length polymorphism (RFLP)
せいこ　精粉【コンニャク】　refined flour
せいこうさいばい　清耕栽培　clean culture
せいこうさくもつ　清耕作物　cleaning crop
せいごうせい　生合成　biosynthesis
せいこうはい　正交配　straight cross
せいさいぼう　性細胞　sexual cell
せいさん　生産, 生産高, 生産量　production, output
せいさんかじょう　生産過剰　over production
せいさんけいしつ　生産形質　production trait
せいさんこうぞう　生産構造　productive structure
せいさんしゃ　生産者　producer
せいさんしゃべいか　生産者米価　producer rice price
せいさんだか　生産高, 生産, 生産量　production, output
せいさんち　生産地　producing area
せいさんちゅうどく　青酸中毒　cyanide toxicity
せいさんちょうせい　生産調整　production adjustment, production control
せいさんひ　生産費　production cost
せいさんぶつそがい　生産物阻害　product inhibition
せいさんりょう　生産量, 生産, 生産高　production, output
せいさんりょく　生産力　productivity, yield potential

せいさんりょくけんてい [しけん] 生産力検定 [試験] performance test, yield trial

せいさんりょくけんていよびしけん 生産力検定予備試験 preliminary performance test, preliminary yield trial

せいし 整枝 1) training, trimming 2) skiffing【チャ】

せいしき 静止期【細胞分裂】 resting stage, quiescent state

せいしちゅうしん 静止中心 quiescent center

せいじゅく 成熟 1) maturation, maturity 2) ripening【果実】

せいじゅんそうかん 正準相関 canonical correlation

せいじゅんはんべつぶんせき 正準判別分析 canonical discriminant analysis

せいじゅんぶんせき 正準分析⇒正準判別分析

せいじょう せい(臍)条 raphe

せいじょううえ 正条植え regular planting

せいじょうたい 星状体 aster

せいしょく 生殖 reproduction

せいしょくかく 生殖核 germ nucleus

せいしょくきかん 生殖器官 reproductive organ

せいしょくさいぼう 生殖細胞 reproductive cell

せいしょくしつ 生殖質 germplasm

せいしょくしつほぞん 生殖質保存 germplasm preservation

せいしょくせいちょう 生殖成長 reproductive growth

せいしょくせいちょうき 生殖成長期 reproductive stage

せいしょくそう 生殖相 reproductive phase

せいしりょうさくもつ 製紙料作物 paper-making crop

せいせい 精製 purification

せいせいとう 精製糖 refined sugar

せいせっかい 生石灰 calcium oxide, calx, quicklime

せいせんしょくたい 性染色体 sex chromosome

せいそう 生草, 青刈り飼料 fresh forage, green forage, soilage, green chop

せいそうしゅうりょう 生草収量 fresh herbage yield, fresh forage yield

せいぞんきょうそう 生存競争 struggle for existence

せいぞんねんげん 生存年限 persistency, longevity

せいぞんりつ 生存率 viability, survival rate

せいたいがいで 生体外で, インビトロ in vitro

せいたいがく 生態学 ecology

せいたいがた 生態型 ecotype

せいたいけい 成体形 adult form

せいたいけい 生態系 ecosystem

せいたいこうがく 生体工学, バイオニクス bionics

せいたいしゅ 生態種 ecospecies

せいたいじゅう 生体重 1) fresh weight 2) live weight (LW)【家畜】

せいたいせいりがく 生態生理学 ecophysiology

せいたいせんしょく 生体染色 vital staining

せいたいてきかくり 生態的隔離 ecological isolation

せいたいてきぼうじょ 生態的防除 ecological [pest] control

せいたいないで 生体内で, インビボ in vivo

せいたいまく 生体膜 biomembrane, biological membrane

セイタカアワダチソウ tall goldenrod, *Solidago altissima* L.

セイタカウコギ→アメリカセンダングサ

ぜいたくきゅうしゅう ぜいたく吸収

luxury absorption
せいち　整地　land grading, ground making
せいちゃ　製茶　tea manufacturing
せいちゅう　成虫　adult
せいちゅんだん　正中断【形態】　median section
せいちょう　成長　growth
せいちょういんし　成長因子, 成長要因　growth factor
せいちょううんどう　成長運動　growth movement
せいちょうかいせき　成長解析　growth analysis
せいちょうきょくせん　成長曲線　growth curve
せいちょうこうりつ　成長効率　growth efficiency
せいちょうこきゅう　成長呼吸, 構成呼吸　growth respiration
せいちょうしゅうせい　成長習性　growth habit
せいちょうそうかん　成長相関　growth correlation
せいちょうそがいぶっしつ　成長阻害物質　growth inhibitor
せいちょうそくど　成長速度, 成長率　growth rate
せいちょうちょうせつ　成長調節　growth regulation
せいちょうちょうせつぶっしつ　成長調節物質　growth regulator
せいちょうてん　成長点　growing point
せいちょうてんばいよう　成長点培養　growing point culture
せいちょうぶっしつ　成長物質　growth substance
せいちょうホルモン　成長ホルモン　growth hormone
せいちょうモデル　成長モデル　growth model
せいちょうよういん　成長要因, 成長因子　growth factor
せいちょうよくせいざい　成長抑制剤, 成長抑制物質, わい(矮)化剤　growth retardant
せいちょうよくせいぶっしつ　成長抑制物質, 成長抑制剤, わい(矮)化剤　growth retardant
せいちょうりつ　成長率, 成長速度　growth rate
せいてきかくり　性的隔離　sexual isolation
せいてきしんわせい　性的親和性　sexual affinity
せいでんようりょうがたしつどけい　静電容量型湿度計　capacitance humidity sensor
せいのうけんてい　性能検定　performance test
せいはく　精白　milling, polishing
せいばく　精麦　1) pearled barley 2) pearling of barley
せいはくまい　精白米⇒精米　milled rice, polished rice
せいぱんてきせい　製パン適性　baking quality
セイバンモロコシ→ジョンソングラス
せいひ　性比　sex ratio
せいびょう　成苗　mature seedling with 5.0-7.0 leaf stage
せいびょうりつ　成苗率【チャ】　ratio of good nursery plant
せいぶつがくてきしゅうりょう　生物学的収量　biological yield
せいぶつきかいがく　生物機械学　biomechanics
せいぶつきせつがく　生物季節学　phenology
せいぶつけん　生物圏　biosphere
せいぶつけんてい　生物検定, バイオアッセイ　bioassay
せいぶつこうがく　生物工学　biological engineering,

bioengineering
せいぶつそくていがく　生物測定学, 計量生物学　biometrics, biometry
せいぶつ[たい]りょう　生物[体]量, バイオマス, 現存量　biomass, standing crop
せいぶつたようせい　生物多様性　biodiversity, biological diversity
せいぶつてきじょうか　生物的浄化, バイオレメディエーション　bioremediation
せいぶつてきじょそう　生物的除草　biological weed control
せいぶつてきふうじこめ　生物的封じ込め　biological containment
せいぶつてきぼうじょ　生物的防除　biological [pest] control
せいぶつとうけいがく　生物統計学　biostatistics
せいぶつどけい　生物時計　biological clock
せいぶつのうやく　生物農薬　biocide, biopesticide, biotic pesticide
せいぶつぶつり[がく]　生物物理[学]　biophysics
せいぶつリズム　生物リズム, バイオリズム　biorhythm
せいふん　製粉　milling
せいふんき　製粉機　flour mill
せいふんしけん　製粉試験　milling test
せいふんせい　製粉性　milling quality
せいふんぶあい　製粉歩合, 製粉歩留り　flour milling percentage, flour yield
せいふんぶどまり　製粉歩留り, 製粉歩合　flour yield, flour milling percentage
せいぶんりょう　成分量　quantity of element
せいほうけいうえ　正方形植え　square planting
せいぼく　成木　adult tree
せいまい　精米　1) milled rice, polished rice　2) rice milling
せいまいき　精米機　rice mill, rice milling machine
せいみつさいばいしけん　精密栽培試験　precise culture experiment
せいみつほじょうかんり　精密圃場管理　precision farming
せいめいかがく　生命科学　life science
せいもみ　精籾　winnowed rough rice
せいやくじょうけん　制約条件　constraint condition
せいゆ　精油　essential oil
せいよう　成葉　mature leaf
せいよう　生葉　fresh leaf, green leaf
セイヨウアカネ(西洋茜)　common madder, madder, *Rubia tinctorum* L.
セイヨウアブラナ→洋種ナタネ
セイヨウカボチャ　pumpkin, winter squash, *Cucurbita maxima* Duch. ex Lam.
セイヨウカラシナ　brown mustard, *Brassica juncea* (L.) Czern. et Cross.
セイヨウグリ　European chestnut, *Castanea sativa* Mill.
セイヨウタンポポ　dandelion, *Taraxacum officinale* Weber
セイヨウナシ　pear, *Pyrus communis* L. var. *sativa* DC.
セイヨウネズ→セイヨウビャクシン
セイヨウハッカ→ペパーミント
セイヨウビャクシン(洋種杜松)　common juniper, *Juniperus communis* L.
セイヨウミヤコグサ→バーズフット・トレフォイル
セイヨウワサビ(西洋山葵)→ワサビダイコン
せいりがく　生理学　physiology
せいりかっせいぶっしつ　生理活性物質　physiological active substance
せいりしょうがい　生理障害　physiological disorder
せいりせいたいがく　生理生態学　physiological ecology

せいりてきいんし　生理的因子　physiological factor

せいり[てき]しょくえんすい　生理[的]食塩水　physiological salt solution

せいりてきはんてんびょう　生理的斑点病【タバコ】　physiological leaf spot, weather fleck

せいりびょう　生理病　physiological disease

せいりひんしゅ　生理品種, 生理変種　physiological race

せいりへんしゅ　生理変種, 生理品種　physiological race

せいりゅう　整粒　whole grain

せいれん　精練【アサ】　retting

セイロンシトロネラソウ　Ceylon citronella grass, *Cymbopogon nardus* (L.) Rendle

セイロンニッケイ→シナモン

セインフォイン　sainfoin, *Onobrychis viciifolia* Scop.

セージ　sage, *Salvia officinalis* L.

ゼオカルパマメ　geocarpa bean, *Macrotyloma geocarpum* (Harms) Maréchal et Baudet (= *Kerstingiella geocarpa* Harms)

せがわの　背側の　dorsal

せきか　石果　⇒核果　drupe

せきがいせん　赤外線　infrared radiation

せきがいせんガスぶんせきけい　赤外線ガス分析計　infrared gas analyzer

せきさいぼう　石細胞　stone cell

せきさんうりょうけい　積算雨量計　totalizing rain gauge

せきさんおんど　積算温度　cumulative temperature, accumulated temperature

せきしょくど　赤色土　Red soil

せきせつ　積雪　snow cover

せきどうばん　赤道板【細胞】　equatorial plate

せきどうめん　赤道面【細胞】　equatorial plane

せきりつ　積率【統計】, モーメント　moment

せきわ　積和　sum of products

せじろまい　背白米　white-back rice

せだい　世代　generation

せだいかんかく　世代間隔　generation interval

せだいこうたい　世代交代, 世代交番　alternation of generation

せだいこうばん　世代交番, 世代交代　alternation of generation

せだいそくしん　世代促進　accelerated generation advancement, 世代短縮　shortening of breeding cycle

せだいたんしゅく　世代短縮　shortening of breeding cycle, 世代促進　accelerated generation advancement

せつ　節　1) node　2) section【分類】

せつい　節位　node order

せつえいほう　切えい(穎)法　clipping method

せっかい　石灰　lime

せつがい　雪害　snow damage (injury)

せっかいいおうごうざい　石灰硫黄合剤　lime sulfur

せっかいさんぷき　石灰散布機　lime sower, lime spreader

せっかいしつどじょう　石灰質土壌　calcareous soil

せっかいちっそ　石灰窒素　calcium cyanamide

せっかん　節間　internode

せっかんしんちょう　節間伸長　internode elongation

せっかんしんちょうき　節間伸長期　internode elongation stage

せつがんレンズ　接眼レンズ【顕微鏡】　eyepiece, ocular

せっきょくてききゅうしゅう　積極的吸収　active absorption

せつごう　接合【染色体】　synapsis (*pl.*

synapses), syndesis (*pl.* syndeses)
せつごうし 接合子, 接合体 zygote
せつごうたいちし 接合体致死 zygotic lethal
せつごうたいひ 接合体比 zygotic ratio
せっこん 節根 nodal root
せっしゅ 接種 inoculation
せっしゅげん 接種原 inoculum
せっしゅ[りょう] 摂取[量] intake
せつじょ 切除 clipping
せっしょくけいたいけいせい 接触形態形成 thigmomorphogenesis
せっしょくざい 接触剤 1) contact herbicide【雑草】 2) contact insecticide【害虫】
せっしょくせいじょそうざい 接触性除草剤 contact herbicide
せっしょくでんせん 接触伝染 contact transmission
せっすいさいばい 節水栽培 water-saving culture
せっせんぶんれつ 接線分裂【細胞分裂】 tangential division
ぜったいおんど 絶対温度 absolute temperature
ぜったいしつど 絶対湿度 absolute humidity
ぜったいちがた 絶対値型【ガス分析】 absolute type
せつだん 切断【染色体】 fragmentation, breakage
せつだんがたせんばつ 切断型選抜 selection by truncation, truncation selection
せつだんぶんぷ 切断分布 truncated distribution
せつだんよう 切断葉 detached leaf
せっちきこう 接地気候 climate near the ground
せっちゅうなわしろ 折衷苗代 semi-irrigated rice nursery
ゼットスキーム【光合成】 Z scheme

せつめいへんすう 説明変数【回帰分析】 explanatory variable
セティゲルムケシ *Papaver setigerum* DC.
セネガ senega, seneca, snake root, *Polygala senega* L.
せひ 施肥 fertilization, fertilizer application, manuring
せひき 施肥機 fertilizer applicator, fertilizing machine
せひきじゅん 施肥基準 standard application rate of fertilizer
せひこう 施肥溝 dressing furrow
せひじき 施肥時期 time of fertilizer application
せひせっけい 施肥設計 design for fertilizer application
せひはしゅき 施肥播種機 drill, fertilizer drill
せひほう 施肥法 method of fertilizer application
せひりょう 施肥量 amount of fertilizer application
セファデックス Sephadex
セライト Celite
ゼラチン gelatin
セラデラ serradella, *Ornithopus sativus* Brot.
ゼラニウム geranium, *Pelargonium* spp.【*P. graveolens* L'Her., *P. radura* L'Her. など】
セリ dropwort, *Oenanthe javanica* (Blume) DC.
セリン serine (Ser)
セルラーゼ cellulase
セルロース cellulose
セロイジンほう セロイジン法 celloidin method
せん 腺 gland
せんい 繊維 fiber
せんい 遷移 succession
せんいけいれつ 遷移系列 sere
ぜんがり 全刈り whole sampling

ぜんき　前期【細胞分裂】　prophase
ぜんきぜんびしょうかんそく　前期前微小管束　preprophase band (PPB)
センキュウ(川芎)　*Cnidium officinale* Makino
ぜんきゅうそくいシステム　全球測位システム　global positioning system (GPS)
せんぎょうのうか　専業農家　full-time farmer
ぜんくぶっしつ　前駆物質　precursor
せんけいかいき　線形回帰　linear regression
せんけいけいかくほう　線形計画法　linear programming
せんけいけつごう　線形結合　linear combination
せんけいせい　線形性　linearity
ぜんけいせいそう　前形成層　procambium
せんけいへんかん　線形変換　linear transformation
せんけいもけい　線形模型　linear model
せんこう　せん(穿)孔　perforation
せんこう　浅耕　shallow tillage
せんこうせいがいちゅう　せん(穿)孔性害虫　borer
せんこうばん　せん(穿)孔板　perforation plate
ぜんこくへいきん　全国平均　national average
せんこん　せん(剪)根, 断根　root pruning
せんこんせいさくもつ　浅根性作物　shallow-rooted crop
せんこんせいの　浅根性の　shallow-rooted
センサー　sensor
せんざいちりょく　潜在地力　potential soil fertility, potential soil productivity
ぜんさく　前作　preceding cropping

ぜんさく[もつ]　前作[物]　preceding crop
せんし　せん(剪)枝【チャ】　pruning
せんしゅ　選種　seed grading, seed selection
ぜんしゅつよう　前出葉, 前葉　prophyll
せんじょうち　扇状地　alluvial fan
せんしょく　染色　staining
せんしょくしつ　染色質, クロマチン　chromatin
せんしょくたい　染色体　chromosome
せんしょくたいいじょう　染色体異常　chromosome aberration
せんしょくたいちかん　染色体置換　chromosome substitution
せんしょくたいちず　染色体地図　chromosome map
せんしょくたいとつぜんへんい　染色体突然変異　chromosomal mutation
せんしょくたいばいか　染色体倍加　doubling of chromosome, chromosome doubling
ぜんしょり　前処理　pretreatment
ぜんそうせひ　全層施肥　uniform application of fertilizer to top soil
ぜんそうまき　全層播き　broadcasting with rotary cultivation, broadcasting with mixing-in soil layer
せんぞがえり　先祖返り　atavism, reversion
せんたく　選択, 選抜　selection
せんたくかぶ　選択株　elite plant
せんたくきゅうしゅう　選択吸収　selective absorption
せんたくじゅせい　選択受精　selective fertilization
せんたくせいじょそうざい　選択性除草剤　selective herbicide
せんたくてきとうかせい　選択的透過性　selective permeability
せんだつ　洗脱【土壌】⇒溶脱　leaching, eluviation

センダングサ *Bidens biternata* (Lour.) Merr. et Sherff.
ぜんちっそ　全窒素　total nitrogen
せんちゃ　煎茶　Sencha
せんちゅう　線虫　nematode
せんてい　せん(剪)定　pruning
ぜんていじょうけん　前提条件　assumption
セントロ　centro, *Centrosema pubescens* Benth.
センナ　senna, *Senna angustifolia* Batka および *S. alexandrina* Mill. (= *Cassia senna* L.)
センニンコク　grain amaranth, *Amaranthus caudatus* L. (= *A. edulis* Spegazzini)
せんねつ　潜熱　latent heat
ぜんのうせい　全能性, 分化全能性　totipotency
ぜんはい　前胚　proembryo
せんば[こき]　千歯[こ(扱)き]　comb thresher, threshing comb
せんばつ　選抜, 選択　selection
せんばつ(せんたく)あつ　選抜(選択)圧　selection pressure
せんばつ(せんたく)きょうど　選抜(選択)強度　selection intensity
せんばつ(せんたく)しすう　選抜(選択)指数　selection index
せんばつ(せんたく)じっけん　選抜(選択)実験　selection experiment
せんばつ(せんたく)マーカー　選抜(選択)マーカー　selection marker
せんばつ(せんたく)もくひょう　選抜(選択)目標　selection objective
せんばつ(せんたく)りつ　選抜(選択)率　percentage of selection, selection rate
ぜん(げん)ひょうひ　前(原)表皮　protoderm
せんぷくき[かん]　潜伏期[間]　incubation period, latent period
ぜんぶんれつそしき　前分裂組織　promeristem
せんべい(煎餅)　rice flake, rice cracker
せんべつ　選別　1) grading, sorting 2) sorting【タバコ】
せんべつき　選別機　grader, sorter, sizer
ぜんめんさんぷ　全面散布　broadcasting
せんもう　腺毛　glandular hair, glandular trichome
せんよう　せん(剪)葉　1) leaf cutting 2) defoliation【草地】
ぜんよう　前葉, 前出葉　prophyll
せんりゅうじゅう　千粒重　1000-grain weight, one-thousand-grain weight
せんりょうさくもつ　染料作物　dye [stuff] crop
ぜんりん　前鱗　ventral scale

[そ]

そう　層【標本抽出】　stratum (*pl.* strata)
そう　相　phase
そういちじせいさんりょく　総一次生産力　gross primary productivity
そうえん　桑園　mulberry field
ぞうえんがく　造園学, 緑地学　landscape architecture
そうおう　挿秧⇒田植え　rice transplanting
そうか　そう(痩)果　achene
そうがい　霜害, 凍霜害　frost damage (injury)
そうかこうか　相加効果　additive effect
そうかてき　相加的　additive
そうか[てき]いでんこうか　相加[的]遺伝効果　additive genetic effect
そうかびょう　そうか(瘡痂)病【ジャガイモ】　scab
そうか[ひょうほん]ちゅうしゅつ

層化[標本]抽出　stratified sampling
そうかん　草冠　1) canopy　2) sward canopy【草地】
そうかん　相関　correlation
そうかんぎょうれつ　相関行列　correlation matrix
そうかんけいすう　相関係数　correlation coefficient, coefficient of correlation
そうかんこうぞう　草冠構造, 個体群構造　canopy structure, canopy architecture
そうかんひょう　相関表　correlation table
そうきさいばい　早期栽培　early-season culture
そうきちゅうだい　早期抽だい(苔)　premature bolting
そうきょくせん　双曲線　hyperbola
そうげん　草原　grassland
そうこう　草高【≠草丈　plant length】1) plant height　2) sward height【草地】
そうこうごうせい　総光合成　gross photosynthesis, 真の光合成　true photosynthesis
そうごうてきびょうがいちゅうかんり　総合的病害虫管理　integrated pest management (IPM)
そうごうぼうじょ　総合防除　integrated control
そうごさよう　相互作用　interactin
そうごしゃへい　相互遮へい　mutual shading
そうごてんざ　相互転座　reciprocal translocation, mutual translocation
ゾウコンニャク　elephant yam, *Amorphophallus paeoniifolius* (Dennst.) Nicolson (= *A. campanulatus* Bl.)
そうさがたでんしけんびきょう　走査型電子顕微鏡　scanning electron microscope (SEM)

そうじ　相似　1) analogy【形態】2) similarity【図形】
そうじがり　掃除刈り【牧草】trimming cut
そうじきかん　相似器官　analogous organ
そうしゅ　草種【牧草】forage species
ぞうしゅう　増収　yield increase
そうしゅうにゅう　総収入, 粗収入　gross income, gross revenue
そうじゅく　早熟　early maturation
そうじゅくせい　早熟性　early maturing habit
そうしゅこうせい　草種構成　sward composition
そうしょう　相称, 対称　symmetry
そうじょうかじょ　総状花序　raceme
そうじょうこうか　相乗効果　multiplicative effect
そうじょうさよう　相乗作用　synergism
そうしようしょくぶつ　双子葉植物　dicotyledon, dicot
そうじょうてき　相乗的　multiplicative
そうしようの　双子葉の　dicotyledonous
そうしょく　草食　herbivory
ぞうしょく　増殖　multiplication, proliferation, propagation
そうしょくどうぶつ　草食動物　herbivore
そうじょこうか　相助効果　synergistic effect
そうじょてき　相助的　synergistic
そうせい　そう(叢)生　tussock
そうせい　走性　taxis
そうせいさいばい　草生栽培　sod culture
そうせいさん　総生産, 粗生産　gross production
そうせいそう　そう(叢)生草　bunch-type grass
ぞうせいそうち　造成草地　tame pasture

そうせいの　そう(叢)生の, 束生の　fasciculate
そうせいほう　草生法　sod culture system
そうせいマルチほう　草生マルチ法　sod-mulch system
そうたいがんすいりょう　相対含水量　relative water content
そうたいしつど　相対湿度　relative humidity
そうたいせいちょう　相対成長　relative growth, アロメトリー　allometry
そうたいせいちょうけいすう　相対成長係数　allometric coefficient
そうたいせいちょうりつ　相対成長率　relative growth rate (RGR)
そうたいひんど　相対頻度　relative frequency
ぞうたい[りょう]　増体[量]　liveweight gain (LWG), body weight gain
そうち　草地　grassland, pasture, meadow, sward
そうちがく　草地学　grassland science
そうちかんり　草地管理　sward management
そうちこうしん　草地更新　sward renovation
そうちしけんじょう　草地試験場　National Grassland Research Institute
そうちぞうせい　草地造成　sward establishment, forage establishment
そうちちくさん　草地畜産, 草地農業　grassland farming (agriculture)
そうどうきかん　相同器官　homologous organ
そうどう[せい]　相同[性]　homology
そうどうせんしょくたい　相同染色体　homologous chromosome
そうはんこうざつ　相反交雑, 正逆交雑　reciprocal crosses (crossing)
そうはんじゅんかんせんばつ　相反循環選抜　reciprocal recurrent selection
そうばんせい　早晩性　earliness
そうびょう　挿苗, 苗挿し【サツマイモ】　sprouted vine planting
そうふうしきふんむき　送風式噴霧機, ミスト機　mist blower
そうべつかりとりほう　層別刈取法　stratified clip method
そうほう　総苞　involucre
そうほうきょくせん　双峰曲線⇒二頂曲線　bimodal curve
ぞうほうたい　造胞体, 胞子体　sporophyte
そうぼうぼう　僧帽芒, 三叉芒　hooded awn
そうほてきディーエヌエー　相補的DNA　complementary DNA (cDNA)
そうほん　草本　herb, herbaceous plant
そうほんさくもつ　草本作物　herbaceous crop
そうほんの　草本の　herbaceous
そうめん　繰綿　1) ginned cotton, lint【製品】　2) ginning【工程】
そうめん　草棉　herbaceous cotton
そうめんしすう　繰綿指数　lint index
そうめんぶあい　繰綿歩合　ginning percentage, lint percentage, ginning outturn (GOT)
そうもくかい　草木灰　plant and wood ashes
そうりゅう　層流【気象】　laminar flow
そうりょう　草量　herbage mass
ぞうりょうざい　増量剤　1) dust diluent【農薬】　2) extender【授粉】
そうりんきゅうかん　そう(叢)林休閑　bush fallow
そうるい　藻類　algae (*sing.* alga)
そがい　阻害, 抑制　inhibition
そがいざい　阻害剤　inhibitor
そかいぶん　粗灰分　crude ash
そぎつぎ　そぎ接ぎ　whittle grafting, crown grafting
ぞく　1) 属　genus　2) 族　tribe

そくが 側芽 lateral bud
ぞくかんこうざつ 属間交雑 intergeneric crossing, genus cross
ぞくかんざっしゅ 属間雑種 intergeneric hybrid, genus hybrid
そくし 側枝 lateral branch
そくじょうせひ 側条施肥 side dressing
そくしんざい 促進剤 accelerator
そくせい 促成 forcing
そくせい 属性 attribute
そくせいか 側生花 lateral flower
そくせいさいばい 促成栽培 forcing culture
ぞくせいじょうほう 属性情報 attribute information
そくせいたいひ 速成堆肥 artificial [farmyard] manure
そくせいの 束生の, そう(叢)生の fasciculate
そくてい 測定 measurement
そくていち 測定値 1) measured value 2) measurement【統計】
そくみゃく 側脈 lateral vein
ソケイ(素馨), ツルマツリ(蔓茉莉) poet's jasmine, common white jasmine, *Jasminum officinale* L. f.
そさい そ(蔬)菜⇒野菜 vegetable [crop]
そさいえんげい そ(蔬)菜園芸⇒野菜園芸 vegetable gardening, olericulture
そしき 組織 tissue
そしきかがく 組織化学 histochemistry
そしきがく 組織学 histology
そしきけいせい 組織形成 histogenesis
そしきとくいてきいでんしはつげん 組織特異的遺伝子発現 tissue-specific gene expression
そしきばいよう 組織培養 tissue culture

そしつ 礎質【細胞】 matrix
そしぼう 粗脂肪 crude fat
そしゅうにゅう 粗収入, 総収入 gross income, gross revenue
そしょく 疎植 sparse planting, planting with wide spacing
そしりょう 粗飼料 roughage
ソージン【ワタ】 saw gin
ソース source
そすい 疎穂 lax head
そすいせいの 疎水性の hydrophobic
そせい 塑性, 可塑性 plasticity
そせいぐんらく 疎生群落 open community
そせいさん 粗生産, 総生産 gross production
そせんい 粗繊維 crude fiber
そせんけい 祖先型 ancestral form
そせんしゅ 祖先種 ancestral species
そタンパクしつ 粗タンパク質 crude protein (CP)
そっこうせいひりょう 速効性肥料 quick acting fertilizer, readily available fertilizer
そっこん 側根 lateral root, 分枝根 branch root
ソテツ(蘇鉄) 1) cycad, *Cycas* spp.【広義】 2) Japanese sago palm, *Cycas revoluta* Bedd. (= *C. revoluta* Thunb.)【狭義】
そとう 粗糖 raw sugar
そとがえしこう 外返し耕 casting plowing
ソドマルチ sod mulch
そは 疎播, 薄播き sparse sowing (seeding), thin sowing (seeding)
ソバ(蕎麦) buckwheat, *Fagopyrum esculentum* Moench
そば(蕎麦), そば切り soba, buckwheat noodle
ソバがら ソバ殻 buckwheat husk
そばきり そば切り, そば(蕎麦) soba, buckwheat noodle

ソバコ ソバ粉 buckwheat flour
そひ 粗皮【アサ】 bark
ソフトウェア software
ソフトコーン soft corn, *Zea mays* L. var. *amylacea* Sturt.
そへんりょうさくもつ 組編料作物 rough-weaving crop
そほうさいばい 粗放栽培 extensive cultivation
そほうさくもつ 粗放作物 extensive crop
ソマクローン，体細胞由来繁殖系 somaclone
そめん 梳綿 carding
そめんしょうほうたい 粗面小胞体 rough-surfaced endoplasmic reticulum, granular endoplasmic reticulum
ソラニン solanine
ソラマメ(蚕豆) broad bean, *Vicia faba* L.
そりゅうこくもつ 粗粒穀物 coarse grain
ゾル sol
ソルガム→モロコシ
ソルビトール sorbitol
ソロネッツ Solonetz
ソロンチャク Solonchak
そんしつかんすう 損失関数 loss function

[た]

ダイアフラムポンプ diaphragm pump
たいあるかりせい 耐アルカリ性 alkari tolerance
ダイアレル交配 diallel cross
だいいかんそく 大維管束 large vascular bundle
だいいちよう 第一葉 the first leaf
だいいっしゅのかご 第一種の過誤【統計】 error of the first kind
たいいんせい 耐陰性 shade tolerance

ダイウイキョウ(大茴香・八角茴香) star anise, *Illicium verum* Hook. f.
たいえんせい 耐塩性 salt (salinity) tolerance
ダイオウ(大黄) medicinal rhubarb, *Rheum officinale* Baill.
たいか 帯化 fasciation
たいか 退化 degeneration
たいかかふん 退化花粉 abortive pollen
たいかきかん 退化器官 reduced organ
たいかこんせき 退化痕跡 vestige
だいがち 台勝ち overgrowth of rootstock
だいがり 台刈り pruning at the base
たいかんせい 耐旱(干)性⇒耐乾性 drought resistance (tolerance)
たいかんせい 耐寒性 cold hardiness, cold resistance
たいき 大気 atmosphere
だいぎ 台木 rootstock, stock
たいきおせん 大気汚染 air pollution
たいきおせんぶっしつ 大気汚染物質 air pollutant
たいきぶんれつそしき 待機分裂組織 waiting meristem
たいきほせい 大気補正 atmospheric correction
たいきゅうひ 堆厩肥 stable manure
だいぎり 台切り【チャ】 collar pruning
たいごう 対合【染色体】 pairing, synapsis, syndesis
タイコウイ(太甲藺) bulrush, *Schoenoplectus triqueter* (L.) Palla (= *Scirpus triqueter* L.)
たいこうさよう 対抗作用⇒拮抗作用 antagonism
たいこうせんい 退行遷移 retrogressive succession
ダイコン(大根) radish, *Raphanus sativus* L.

たいざ 胎座 placenta
たいさいぼう 体細胞 somatic cell
たいさいぼう[せい]へんい 体細胞[性]変異 somaclonal variation
たいさいぼうとつぜんへんい 体細胞突然変異 somatic mutation
たいさいぼうふていはいけいせい 体細胞不定胚形成 somatic embryogenesis
たいさいぼうゆらいはんしょくけい 体細胞由来繁殖系, ソマクローン somaclone
だいさく 代作 substitute cropping
だいさく[もつ] 代作[物] substitute crop
たいさんせい 耐酸性 acid tolerance
たいしつせい 耐湿性 wet endurance, excess water tolerance
たいしゃ 代謝 metabolism
たいしゃけいろ 代謝経路 metabolic pathway
たいしゃさんぶつ 代謝産物 metabolite
たいしゃそがいざい 代謝阻害剤 metabolic inhibitor
たいじゅう 体重 body weight (BW)
ダイジョ(大薯) greater yam, water yam, winged yam, *Dioscorea alata* L.
たいしょう 対称, 相称 symmetry

たいしょう 対照 control
たいしょう[しけん]く 対照[試験]区, 標準[試験]区 control plot
たいしょうひんしゅ 対照品種, 比較品種 check variety
たいしょくりゅう 退色粒【ムギ類】 bleached kernel, bleached wheat
たいしょせい 耐暑性, 耐熱性 heat tolerance
ダイズ(大豆) soybean, *Glycine max* (L.) Merr.
たいすうせいきぶんぷ 対数正規分布 lognormal distribution

たいすうのほうそく 大数の法則 law of large numbers
たいすうへんかん 対数変換 logarithmic (log) transformation
ダイズかす 大豆粕 soybean meal
ダイズこ 大豆粉 soybean flour
ダイズサヤタマバエ soybean pod gall midge
ダイズゆ 大豆油 soybean oil
たいせい 胎生 vivipary, viviparity
たいせい 体制 organization
たいせい 耐性 tolerance
たいせいしゅし 胎生種子 viviparous seed
たいせいの 対生の opposite
たいせき 堆積 sedimentation
たいせつせい 耐雪性 snow endurance
たいそうせい 耐霜性 frost hardiness (resistance)
たいちゅうせい 耐虫性 insect resistance
だいちょうきん 大腸菌 *Escherichia coli*
たいとうせい 耐冬性 winter hardiness
たいとうせい 耐凍性 freezing hardiness
たいとうふくせい 耐倒伏性 lodging resistance
だいにしゅのかご 第二種の過誤【統計】 error of the second kind
タイヌビエ barnyard grass, *Echinochloa crus-galli* (L.) Beauv. var. *oryzicola* (Vasing.) Ohwi
たいねつせい 耐熱性, 耐暑性 heat tolerance
たいひ 堆肥 compost
たいひさんぷき 堆肥散布機, マニュアスプレッダ manure spreader
たいひしゃ 堆肥舎 compost depot
たいひせい 耐肥性 adaptability for heavy manuring
たいびょうせい 耐病性, 病害抵抗性

disease resistance
たいびょうせいけんていほ　耐病性検定圃　disease garden, test field for disease-tolerance [evaluation]
たいびょうせいひんしゅ　耐病性品種　disease tolerant variety
たいふう　台風　typhoon
たいふうがい　台風害　typhoon damage
たいふうせい　耐風性　wind tolerance
たいぶつレンズ　対物レンズ　objective
タイマ（大麻）　hemp, *Cannabis sativa* L.
だいまけ　台負け　overgrowth of the scion
タイマしゅ　大麻子油, 麻実油　hemp seed oil
タイム　common thyme, garden thyme, *Thymus vulgaris* L.
だいめ　台芽　sucker
たいようエネルギーりようこうりつ　太陽エネルギー利用効率　efficiency of solar energy utilization
たいようこうど　太陽高度　solar altitude
たいようじょうすう　太陽常数　solar constant
たいようねんすう　耐用年数　durable years
たいようほうしゃ　太陽放射　solar radiation
たいりついでんし　対立遺伝子　allele
たいりつかせつ　対立仮説　alternative hypothesis
たいりつけいしつ　対立形質　allelomorph
たいりゅう　対流　convection
だいりゅうひんしゅ　大粒品種　large grain variety
たいれいすいせい　耐冷水性　cold water tolerance
たいれいせい　耐冷性　cold weather resistance
タイワンアシカキ　bareet grass, southern cutgrass, tiger's-tongue grass, *Leersia hexandra* Sw.
タイワンツナソ→シマツナソ
ダーウィニズム, ダーウィン説　Darwinism
たうえ　田植え　rice transplanting
たうえき　田植機　rice transplanter
たうえづな　田植綱　rice planting rope
たうえわく　田植枠　frame rule for transplanting of rice seedling, rice transplanting frame
タウコギ　bur beggarticks, *Bidens tripartita* L.
だえん　楕円　ellipse
たかうね　高畝　high ridge, high bed
たかか　多花果, 集合果　multiple fruit, aggregate fruit, syncarp
たかがり　高刈り　high-level cutting
タカキビ→モロコシ
タカサブロウ　eclipta, false daisy, *Eclipta prostrata* L. (= *E. alba* L.)
たかじゅせい　他家受（授）精　cross fertilization
たかじゅふん　他家受（授）粉　cross pollination
たかせいしょく　他家生殖, 他殖　outcrossing, allogamy
たかせんしょくたい　多価染色体　multivalent chromosome, polyvalent chromosome
たがたい　多芽体　multiple shoot
たかつぎ　高接ぎ　top grafting, top working
たかつくり　高作り　tall training
たかねがりしたて　高根刈り仕立て【クワ】　semi-low cut training
たかぶあなまき　多株穴播き　dense dibbling
タガラシ　crowfoot, buttercup, *Ranunculus sceleratus* L.
たかんさよう　他感作用, アレロパシー

allelopathy
たくよう 托葉 stipule
タケアズキ rice bean, *Vigna umbellata* (Thunb.) Ohwi et Ohashi (= *Phaseolus calcaratus* Roxb.)
たけいこうざつ 多系交雑 multiple crosses (crossings)
たけいひんしゅ 多系品種 multiline cultivar (variety)
タケニグサ *Macleaya cordata* (Willd.) R. Br.
タケノコ bamboo shoot
たげんけい 多原型 polyarch
たげんせいの 多元性の polyphyletic
たこうしきかいき 多項式回帰 polynomial regression
たこうはい 多交配 polycross[ing]
たこうぶんぷ 多項分布 multinomial distribution
たごしかんがい 田越灌漑 plot-to-plot irrigation
タコノキ screw pine, *Pandanus* spp. 【*P. tectorius* Sol. ex Parkins. 他多種】
タゴボウ→チョウジタデ
たさいぼうせいぶつ 多細胞生物 multicellular organism
たじげんしゃくどほう 多次元尺度法 multi-dimensional scaling method
たしゅうかくさいばい 多収穫栽培 high-yielding culture
たしゅうせい 多収性 high-yielding ability
たしゅうせいひんしゅ 多収性品種, 高収性品種 high-yielding variety
たじゅうひかく 多重比較 multiple comparisons
たしょう 多照 abundant sunshine
たじょうまき 多条播き dense drilling
たしょく 他殖, 他家生殖 outcrossing, allogamy
たしょくせいしょくぶつ 他殖性植物 allogamous plant
タシロイモ East Indian arrowroot, Tahiti arrowroot, *Tacca leontopetaloides* (L.) Kuntze (= *T. pinnatifida* Forst.)
たせいざっしゅ 多性雑種 polyhybrid, multihybrid
たそうさく 多層作 multi-layered cropping
ただん[ひょうほん]ちゅうしゅつ 多段[標本]抽出 multistage sampling
タチイヌノフグリ corn speedwell, *Veronica arvensis* L.
たちがれ 立枯れ damping-off
たちがれそう 立枯れ草【飼料】 foggage
たちがれびょう 立枯病【ムギ類】 take-all
タチジャコウソウ→タイム
たちせいの 立性の upright
タチナタマメ(立刀豆) Jack bean, *Canavalia ensiformis* (L.) DC.
たちほ 立ち穂 erect ear
だついおんすい 脱イオン水, 脱塩水 deionized water
だつえんすい 脱塩水, 脱イオン水 deionized water
タックポリメラーゼ *Taq* ポリメラーゼ *Taq* polymerase
だっこく 脱穀 threshing
だっこくき 脱穀機, スレッシャ thresher
だっしダイズ 脱脂大豆 defatted soybean
だっしめん 脱脂綿 absorbent cotton
だっすい 脱水 dehydration, desiccation
だっすいそこうそ 脱水素酵素 dehydrogenase
だつたんさんこうそ 脱炭酸酵素 decarboxylase
ダッタンソバ(韃靼蕎麦) Tartary buckwheat, Kangra buckwheat, *Fagopyrum tataricum* (L.) Gaertn.

だっちつ [さよう] 脱窒 [作用] denitrification
タツノツメガヤ crowfootgrass, *Dactyloctenium aegyptium* (L.) P. Beauv.
タッピング, 切付け【ゴム・ウルシなど】 tapping
だっぷ 脱ぷ (稃) dehulling, hulling, husking
だっぷき 脱ぷ (稃) 機 huller
だつぶんか 脱分化 dedifferentiation
だつぼう 脱芒 awning
だつりゅうせい 脱粒性 shattering habit, shedding habit, threshability
タデアイ (蓼藍) →アイ
たてじく 縦軸 ordinate
たてせいちょう 縦成長 longitudinal growth
たてたば 立束 shock
たてぶんれつ 縦分裂 longitudinal division
たてぼし 立干し shocking
たとうるい 多糖類 polysaccharide
たなじたて 棚仕立て trellis training
たなだ 棚田 rice terrace, terrace paddy field
たにくか 多肉果, 液果 sap fruit, succulent fruit, berry
たにくけい 多肉茎 succulent stem
たにくこん 多肉根 succulent root
たにくしょくぶつ 多肉植物 succulent plant
たにくの 多肉の succulent
たねイモ 種イモ seed tuber
たねごえ 種肥 seed impregnation
タネツケバナ flexuous bittercress, *Cardamine flexuosa* With.
たねば 種場 seed home
たねばた 種畑 field for seed production
たねもの 種物 seeds
たねもみ 種籾 rice seed, seed rice

たねんせい [の] 多年生 [の] perennial
たねんせいさくもつ 多年生作物 perennial crop
たねんせいざっそう 多年生雑草 perennial weed
たねんせいしょくぶつ 多年生植物 perennial [plant]
たねんせいぼくそう 多年生牧草 perennial forage
たねんそう 多年草 perennial herb (grass)
たはいげんしょう 多胚現象 polyembryony
たはいしゅし 多胚種子 multigerm seed
タバコ (煙草) tobacco, *Nicotiana tabacum* L.
タピオカ tapioca
たひさいばい 多肥栽培 heavy manuring culture
タビラコ→コオニタビラコ
たぶね 田舟 paddy field boat
タペートさいぼう タペート細胞 tapetal cell, tapete cell
タペート [そしき] タペート [組織] tapetum
たへんりょうかいせき 多変量解析 multivariate analysis
たへんりょうせいきぶんぷ 多変量正規分布 multivariate normal distribution
たほうぶんぷ 多峰分布 multimodal distribution
タマガヤツリ smallflower umbrella sedge, *Cyperus difformis* L.
タマザキツヅラフジ *Stephania cepharantha* Hayata
ダマスクバラ damask rose, *Rosa damascena* Mill.
タマネギ onion, *Allium cepa* L.
タマリンド tamarind, *Tamarindus indica* L.

たまわれ 玉割れ cracking of fruit
ターミネーター terminator
ためんさよう 多面作用 pleiotropy
ためんはつげん 多面発現 pleiotropism
たもうさく 多毛作 multiple cropping
タラゴン tarragon, estragon, *Artemisia dracunculus* L.
ダリスグラス dallisgrass, *Paspalum dilatatum* Poir.
たりょうげんそ 多量元素, 多量要素 macroelement, macronutrient, major element
たりょうようそ 多量要素, 多量元素 macroelement, macronutrient, major element
タルホコムギ *Aegilops squarrosa* L.
たれほ 垂れ穂 nodding ear
タロイモ taro, dasheen, *Colocasia esculenta* (L.) Schott var. *esculenta* Hubbard & Rehder
たわら 俵 straw bag, bale
たんいけっか 単為結果 parthenocarpy
たんいせいしょく 単為生殖, 単為発生, 処女生殖 parthenogenesis
たんいつさいばい 単一栽培 monoculture
たんいはっせい 単為発生, 単為生殖, 処女生殖 parthenogenesis
たんいまく 単位膜 unit membrane
たんえきすいこう 湛液水耕 deep flow technique (DFT)
だんおんたい 暖温帯 warm temperate zone
だんかい 段階【標本抽出】 stage
たんかすいそ 炭化水素 hydrocarbon
たんかちゅうか 短花柱花, 短柱花 short-styled flower, thrum flower
たんかまい 炭化米 carbonized rice grain
たんカル 炭カル⇒炭酸カルシウム calcium carbonate
だんがんあんきょ 弾丸暗きょ(渠), モグラ暗きょ(渠) mole drain
ダンカンのたじゅうけんてい ダンカンの多重検定 Duncan's multiple range test
たんかんひんしゅ 短稈品種 short-culmed variety
たんきさくもつ 短期作物 short-season crop
たんきりんさく 短期輪作 short-term rotation
たんこうざつ 単交雑 single cross
だんごひりょう 団子肥料 ball fertilizer
だんこん 断根, せん(剪)根 root pruning
たんさいぼうせいぶつ 単細胞生物 unicellular organism
たんさく 単作 single cropping, monoculture
たんざくなわしろ 短冊苗代 rectangular nursery
たんさんガス 炭酸ガス⇒二酸化炭素 carbon dioxide (CO_2)
たんさんガスせひ 炭酸ガス施肥 carbon dioxide fertilization (enrichment)
たんさんカルシウム 炭酸カルシウム calcium carbonate
たんさんこてい 炭酸固定 carbon dioxide fixation
たんさんだっすいこうそ 炭酸脱水酵素, カーボニックアンヒドラーゼ carbonic anhydrase
たんさんどうか 炭酸同化 carbon dioxide assimilation
たんじくぶんし 単軸分枝 monopodial branching
たんじつ 短日 short day
たんじつしゅんかせい 短日春化性 short-day vernalization
たんじつしょくぶつ 短日植物 short-day plant
たんじつしょり 短日処理 short-day

treatment
たんじゅんはいれつちょうたけい　単純配列長多型　simple sequence length polymorphism (SSLP)
たんしようしょくぶつ　単子葉植物　monocotyledon, monocot
たんしようの　単子葉の　monocotyledonous
たんすい　湛水　flooding, waterlogging
たんすいかぶつ　炭水化物　carbohydrate
たんすいかぶつたいしゃ　炭水化物代謝　carbohydrate metabolism
たんすいかんがい　湛水灌漑　flooded irrigation
たんすいちょくはん (ちょくは)　湛水直播　direct sowing (seeding) in flooded paddy field
たんすいでん　湛水田　flooded paddy field, flooded ricefield
たんすいどじょう　湛水土壌　flooded soil, submerged soil, waterlogged soil
たんせいか　単性花　unisexual flower
たんそ　炭素　carbon (C)
たんそう　短草　short grass
たんそうかん　単相関　simple correlation
たんそう[たい]　単相[体], 半数体　haploid, haplont
たんそげん　炭素源　carbon source
たんそしゅうし　炭素収支　carbon balance
たんそどういたいぶんべつ　炭素同位体分別　carbon isotope discrimination
たんそどうか　炭素同化　carbon assimilation
たんそびょう　炭そ(疽)病　anthracnose
たんたい　担体, キャリヤー　carrier
だんちがたぼくそう　暖地型牧草　tropical forages
たんちゃ　たん(磚)茶　brick tea

たんちゅうか　短柱花, 短花柱花　short-styled flower, thrum flower
たんとうしゅうりょう　反当収量　10-are yield
たんとう　単糖　monosaccharide
たんどくけいとう　単独系統　single pedigree
ダンドボロギク　American burnweed, *Erechtites hieracifolia* (L.) Raf. ex DC.
タンニン　tannin
タンニンさいぼう　タンニン細胞　tannin cell
タンニンさん　タンニン酸　tannic acid
タンニンりょうさくもつ　タンニン料作物　tannin crop
たんぱいしゅし　単胚種子　monogerm seed
タンパクしつ　タンパク質　protein
タンパクしつたいしゃ　タンパク質代謝　protein metabolism
タンパク[しつ]ぶんかいこうそ　タンパク[質]分解酵素, プロテアーゼ　protease
タンパクたいちっそ　タンパク態窒素　protein nitrogen
タンパクりゅう　タンパク粒　protein grain
だんばた　段畑　terrace field
たんぱほうしゃ　短波放射　short-wave radiation
たんぱ[ん]そうち　単播草地　pure sward
たんぴ　単肥　straight fertilizer
たんぺき　端壁　end wall
だんぺんせんしょくたい　断片染色体, 破片染色体　fragment chromosome
たんぽうぶんぷ　単峰分布　unimodal distribution
たんめいしゅし　短命種子　microbiotic seed, short-lived seed
たんめいの　短命の　ephemeral
だんめんせき　断面積　cross sectional

area
たんもう　短毛, 地毛(じもう)【ワタ】　fuzz, linters
たんよう　単葉　simple leaf
たんり　単離　isolation
だんりゅう　団粒　aggregate
たんりゅうけいとうほう　単粒系統法　single seed descent method (SSD method)
たんりゅうこうぞう　単粒構造　single grained structure
だんりゅうこうぞう　団粒構造　aggregated structure
たんりゅうひんしゅ　短粒品種　short-grained variety

[ち]

ちいききしょうかんそくシステム　地域気象観測システム, アメダス　Automated Meteorological Data Acquisition System (AMeDAS)
ちいきせい　地域性　locality
チェリモヤ　cherimoya, *Annona cherimola* Mill.
チェルノーゼム　Chernozem
チェレー【イネの生態型】　tjereh
ちえんがたれいがい　遅延型冷害　cool summer damage due to delayed growth
ちえんじゅふん　遅延受粉　delayed pollination
ちおん　地温　soil temperature
ちかかんがい　地下灌漑　subirrigation
ちかく　知覚　perception
ちかけい　地下茎　subterranean stem
ちかけつじつ　地下結実　geocarpy
ちかしよう　地下子葉　hypogeal cotyledon
ちかすい　地下水　ground water, groundwater
ちかすいい　地下水位　groundwater level
ちかすいおせん　地下水汚染　groundwater contamination
ちかすいめん　地下水面　groundwater table
ちかはいすい　地下排水　subsurface drainage
ちかぶ　地下部　subterranean part, underground part
チガヤ(茅)　cogongrass, *Imperata cylindrica* (L.) P. Beauv.
チカラシバ　Chinese pennisetum, *Pennisetum alopecuroides* (L.) Spreng.
ちかんさんど　置換酸度⇒交換酸度　exchange acidity
ちかんせいえんき　置換性塩基⇒交換性塩基　exchangeable base
ちかんようりょう　置換容量⇒交換容量　exchange capacity
ちきゅうかがく　地球科学　earth science, geoscience
ちくさん　畜産　animal industry, livestock industry, animal husbandry
ちくさんしけんじょう　畜産試験場　National Institute of Animal Industry
ちくじけんてい　逐次検定　sequential test
ちくしゃ　畜舎　barn, stable
チクル【サポジラ由来のゴム状物質】　chicle
ちけい　地形　topography
ちけいず　地形図　topographic map, relief map
ちこうせいひりょう　遅効性肥料⇒緩効性肥料　controlled release fertilizer, slow release fertilizer
チゴザサ　*Isachne globosa* (Thunb.) O. Kuntze
ちしいでんし　致死遺伝子　lethal gene
ちしいんし　致死因子　lethal factor
ちしのうど　致死濃度　lethal concentration
チシャ→レタス

ちじょうけい　地上茎　terrestrial stem
ちじょうさんぷ　地上散布　ground application
ちじょうしよう　地上子葉　epigeal cotyledon
ちじょうぶ　地上部　top, aboveground part, aerial part
チーゼル　teasel, *Dipsacus fullonum* L.
チチコグサ　Japanese cudweed, *Gnaphalium japonicum* Thunb.
チチコグサモドキ　wandering cudweed, *Gnaphalium pensylvanicum* Willd. (= *G. purpureum* L. var. *stathulatum* (Lam.) Baker)
ちっそ　窒素　nitrogen (N)
ちっそかじょう　窒素過剰　nitrogen excess
ちっそきが　窒素飢餓　nitrogen starvation
ちっそきゅうしゅうけいすう　窒素吸収係数　nitrogen absorption coefficient
ちっそげん　窒素源　nitrogen source
ちっそこうりつ　窒素効率　nitrogen efficiency
ちっそこてい　窒素固定　nitrogen fixation
ちっそこていいでんし　窒素固定遺伝子　nitrogen fixation (*nif*) gene
ちっそこていきん　窒素固定菌　nitrogen fixing bacterium, nitrogen fixer
ちっそさんかぶつ　窒素酸化物　nitrogen oxides (NO_x)
ちっそじゅんかん　窒素循環　nitrogen cycle, cycling of nitrogen
ちっそたいしゃ　窒素代謝　nitrogen metabolism
ちっそどうか　窒素同化　nitrogen assimilation
ちっそひりょう　窒素肥料　nitrogen fertilizer, nitrogenous fertilizer
ちっそりようこうりつ　窒素利用効率　nitrogen use efficiency

チトクロム, シトクロム　cytochrome
チトセラン→サンセベリア
チドメグサ(血止草)　lawn pennywort, *Hydrocotyle sibthorpioides* Lam.
ちどりうえ　千鳥植え　zigzag planting
ちびょう　稚苗　young seedling with 3.0-3.5 leaf stage
ちひょうかんがい　地表灌漑【イネ】　surface irrigation
ちひょうはいすい　地表排水　surface drainage
ちほうひんしゅ　地方品種　local variety
ちほうめい　地方名　local name
チミン　thymine (T)
チモシー　timothy, *Phleum pratense* L.
チモフェービけいコムギ　チモフェービ系コムギ　timopheevi wheat
チモフェービコムギ　timopheevi wheat, *Triticum timopheevi* Zhuk.
チャ(茶)　tea, *Camellia sinensis* (L.) O. Kuntze
チャえん　茶園　tea field, tea garden
ちゃくか　着花　flower setting, flower bearing
ちゃくしょくフィルム　着色フィルム　colored film
ちゃくしょくりゅう　着色粒　colored grain
ちゃくよういち　着葉位置【タバコ】　stalk position
ちゃくらいき　着らい(蕾)期　flower bud appearing stage
チャじつゆ　茶実油　tea seed oil
チャじゅ　茶樹　tea plant, tea bush
ちゃまい　茶米　rusty rice [kernel]
ちゅうえい　虫えい(癭)　gall
ちゅうおうしょりそうち　中央処理装置　central processing unit (CPU)
ちゅうおうち　中央値, メディアン　median
ちゅうおうみゃく　中央脈, 主脈　main vein

ちゅうがい　虫害　insect injury
ちゅうかひ　中果皮　mesocarp
ちゅうがりしたて　中刈り仕立て, 中幹仕立て【クワ】　medium cut training
ちゅうかんおや　中間親, 両親平均　midparent [value]
ちゅうかんがた　中間型【ソバ】　intermediate ecotype
ちゅうかんき　中間期, 間期【減数分裂】　interphase, interkinesis
ちゅうかんざっしゅ　中間雑種　intermediate hybrid
ちゅうかんしたて　中幹仕立て, 中刈り仕立て【クワ】　medium cut training
ちゅうかんしゅくしゅ　中間宿主　intermediate host
ちゅうかんしょくぶつ　中間植物　intermediate plant
ちゅうかんだいぎ　中間台木　interstock, intermediate stock
ちゅうかんたいしゃぶっしつ　中間代謝物質　metabolic intermediate
ちゅうき　中期【細胞分裂】　metaphase
ちゅうぎり　中切り【チャ】　medium pruning
ちゅうけい　中茎, 中胚軸　mesocotyl
ちゅうけいこん　中茎根　mesocotylar root
ちゅうこう　中耕　intertillage
ちゅうこうさくもつ　中耕作物　intertillage crop
ちゅうこうじょそうき　中耕除草機　cultivator
ちゅうごくのうぎょうしけんじょう　中国農業試験場　Chugoku National Agricultural Experiment Station
ちゅうこつ　中骨【タバコ】　stem, midrib
ちゅうこつかんそうき　中骨乾燥期【タバコ】　stem drying stage
ちゅうこん　中根【チャ】　medium root
ちゅうさんかんちいき　中山間地域　hilly and mountainous area

ちゅうじく　中軸　rachis
ちゅうじつしょくぶつ　中日植物⇒中性植物　day-neutral plant, neutral plant
ちゅうしゅつ　抽出　1) emergence, unfolding【発育】　2) extraction【化学】
ちゅうしゅつ　抽出【統計】⇒標本抽出　sampling
ちゅうしゅつけんさ　抽出検査⇒抜取り検査　sampling inspection
ちゅうしゅつたんい　抽出単位【標本抽出】　sampling unit
ちゅうじゅん　中旬　the second ten-days [of a month]
ちゅうしんくうどう　中心空洞　hollow heart
ちゅうしんし　中心子, 中心小体　centriole
ちゅうしんしょうたい　中心小体, 中心子　centriole
ちゅうしんたい　中心体　centrosome
ちゅうしんちゅう　中心柱　stele, central cylinder
ちゅうすいざっそう　抽水雑草　emersed weed, emergent weed
ちゅうせいアミノさん　中性アミノ酸　neutral amino acid
ちゅうせいざっそう　中生雑草　mesophytic weed
ちゅうせいしょくぶつ　中性植物　day-neutral plant, neutral plant
ちゅうせいしょくぶつ　中生植物　mesophyte
ちゅうせいデタージェントせんい　中性デタージェント繊維　neutral detergent fiber (NDF)
ちゅうせい(なかて)ひんしゅ　中生品種　medium [maturing] variety
ちゅうせきそう　沖積層　alluvium
ちゅうせきち　沖積地　alluvial land
ちゅうせきど　沖積土　Alluvial soil
ちゅうせきどじょう　沖積土壌　alluvial soil

ちゅうそう　中層, 中葉　intercellular layer, middle lamella
ちゅうだい　抽だい(苔)　bolting, flower stalk development
ちゅうとう　柱頭　stigma
ちゅうは　中葉【タバコ】　lugs
ちゅうばい　虫媒　entomophily, insect pollination
ちゅうばいか　虫媒花　entomophilous flower
ちゅうはいじく　中胚軸, 中茎　mesocotyl
ちゅうばいしょくぶつ　虫媒植物　entomophilae, entomophilous plant
ちゅうびょう　中苗　middle seedling with 3.5-5.0 leaf stage
チューブかんすい　チューブ灌水　plastic-tube watering
チューブリン　tubulin
ちゅうよう　中葉, 中層　intercellular layer, middle lamella
ちゅうりゅうひんしゅ　中粒品種　medium-grained variety
ちゅうろく　中肋　midrib
ちゅうわ　中和　neutralization
ちょうえいとう　長穎稲　long glume rice
ちょうえついくしゅほう　超越育種法　transgression breeding
ちょうえつぶんり　超越分離　transgressive segregation
ちょうえんしんき　超遠心機　ultracentrifuge
ちょうおんぱしょり　超音波処理　ultrasonication
ちょうか　頂花　terminal flower
ちょうが　頂芽　apical bud, terminal bud
ちょうがい　鳥害　crop damage due to bird, bird damage
ちょうかちゅうか　長花柱花, 長柱花　long-styled flower, pin flower
ちょうがゆうせい　頂芽優性, 頂部優性　apical dominance
ちょうかんひんしゅ　長稈品種　long-culmed variety
ちょうきちょぞう　長期貯蔵　long-term storage
ちょうけいか　蝶形花　papilionaceous flower
チョウジ (丁子, 丁字)　clove, *Syzygium aromaticum* (L.) Merr. et Perry (= *Eugenia caryophyllata* Thunb., *E. caryophyllus* Spreng., *E. aromatica* Kuntze)
チョウジタデ　*Ludwigia epilobioides* Maxim.
ちょうじつ　長日　long day
ちょうじつしょくぶつ　長日植物　long-day plant
ちょうじつしょり　長日処理　long-day treatment
ちょうじょうせい　頂上性　ascending habit
ちょうせい　調製　preparation, processing
ちょうせいの　頂生の　terminal
ちょうせいついでんし　調節遺伝子　regulator gene, regulatory gene
チョウセンアサガオ　white datura, *Datura metel* L. (= *D. alba* Nees)
チョウセンニンジン→ヤクヨウニンジン
ちょうそう　長草　tall grass
ちょうたんさいぼう　頂端細胞　apical cell
ちょうたんせいちょう　頂端成長　apical growth
ちょうたんぶんれつそしき　頂端分裂組織　apical meristem
ちょうちゅうか　長柱花, 長花柱花　long-styled flower, pin flower
ちょうちんほ　提灯穂　open-floret panicle
ちょうていおんこ　超低温庫, 冷凍冷蔵庫　deep freezer
ちょうはくせっぺん　超薄切片

ultrathin section
ちょうはほうしゃ　長波放射　long-wave radiation
ちょうぶ　頂部　apex, tip
ちょうふうがい　潮風害　salty wind damage (injury)
ちょうふくいでんし　重複遺伝子　duplicate gene
ちょうふくじゅせい　重複受精　double fertilization
ちょうふくせんしょくたい　重複染色体　duplicated chromosome
ちょうぶじょきょ　頂部除去　decapitation
ちょうぶゆうせい　頂部優性, 頂芽優性　apical dominance
ちょうほうけいうえ　長方形植え　rectangular planting
ちょうめいしゅし　長命種子　macrobiotic seed, long-lived seed
ちょうゆうせい　超優性　overdominance
ちょうりゅうひんしゅ　長粒品種　long-grained variety
ちょうりょく　張力　tension
ちょうわへいきん　調和平均　harmonic mean
ちょくせついでんしどうにゅう　直接遺伝子導入　direct gene transfer, direct transformation
ちょくせんかいき　直線回帰⇒線形回帰　linear regression
ちょくたつこう　直達光　direct light (sunlight)
ちょくたつたいようほうしゃ　直達太陽放射, 直達日射　direct solar radiation
ちょくたつにっしゃ　直達日射, 直達太陽放射　direct solar radiation
ちょくはん（ちょくは）　1) 直播, 直播き　direct sowing (seeding)　2) 直播【サツマイモ】　direct planting
ちょくりつうえ　直立植え, 直立挿し【サツマイモ】　upright planting
ちょくりつがた　直立型　erect type, upright type
ちょくりつけい　直立茎　erect stem, upright stem
ちょくりつざし　直立挿し, 直立植え【サツマイモ】　upright planting
ちょくりつせい　直立性　upright habit
ちょくりつよう　直立葉　erect leaf
ちょくりつようがたひんしゅ　直立葉型品種　erect-leaved variety
ちょすいち　貯水地　reservoir
ちょぞう　貯蔵　storage
ちょぞうきかん　貯蔵器官　storage organ
ちょぞうきゅうよ　貯蔵給与, 貯蔵利用【飼料】　storage feeding
ちょぞうこん　貯蔵根　storage root
ちょぞうせい　貯蔵性　storage ability
ちょぞうそしき　貯蔵組織　storage tissue
ちょぞうたんすいかぶつ　貯蔵炭水化物　reserve carbohydrate
ちょぞうタンパクしつ　貯蔵タンパク質　reserve protein, storage protein
ちょぞうデンプン　貯蔵デンプン　reserve starch, storage starch
ちょぞうぶっしつ　貯蔵物質　reserve substance
ちょぞうまい　貯蔵米　stored rice
ちょぞうようぶん　貯蔵養分⇒貯蔵物質　reserve substance
ちょぞうりよう　貯蔵利用, 貯蔵給与【飼料】　storage feeding
ちょっけい　直径　diameter
ちょっこうせっけい　直交設計　orthogonal design
ちょっこうたこうしき　直交多項式　orthogonal polynomial
ちょっこん　直根　taproot
チョマ（苧麻）　ramie, China grass, *Boehmeria nivea* (L.) Gaud.
チョロギ　chorogi, *Stachys sieboldii* Miq. (= *S. affinis* Fresen)

チラコイド　thylakoid
ちりじょうほう　地理情報　geographic information
ちりじょうほうシステム　地理情報システム　geographic information system (GIS)
ちりてきぶんぷ　地理的分布　geographical distribution
チリメンシソ→シソ
ちりょく　地力⇒土壌肥沃度　soil fertility
チロシン　tyrosine (Tyr)
ちんあつ　鎮圧　1) compaction【農作業】2) packing, tamping【土木】
ちんあつき　鎮圧機，カルチパッカ　culti-packer
ちんこう　沈降　sedimentation
ちんすいざっそう　沈水雑草　submerged weed, submersed weed
ちんでんぶつ　沈殿物【遠心の】　pellet

[つ]

ついじゅく　追熟⇒後熟(こうじゅく)　afterripening
ついひ　追肥　topdressing, supplement application
ついひかく　対比較　pair comparison
つうかさいぼう　通過細胞　passage cell
つうき　通気　aeration
つうきせい　通気性　air permeability
つうきそしき　通気組織　aerenchyma
つうすいていこう　通水抵抗　resistance to water flow
つうどうこうりつ　通導効率　hydraulic conductance
つうどうそしき　通導組織　conductive tissue
つうどうていこう　通導抵抗　hydraulic resistance, resistance to water flow
つうふうかんそうき　通風乾燥機　forced air dryer, power dryer

つぎき　接ぎ木　grafting, graftage
つぎきざっしゅ　接ぎ木雑種　graft hybrid
つぎきしんわせい　接ぎ木親和性　graft compatibility, graft affinity
つぎきふしんわせい　接ぎ木不親和性　graft incompatibility
つきべり　つ(搗)き減り　milling loss
つぎほ　接ぎ穂，穂木　scion
つちいれ　土入れ　topsoiling
つちなしさいばい　土なし栽培⇒無土壌栽培　soilless culture
つちよせ　土寄せ，培土　earthing up, ridging, molding
つちよせき　土寄機　ridger
つなぎさく　つなぎ作　relay [inter]cropping
ツナソ→ジュート
ツノアイアシ　itch grass, *Rottboellia exaltata* (L.) L. f.
ツノウマゴヤシ→セラデラ
つぼがり　坪刈り　unit area sampling, quadrat sampling
ツボクサ　Indian pennywort, *Centella asiatica* (L.) Urban
つぼみじゅふん　つぼみ(蕾)受粉　bud pollination
つみがり　摘刈り　picking, plucking
つみごえ　積み肥⇒堆肥(たいひ)　compost
ツメクサ　Japanese pearlwort, *Sagina japonica* (Sw.) Ohwi
つゆ　露　dew
つゆあけ　つゆ(梅雨)明け　end of bai-u
つゆいり　つゆ(梅雨)入り，入梅　beginning of bai-u
ツユクサ(露草)　dayflower, Asiatic dayflower, *Commelina communis* L.
つりぐされ　吊り腐れ【タバコ】　shed burn
つりばりうえ　釣針植え，釣針挿し【サツマイモ】　hooked planting

つりばりざし　釣針挿し，釣針植え【サツマイモ】　hooked planting
つる（蔓）　vine, liana
つるあげ　つる（蔓）揚げ　lifting up of vines
ツルアズキ（蔓小豆）　*Vigna umbellata* (Thunb.) Ohwi et Ohashi (= *Phaseolus pendulus* Makino)
ツルカ　turka, ケンディル　kendyr, *Apocynum sibiricum* Jacq.
つるがえし　つる返し【サツマイモ】　turning of vines
つるしょくぶつ　つる（蔓）植物　climbing plant, vine
つるせいの　つる（蔓）性の　viny, pole climbing
ツルナシインゲン（蔓無隠元）　dwarf bean, bush bean →インゲンマメ
ツルノゲイトウ　sessile joy-weed, *Alternanthera sessilis* (L.) R. Br. ex Roem. et Schult.
つるぼけ　つる（蔓）ぼけ　excessive vine growth
ツルマツリ（蔓茉莉），ソケイ（素馨）　poet's jasmine, common white jasmine, *Jasminum officinale* L. f.
ツルマメ（蔓豆）　wild soybean, *Glycine max* (L.) Merr. ssp. *soja* (Sieb. et Zucc.) H. Ohashi
ツルメヒシバ →カーペットグラス
ツルレイシ →ニガウリ
つるわれびょう　つる割病【サツマイモ】　stem rot
ツンドラきこう　ツンドラ気候　tundra climate

[て]

ティーアイプラスミド　Tiプラスミド　Ti plasmid
ディアキネシスき　ディアキネシス期，移動期【減数分裂】　diakinesis stage
ティーアールりつ　TR率　top-root ratio (T-R ratio)
ていいしゅうかくでん　低位収穫田　low-yielding paddy field
ていいせいさんち　低位生産地　low-productivity land (area)
ていいぶんげつ　低位分げつ　lower nodal tiller
ディーエヌエーポリメラーゼ　DNAポリメラーゼ　DNA polymerase
ディーエヌエーリガーゼ　DNAリガーゼ，ポリデオキシリボヌクレオチドシンターゼ　polydeoxyribonucleotide synthase
ディーエーへんかん　D/A変換　digital-analog (D/A) conversion
ていおんき　定温器, 恒温器　incubator
ていおんしょうがい　低温障害　low-temperature injury, chilling injury
ていおんしょり　低温処理　low-temperature treatment
ていおんそうこ　低温倉庫　low-temperature warehouse
ていおんちょぞう　低温貯蔵　low-temperature storage
ていおんはつがせい　低温発芽性　low-temperature germinability
ていが　定芽　definite bud
ていかんしたて　低幹仕立て，根刈り仕立て【クワ】　low cut training
ていきあつ　低気圧　depression, low, cyclone
ディクチオゾーム　dictyosome
ティーけんてい　t検定　t-test
ていこうおんどけい　抵抗温度計　resistance thermometer
ていこうせい　抵抗性　resistance
ていこうせいひんしゅ　抵抗性品種　resistant variety
ていこうほう　蹄耕法　hoof cultivation
ていさんそじょうたい　低酸素状態　hypoxia
ていし　底刺【オオムギ】　basal bristle

ティーシーエー(トリカルボンさん)かいろ　TCA(トリカルボン酸)回路　TCA (tricarboxylic acid) cycle, クレブス回路　Krebs cycle, クエン酸回路　citric acid cycle

ティーじかん　T字管　T-shaped tubing connector

ディジタイザー　digitizer

ていしつち　低湿地　swamp, marsh

ていじぶんげつ　低次分げつ　lower order tiller

ていじょうじょうたい　定常状態　steady state

ていしょく　定植　setting, planting

ていしょくき　定植期　planting time

ていすう[こうか]もけい　定数[効果]模型, 固定[効果]模型　fixed [effects] model

ディスクプラウ　disk plow

ていせいぶんせき　定性分析　qualitative analysis

でいたん　泥炭　peat

でいたんち　泥炭地　bog, moor

でいたんど　泥炭土　Peat soil, Bog soil

ていタンパクしつまい　低タンパク質米　low protein rice

ていちのうぎょう　低地農業　lowland agriculture

ていちゃく　定着, 土着　establishment, ecesis

ていちゃく　定着【写真】　fixation

ていちゃくえき　定着液　fixative

ティーディーアールほう　TDR法　time domain reflectometry

ティーディーエヌエー　T-DNA (transferred DNA)

ていとうにゅうじぞくがたのうぎょう　低投入持続型農業　low-input sustainable agriculture (LISA)

ティーひょう　t表　t-table

ディファレンシャルディスプレイ　differential display

ディプロテンき　ディプロテン期, 複糸期【減数分裂】　diplotene stage

ティーぶんぷ　t分布　t-distribution

ていぼく　低木　shrub, bush

ていりょうぶんせき　定量分析　quantitative analysis

ディル, イノンド　dill, *Anethum graveolens* L.

デオキシリボかくさん　デオキシリボ核酸　deoxyribonucleic acid (DNA)

テオシント　teosinte, *Euchlaena mexicana* Schrad.

てがり　手刈り　hand cutting

てきおう　適応　adaptation

てきおうけいしつ　適応形質　adaptive character

てきおうせい　適応性　adaptability

てきおうせんりゃく　適応戦略　adaptive strategy

てきおうど　適応度　fitness

てきおん　適温, 最適温度　optimum temperature

てきか　摘果　fruit thinning

てきか　摘花　defloration, flower picking, flower thinning

てきが　摘芽　disbudding, bud picking

てきごうど　適合度, 群落適合度【生態】　fidelity

てきごうど　適合度【統計】　goodness of fit

てきごうどけんてい　適合度検定　test of goodness of fit

てきさい　摘採　1) picking【ワタなど】2) plucking【チャなど】

てきさいき　摘採機　1) picker, plucker 2) tea plucker, plucking machine【チャ】

てきさいてきき　摘採適期　1) optimum picking time【ワタなど】2) optimum plucking time【チャなど】

てきさいめん　摘採棉(綿)　picked cotton

てきさいめん　摘採面【チャ】　plucking surface

てきしん　摘心　1) topping, pinching, top pruning　2) 摘心【タバコ】topping
デキストリン　dextrin
てきちてきさく　適地適作　right crop for right land
てきてい　滴定　titration
てきは　摘播　nest sowing (seeding)
てきびん　摘びん(瓶)　dropping bottle
てきよう　摘葉　defoliation
テクスチャー　texture
テクスチュロメータ　texturometer
デシケーター　desiccator
デジタル　digital
デスモしょうかん　デスモ小管　desmotuble
デスモディウム　desmodium, *Desmodium* spp.
データベース　database
てつ　鉄　iron (Fe)
てづみ　手摘み　1) hand picking【ワタなど】2) hand plucking【チャなど】
てどりじょそう　手取り除草　hand weeding
テパリビーン　tepary bean, *Phaseolus acutifolius* A. Gray var. *latifolius* Freem.
でびらきが　出開き芽【チャ】*banjhi* shoot
でびらきど　出開き度【チャ】percentage of banjhi shoots to the total (P. B. S.), ratio of banjhi shoot
テフ　teff [grass], *Eragrostis abyssinica* (Jacq.) Link (= *E. tef* Trotter)
てまき　手播き　hand sowing (seeding)
デュラムコムギ　durum wheat, マカロニコムギ　macaroni wheat, *Triticum durum* Desf.
テラスさいばい　テラス栽培, 階段耕作　terrace culture, terrace farming
テラロサ　Terra Rossa
デリス　derris, *Derris* spp.【*D. elliptica* (Roxb.) Benth. など】

デルタ, 三角州　delta
テロームせつ　テローム説　telome theory
てんいアールエヌエー　転移 RNA　transfer RNA (tRNA)
てんいいでんし　転移遺伝子　transfer genes
てんいせいいでんいんし　転移性遺伝因子　transposable genetic element, トランスポゾン　transposon
てんか　転化　1) conversion 2) inversion【糖】
てんか　添加　addition
てんかざい　添加剤, 添加物　additive
てんかとう　転化糖　invert sugar
てんかぶつ　添加物, 添加剤　additive
てんかん[すい]でん　転換[水]田　paddy field converted from upland field
てんかんばた　転換畑　upland field converted from paddy field
てんかんぼくそうち　転換牧草地　temporary grassland
でんきえいどう　電気泳動　electrophoresis
でんきせんこうほう　電気穿孔法, エレクトロポレーション　electroporation
でんきでんどうりつ　電気伝導率　electric conductivity
てんきよほう　天気予報　weather forecasting
てんぐすびょう　天狗巣病　witche's broom
てんざ　転座【染色体】translocation
テンサイ(甜菜)　sugar beet, *Beta vulgaris* L. var. *rapa* Dumort.
テンサイとう　テンサイ(甜菜)糖　beet sugar
テンサイパルプ　テンサイ(甜菜)パルプ ⇒ビートパルプ　beet pulp
でんさくもつ　田作物　lowland crop
てんざぶんせき　転座分析　translocation analysis

テンシオメーター tensiometer
でんしけいさんき 電子計算機, コンピュータ computer
でんしけんびきょう 電子顕微鏡 electron microscope
デンジソウ pepperwort, water clover, *Marsilea quadrifolia* L.
でんしでんたつけい 電子伝達系 electron transport system (chain)
デンシトメーター densitometer
でんじは 電磁波 electromagnetic wave
てんじほ[じょう] 展示圃[場] demonstration farm, demonstration plot
でんしみつど 電子密度 electron density
てんしゃ 転写 transcription
でんしょうさいばい 電照栽培 light culture, cultivation under lightening
てんすいさいばい 天水栽培 rain-fed cultivation, rain-fed culture
てんすいでん 天水田 rain-fed paddy field
でんせんびょう 伝染病 infectious disease
てんそうさいぼう 転送細胞, 輸送細胞 transfer cell
でんたつかんすう 伝達関数 transfer function
てんちがえし 天地返し⇒反転耕 upside down plowing
てんちゃくざい 展着剤 wetting agent
テンツキ forked fringerush, *Fimbristylis dichotoma* (L.) Vahl
てんてき 天敵 natural enemy
てんてきかんがい 点滴灌漑 drip irrigation, trickle irrigation
でんどうど 伝導度, コンダクタンス conductance
でんどうど 電導度⇒電気伝導率 electric conductivity
デントコーン dent corn, *Zea mays* L.

var. *indentata* Bailey
デンドログラム, 樹状図 dendrogram
でんねつおんしょう 電熱温床 electric hotbed
てんねんきょうきゅう[りょう] 天然供給[量] natural supply
てんぱ 天葉【タバコ】 tips
てんぱ 点播 hill sowing (seeding)
でんぱ(でんぱん) 伝播 dissemination
てんぱき 点播機 planter
でんぱたりんかん 田畑輪換 paddy-upland rotation
テンパリング tempering
デンプン starch
デンプンか デンプン価 starch value
デンプンごうせいこうそ デンプン合成酵素 starch synthase
デンプンしつはいにゅう デンプン質胚乳 starchy endosperm
デンプンしゅし デンプン種子 starch seed
デンプンぶどまり デンプン歩留り starch yielding percentage
デンプンよう デンプン葉 starch leaf, amylophyll
デンプンりゅう デンプン粒 starch grain (granule)
デンプン[りょう]さくもつ デンプン[料]作物 starch crop
でんめんすい 田面水 flooded water in paddy field
てんりゅう 転流 translocation
でんれいアールエヌエー 伝令 RNA, メッセンジャー RNA messenger RNA (mRNA)

[と]

トウ(籐) rattan [palm], rotan, cane palm, *Calamus caesius* Blume 他多種
とう(薹)⇒花茎 flower stalk, scape
とう 糖 sugar

どう 銅 copper (Cu)
とうあつ 踏圧 trampling
どういげんそ 同位元素, アイソトープ, 同位体 isotope
どういたい 同位体, アイソトープ, 同位元素 isotope
どういたいきしゃくほう 同位体希釈法 isotope dilution method
どういたいこうか 同位体効果 isotope effect
どういでんしがたこたいぐん 同遺伝子型個体群, バイオタイプ biotype
とうおんせん 等温線 isotherm
とうか 稲架⇒はさ, はざ [paddy] sheaf rack
とうか 冬禾⇒冬穀物 winter cereals
とうか 糖化 saccharification
とうか 豆果 legume
とうか 頭花 caput, capitulum (*pl.* capitula)
とうがい 凍害 freezing damage (injury)
とうかがたでんしけんびきょう 透過型電子顕微鏡 transmission electron microscope (TEM)
どうかきかん 同化器官 assimilation organ
どうか[さよう] 同化[作用] assimilation, anabolism
どうかさんぶつ 同化産物 assimilate, assimilation product
とうかせい 透過性 permeability
どうかそしき 同化組織 assimilation tissue
どうかちゅうか 同花柱花, 同柱花 homomorphous flower
どうかデンプン 同化デンプン assimilation starch
トウガラシ (唐芥子) red pepper, chili, capsicum, *Capsicum annuum* L. (= *C. frutescens* L.)
とうかりつ 透過率 transmittance, transmissivity

どうかん 導管 vessel
トウキ (当帰) *Angelica acutiloba* (Sieb. et Zucc.) Kitagawa
どうぎいでんし 同義遺伝子 multiple genes, polymeric genes
とうきかんがい 冬期灌漑 winter irrigation
とうききゅうかん 冬期(季)休閑 winter fallow
とうきせんてい 冬期せん(剪)定 winter pruning
とうきゅう 等級【玄米など】 grade
とうきゅうづけ 等級付け, 等級分け⇒格付 grading
どうぎれまい 胴切米 notched-belly rice kernel
とうけいいでんがく 統計遺伝学 statistical genetics
とうけいがく 統計学 statistics
どうけいせつごう 同型接合, ホモ接合 homozygosis
どうけいせつごうせい 同型接合性, ホモ接合性 homozygosity
どうけいせつごうたい 同型接合体, ホモ接合体 homozygote
とうけいてきすいろん 統計的推論 statistical inference
とうけい[てき]ぶんせき 統計[的]分析 statistical analysis
とうけいてきほうほう 統計的方法 statistical method
どうけいはんしょく 同系繁殖 inbreeding
どうけいぶんれつ 同型分裂 homotypic division
とうけいりょう 統計量 statistic
とうけつエッチング 凍結エッチング, フリーズエッチング【電顕】 freeze-etching
とうけつかつだん 凍結割断, フリーズフラクチャ【電顕】 freeze-fracture
とうけつかんそう 凍結乾燥 freeze-drying

どうげんたい　動原体　kinetochore
とうこう　冬耕　winter plowing
とうこうせんかんがい　等高線灌漑　contour border irrigation, contour furrow irrigation
とうこうせんさいばい　等高線栽培　contour cropping, contour cultivation, contour farming
トウゴマ→ヒマ
トウシキミ→ダイウイキョウ
とうししつ　糖脂質　glycolipid
どうしついでんしけいとう　同質遺伝子系統　isogenic line
どうしつばいすうせい　同質倍数性　autopolyploidy
とうじゅく　登熟　ripening, grain filling
とうじゅくきかん　登熟期間　ripening period, [grain] filling period
とうじゅくぶあい　登熟歩合　percentage of ripened grains
とうじゅくふりょう　登熟不良　poor ripening
とうじょう　凍上　frost heaving
とうじょうがい　凍上害　frost-heaving damage
とうじょうかじょ　頭状花序　capitulum, caput
とうじょうよう　筒状葉　tubular leaf
どうしょしゅ　同所種　sympatric species
どうしょせい　同所性　sympatry
どうしょせいの　同所性の　sympatric
どうしんせつごう　同親接合, 同親対合【染色体】　autosynapsis, autosyndesis
どうしんたいごう　同親対合, 同親接合【染色体】　autosynapsis, autosyndesis
トウジンビエ（唐人稗）→パールミレット
どうしんぶんげつ　同伸分げつ　synchronously emerging tiller
どうしんよう　同伸葉　synchronously emerging leaf

とうすいせい　透水性　water permeability
とうすいにっすう　到穂日数　days to heading
とうせい　とう(搗)精　milling, pearling
とうせいぶあい　とう(搗)精歩合　milling percentage
とうせき　透析　dialysis
とうそうがい　凍霜害, 霜害　frost damage (injury)
とうた　淘汰⇒選抜　selection
とうたあつ　淘汰圧⇒選抜圧　selection pressure
とうたすいじゅん　淘汰水準　culling level
とうたつかのうしゅうりょう　到達可能収量　attainable yield
とうちせん　等値線　isoline
とうちせんマップ　等値線マップ　isoline map
どうちゅうか　同柱花, 同花柱花　homomorphous flower
とうちょうえき　等張液　isotonic solution
どうちょうせいちょう　同調成長　symplastic growth
どうちょうばいよう　同調培養　synchronous culture, synchronized culture
どうちょうぶんれつ　同調分裂　synchronous division
どうてきへいこう　動的平衡　dynamic equilibrium
とうでんてん　等電点　isoelectric point
とうど　糖度　Brix, 屈折計示度　refractometer index
トゥニカ⇒外衣　tunica
トゥニカ・コーパスせつ　トゥニカ・コーパス説⇒外衣内体説　tunica-corpus theory
とうにゅう　透入, 浸透　penetration
とうにゅう　豆乳　soy milk

どうにゅう　導入　introduction
どうにゅうこうざつ　導入交雑, 移入交雑, 浸透交雑　introgressive hybridization
どうにゅうひんしゅ　導入品種　introduced variety
どうはんさくもつ　同伴作物, 随伴作物　companion crop
とうふ　豆腐　tofu, bean curd
とうふく　倒伏　lodging
とうふくしすう　倒伏指数　lodging index
とうふくせいの　倒伏性の　susceptible to lodging
とうふくていこうせい　倒伏抵抗性　lodging resistance
どうぶつせいさん　動物生産　animal production
とうほくのうぎょうしけんじょう　東北農業試験場　Tohoku National Agricultural Experiment Station
とうみ　唐み(箕)　winnower, winnowing machine
とうみせん　唐み(箕)選　machine winnowing
とうめいか　透明化　clearing
とうめいど　透明度　transparency
トウモロコシ(玉蜀黍)　corn, maize, Indian corn, *Zea mays* L.
トウモロコシゆ　トウモロコシ油　corn oil
とうゆ　桐油　tung oil
とうよう　糖葉　sugar leaf, saccharophyll
どうりょくこううんき　動力耕うん(耘)機　power tiller
どうりょくさんぷんき　動力散粉機　power duster
どうりょくだっこくき　動力脱穀機　power thresher
どうりょくふんむき　動力噴霧機　power sprayer
どうりょくもみすりき　動力籾す(摺)り機　power huller
とうりんさん　糖リン酸　sugar phosphate
とうるい　糖類　sugars
とうろく　登録　registration, registry
とうろくしゅし　登録種子　registered seed
とうろくひんしゅ　登録品種　registered variety
どうわれまい　胴割[れ]米　cracked rice kernel
とおえんこうざつ　遠縁交雑　wide cross, wide hybridization
とおえんしゅ　遠縁種　far-related species
とおしなわしろ　通し苗代　permanent nursery
どかい　土塊　clod
トカドヘチマ　towel gourd, *Luffa acutangula* (L.) Roxb.
ときなしさいばい　時無し栽培　nonseasonal culture
ときなしひんしゅ　時無し品種　nonseasonal variety
トキワハゼ　Japanese mazus, *Mazus pumilus* (Burm.f.) V. Steenis (= *M. japonicus* (Thunb.) O. Kuntze)
トキンソウ(吐金草)　*Centipeda minima* (L.) A. Br. et Aschers.
とくせい　特性　1) characteristic 2) character, trait【統計】
とくせいけんてい　特性検定, 特性試験　test of specific character, test for physiological character
とくせいしけん　特性試験, 特性検定　test of specific character, test for physiological character
どくせい　毒性　toxicity
どくそ　毒素　toxin
どくそう　毒草　poisonous herb
ドクダミ　*Houttuynia cordata* Thunb.
とくていくみあわせのうりょく　特定組合せ能力　specific combining ability

とくのう　篤農　well-experienced farmer
どくぶつ　毒物　poison
ドクムギ (毒麦)　poison ryegrass, darnel, *Lolium temulentum* L.
とくようさくもつ　特用作物　industrial crop
どくりつえいよう　独立栄養　autotrophism, autotrophy
とけいざら　時計皿　watch glass
トゲサゴ [ヤシ] (刺サゴ [椰子])　spiny sago palm, prickly sago palm, *Metroxylon rumphii* Mart.
トゲドコロ　lesser yam, potato yam, *Dioscorea esculenta* (Lour.) Burk.
トゲバンレイシ　sour-sop, sour apple, *Annona muricata* L.
トゲミノキツネノボタン　roughseed buttercup, *Ranunculus muricatus* L.
とこ　床　bed
どこう　土耕　soil culture
とこがえ　床替え　nursery transplanting
とこじめ　床じめ　subsoil compaction
とこつち　床土　[nursery] bed soil
とこねり　床練り　kneading of nursery bed
とこまき　床播き　sowing (seeding) in nursery bed
としゅせっぺん　徒手切片　hand section
どじょう　土壌　soil
どじょうおせん　土壌汚染　soil pollution, soil contamination
どじょうおせんぶっしつ　土壌汚染物質　soil pollutant, soil contaminant
どじょうかいりょう　土壌改良　soil improvement, soil amendment
どじょうかいりょうざい　土壌改良剤　soil conditioner
どじょうかいりょう [し] ざい　土壌改良 [資] 材　inorganic soil amendment
どじょうかがく　土壌化学　soil chemistry
どじょうがく　土壌学　soil science, ペドロジー　pedology
どじょうがた　土壌型　soil type
どじょうかんしょうのう　土壌緩衝能　soil buffer action
どじょうかんり　土壌管理　soil management
どじょうきょうど　土壌強度　soil strength
どじょうけんてい　土壌検定　soil testing
どじょうこうげき　土壌孔げき (隙)　soil pore space
どじょうこうげきりつ　土壌孔げき (隙) 率　soil porosity
どじょうこうぞう　土壌構造　soil structure
どじょうこうど　土壌硬度　soil hardness
どじょうこうぶつ　土壌鉱物　soil mineral
どじょうこきゅう　土壌呼吸　soil respiration
どじょうコロイド　土壌コロイド　soil colloid
どじょうさっきん　土壌殺菌　soil sterilization
どじょうさっきんざい　土壌殺菌剤　soil fungicide, soil disinfectant
どじょうさんせい　土壌酸性　soil acidity
どじょうさんそう　土壌三相　three phases of soil
どじょうざんりゅうせいのうやく　土壌残留性農薬　soil persistent pesticide
どじょうしゅしばんく　土壌種子バンク　soil seed bank
どじょうじょうけん　土壌条件　soil condition
どじょうしょうどく　土壌消毒　soil disinfection
どじょうしょうどくき　土壌消毒機

soil fumigator
どじょう-しょくぶつ-たいきでんたつ　土壌-植物-大気伝達　soil-vegetation-atmosphere transfer (SVAT)
どじょう-しょくぶつ-たいきれんぞくたい　土壌-植物-大気連続体　soil-plant-atmosphere continuum (SPAC)
どじょうしんしゅつえき　土壌浸出液　soil extract
どじょうしんしょく　土壌侵食　soil erosion
どじょうしんだん　土壌診断　soil diagnosis
どじょうず　土壌図　soil map
どじょうすい[ぶん]　土壌水[分]　soil water, soil moisture
どじょうすいぶんがんりょう　土壌水分含量　soil water (moisture) content
どじょうすいぶんストレス　土壌水分ストレス　soil moisture stress
どじょうすいぶんちょうりょく　土壌水分張力　soil water tension
どじょうせいさんりょく　土壌生産力　soil productivity
どじょうせいせい　土壌生成　pedogenesis, soil genesis
どじょうせいぶつ　土壌生物　soil organism
どじょうそうい　土壌層位　soil horizon
どじょうだんめん　土壌断面　soil profile
どじょうちっそ　土壌窒素　soil nitrogen
どじょうちゅうにゅうき　土壌注入機　soil injector
どじょうちょうさ　土壌調査　soil survey
どじょうつうき　土壌通気　soil aeration
どじょうでんせん　土壌伝染　soil transmission
どじょうでんせんびょう　土壌伝染病　soil-borne disease
どじょうとう　土壌統　soil series
どじょうのしめかため　土壌の締固め　soil compaction
どじょう[の]みずポテンシャル　土壌[の]水ポテンシャル　soil water potential
どじょうはんのう　土壌反応　soil reaction
どじょうびさいこうぞう　土壌微細構造　soil microstructure
どじょうびせいぶつ　土壌微生物　soil microorganism
どじょうひふく　土壌被覆　soil mulch
どじょうびょうがい　土壌病害　soil-borne disease
どじょうひよくど　土壌肥沃度　soil fertility
どじょうぶつり[がく]　土壌物理[学]　soil physics
どじょうぶんせき　土壌分析　soil analysis
どじょうほすいざい　土壌保水剤　soil water holding agent
どじょうほぜん　土壌保全　soil conservation
どじょうマルチ　土壌マルチ⇒土壌被覆　soil mulch
どじょうモノリス　土壌モノリス，モノリス　soil monolith
どじょうゆうきぶつ　土壌有機物　soil organic matter
どじょうよういん　土壌要因　edaphic factor
どじょうようえき　土壌溶液　soil solution
どじょうりゅうし　土壌粒子　soil particle
どすう　度数，頻度　frequency
どすうぶんぷ　度数分布，頻度分布　frequency distribution
どせい　土性　soil texture
どそうかいりょう　土層改良　subsoil improvement

とち　土地　land
とちかいりょう　土地改良　land improvement
トチカガミ　*Hydrocharis dubia* (Blume) Backer
とちせいさんせい　土地生産性　land productivity
とちぶんるい　土地分類, 土地分級　land classification
とちぶんきゅう　土地分級, 土地分類　land classification
どちゃく　土着, 定着　establishment, ecesis
とちょう　徒長　spindly growth
とちょうし　徒長枝　water shoot, water sprout
とちりょう　土地利用　land use, land utilization
とっき　突起　protuberance, ingrowth
とつぜんへんい　突然変異　mutation
とつぜんへんいあつ　突然変異圧　mutation pressure
とつぜんへんいいくしゅ［ほう］　突然変異育種［法］　mutation breeding
とつぜんへんいげん　突然変異源　mutagen
とつぜんへんいたい　突然変異体　mutant
とつぜんへんいゆうはつ　突然変異誘発　mutagenesis
トッピング【テンサイ】　topping
トップこうざつ　トップ交雑　top cross
ドデシルりゅうさんナトリウム　ドデシル硫酸ナトリウム　sodium dodecyl sulphate (SDS)
トノプラスト, 液胞膜　tonoplast
トピータンブー→トラフヒメバショウ
とびこ　飛粉【コンニャク】　flying out flour
トマト　tomato, *Lycopersicon esculentum* Mill.
ドメイン【タンパク質】　domain
とめぐさ　止め草　last weeding

とめごえ　止め肥　final dressing, last topdressing
とめば　止葉　flag leaf
とめばのまえのは　止葉の前の葉　penultimate leaf
ともだい　共台　free stock
どようぼし　土用干し⇒中干し(なかぼし)　midseason drainage
トラガカントゴムノキ　tragacanth milkvetch, *Astragalus gummifer* Labill.
トラクタ　tractor
トラフヒメバショウ　topee-tambu, *Calathea allouia* (Aubl.) Lindl
トランスフォーメーション, 形質転換　transformation
トランスポゾン　transposon, 転移性遺伝因子　transposable genetic element
トリアコンタノール　triacontanol
ドリアン　durian, *Durio zibethinus* Murr.
トリオース, 三炭糖　triose
とりおきなえ　取置き苗　pulled out and reserved seedling, stand-by seedlings
トリカルボンさん(ティーシーエー)かいろ　トリカルボン酸(TCA)回路　tricarboxylic acid (TCA) cycle, クレブス回路　Krebs cycle, クエン酸回路　citric acid cycle
とりき　取り木　layering, layerage
トリチウム, 三重水素　tritium
トリッピング【アルファルファ】　tripping
トリプトファン　tryptophan (Trp)
とりまき　取播き　immediate sowing (seeding) after harvest
トリュフ　truffle, *Tuber* spp. 【*T. melanosporum* Vittl. 他】
ドリルはしゅき　ドリル播種機⇒条播(じょうは)機　drill
ドリルまき　ドリル播き⇒条播(じょうは)　drilling, row sowing (seeding)

トルイジンブルー【染色】 toluidine blue
トールオートグラス tall oatgrass, *Arrhenatherum elatius* (L.) K. Presl
トールフェスク tall fescue, tall meadow fescue, *Festuca arundinacea* Schreb.
トレーサー tracer
トレオニン threonine (Thr)
トレンチャ, 溝掘り機 trencher
トレンド trend
トレンドかんすう トレンド関数 trend function
トロロアオイ(黄蜀葵) sunset hibiscus, *Abelmoschus manihot* Medik. (= *Hibiscus manihot* L.)
トンキンウルシ *Rhus succedanea* L. var. *dumoutieri* Pier.
トンキンニッケイ→カシア
トンネルさいばい トンネル栽培 plastic-tunnel culture

[な]

ないえい 内穎 palea
ないかひ 内果皮 endocarp
ないきんこん 内菌根⇒内生菌根 endomycorrhiza, endotrophic mycorrhiza
ないしゅひ 内種皮 inner seed coat, internal seed coat
ないしょう 内鞘 pericycle
ないせいの 内生の endogenous
ないせいオーキシン 内生オーキシン endogenous auxin
ないせいきんこん 内生菌根 endomycorrhiza, endotrophic mycorrhiza
ないせいリズム 内生リズム endogenous rhythm
ないそうほう 内挿法 interpolation
ないたい 内体 corpus
ない[はい]にゅう 内[胚]乳 endosperm
ないひ 内皮 endodermis
ないぶよういん 内部要因 internal factor
ないほうえい 内苞穎 inner glume
ないりくきこう 内陸気候 inland climate
なえ 苗, 実生(みしょう), 幼植物 seedling
なえかんせん 苗感染 seedling infection
なえぎ 苗木 nursery stock
なえざし 苗挿し, 挿苗(そうびょう)【サツマイモ】 sprouted vine planting
なえだち 苗立ち establishment [of seedling]
なえたちがれ[びょう] 苗立枯[病] damping-off
なえだちぶあい 苗立歩合 percentage [of seedling] establishment
なえちょぞう 苗貯蔵 storage of seedling, seedling storage
なえどこ 苗床 nursery [bed]
なえどこいしょく 苗床移植 transplanting in nursery
なえどこけんてい 苗床検定 nursery test
なえとり 苗取り pulling of seedling, uprooting of seedling
なえのそしつ 苗の素質 character of seedling
ナガイモ(薯蕷) Chinese yam, *Dioscorea opposita* Thunb. (= *D. batatas* Decne.)
なかうち 中打ち⇒中耕(ちゅうこう) intertillage
なかしろ 中代 second puddling
なかて(ちゅうせい)ひんしゅ 中生品種 medium [maturing] variety
ナガハグサ→ケンタッキーブルーグラス
ナガバスブタ→スブタ
なかぼし 中干し midseason drainage
なかまき 中播き intersowing,

interseeding
ながれず　流れ図　flow chart (diagram)
ナギナタコウジュ　*Elsholtzia ciliata* (Thunb.) Hylander
ナシ (梨)　Japanese pear, sand pear, *Pyrus pyrifolia* (Burm. f.) Nakai (= *P. serotina* Rehder)
ナシじょうか　ナシ状果，仁果　pomaceous fruit, pome
ナス (茄子)　eggplant, *Solanum melongena* L.
ナズナ　shepherd's-purse, *Capsella bursa-pastoris* (L.) Medik.
ナースばいよう　ナース培養，保護培養　nurse culture
なすりつけほう　なすりつけ法【試料作成】smear method
ナタネ (菜種)　rape, *Brassica napus* L. および *B. campestris* L.
ナタネかす　ナタネ粕　rapeseed meal
ナタネゆ　ナタネ油　rapeseed oil
ナタマメ (刀豆)　sword bean, *Canavalia gladiata* (Jacq.) DC.
なつうえ　夏植え　summer planting
なつがた　夏型【ソバ】summer ecotype
なつがれ　夏枯れ【牧草など】summer depression
なつぎり　夏切【クワ】summer pruning
なつこくもつ　夏穀物　summer cereals
なつさく　夏作　summer cropping
なつさく [もつ]　夏作 [物]　summer crop
なつざっそう　夏雑草　summer weed
なつソバ　夏ソバ　summer buckwheat
なつダイズ　夏ダイズ　summer soybean
なっとう　納豆　fermented soybeans
なつまき　夏播き　summer sowing (seeding)
ナツミカン (夏蜜柑)　Watson pomelo, *Citrus natsudaidai* Hayata

ナツメ (棗)　common jujube, Chinese jujube, *Ziziphus jujuba* Mill.
なつめ　夏芽　summer bud
ナツメグ→ニクズク
ナツメヤシ (棗椰子)　date palm, *Phoenix dactylifera* L.
ナトリウム　sodium (Na)
ななめうえ　斜め植え　oblique planting, 斜め挿し　oblique cutting
ななめざし　斜め挿し　oblique cutting, 斜め植え　oblique planting,
ナフタレンさくさん　ナフタレン酢酸　naphthalene acetic acid (NAA)
なまぐさくろほびょう　なまぐさ黒穂病【コムギ】bunt
なまは　生葉　1) green leaf【タバコ】2) plucked new shoot【チャ】
なまもみ　生籾　fresh paddy
なみがたの　波形の　undulate
なみきうえ　並木植え　row planting
なみせい　並性【オオムギ】normal type
ナメコ　nameko fungus, *Pholiota nameko* S. Ito et Imai
なりくび　なり首, 藷梗【サツマイモ】joint (upper) part of tuberous root
なりどし　成り年　on-year
なわしろ　苗代　nursery bed
なわしろにっすうかんのうせい　苗代日数感応性　sensitivity to nursery days
なわしろようしき　苗代様式　type of nursery bed
ナワシンえき　ナワシン液【固定液】Navashin's fluid, Nawashin fluid
なんか　軟化　softening
なんか　軟化【野菜】⇒軟白　blanching
ナンキンマメ (南京豆)→ラッカセイ
なんしつコムギ　軟質コムギ　soft wheat
なんしつまい　軟質米　soft-textured rice
なんしつりゅう　軟質粒　soft grain
なんだつりゅうせい　難脱粒性　seed

retention [habit]
なんぱく　軟白　blanching
なんぱくさいばい　軟白栽培　blanching culture
ナンバンアイ→キアイ
ナンバンコマツナギ　West Indian indigo, *Indigofera suffruticosa* Mill.
ナンバンルリソウ　Indian heliotrope, *Heliotropium indicum* L.
なんぷ[びょう]　軟腐[病]　soft rot

[に]

にいでんしざっしゅ　二遺伝子雑種　dihybrid
にお, かたい（禾堆）　stack, cock
ニオイイガクサ　wild spikenard, wild basil, *Hyptis suaveolens* Poir.
ニオイタコノキ　screw pine, *Pandanus odorus* Ridl.
においまい　臭米⇒香り米　aromatic rice, scented rice
ニガウリ（苦瓜）　bitter gourd, balsam pear, *Momordica charantia* L.
ニガーシード　niger seed, ramtil, *Guizotia abyssinica* (L.f.) Cass.
ニガーしゅゆ　ニガー種油　niger seed oil
にかせんしょくたい　二価染色体　bivalent [chromosome]
ニガソバ（苦蕎麦）→ダッタンソバ
にかてつ　二価鉄　ferrous iron
ニガナ　*Ixeris dentata* (Thunb. ex Murray) Nakai
にがみ　苦み　bitterness
にきさく　二期作　1) double cropping 2) second crop【作物】
にきしゅ　二基種　digenomic species
にくがんかんてい　肉眼鑑定　eye judgement
にくがんせんばつ　肉眼選抜　visual selection
ニクズク（肉豆蔲）　nutmeg, *Myristica fragrans* Houtt.
に-クロロエチルホスホンさん　2-クロロエチルホスホン酸　2-chloroethyl phosphonic acid (CEPA), エテフォン　ethephon
にげんはいち　二元配置　two-way layout
にこうけいすう　二項係数　binomial coefficient
にこうていり　二項定理　binomial theorem
にこうぶんぷ　二項分布　binomial distribution
ニコチンアミドアデニンジヌクレオチド　nicotinamide adenine dinucleotide (NAD)
ニコチンアミドアデニンジヌクレオチドリンさん　ニコチンアミドアデニンジヌクレオチドリン酸　nicotinamide adenine dinucleotide phosphate (NADP)
ニコチンさん　ニコチン酸　nicotinic acid
にさぶんし　二又分枝, 二叉分枝　dichotomous branching, dichotomy
にさんかイオウ　二酸化イオウ　sulfur dioxide (SO_2)
にさんかたんそ　二酸化炭素　carbon dioxide (CO_2)
にさんかたんそきゅうしゅう　二酸化炭素吸収　carbon dioxide uptake (absorption)
にさんかたんそこうかんそくど　二酸化炭素交換速度　carbon dioxide exchange rate (CER)
にさんかたんそほしょうてん　二酸化炭素補償点　carbon dioxide compensation point
にさんかちっそ　二酸化窒素　nitrogen dioxide (NO_2)
にさんごトリヨードあんそくこうさん　2,3,5-トリヨード安息香酸　2,3,5-triiodobezoic acid (TIBA)

にしアフリカいねかいはつきょうかい 西アフリカ稲開発協会　West Africa Rice Development Association (WARDA)

にじいかんそく　二次維管束　secondary vascular bundle

にしインドアロールート　西インドアロールート→アロールート

にしインドレモングラス　西インドレモングラス　West Indian lemongrass, *Cymbopogon citratus* (D.C. ex Nees) Stapf

ニシキソウ　*Euphorbia humifusa* Willd. (= *E. pseudochamaesyce* Fisch., Mey. et Lallem.)

にじきゅうみん　二次休眠　secondary dormancy

にじきょくせん　二次曲線　quadratic curve

にじこん　二次根　secondary root

にじ[さいぼう]へき　二次[細胞]壁　secondary [cell] wall

にじさくもつ　二次作物　secondary crop

にじしこう　二次枝梗　secondary rachis-branch of panicle, secondary branch of panicle

にじしぶ　二次篩部　secondary phloem

にじせいちょう　二次成長　secondary growth

にじせんい　二次遷移　secondary succession

にじせんばつ　二次選抜　secondary selection

にじそしき　二次組織　secondary tissue

にじたいごう　二次対合　secondary pairing, secondary association

にじたいしゃさんぶつ　二次代謝産物　secondary metabolite

にじばいすうせい　二次倍数性　secondary polyploidy

にじひだい[せいちょう]　二次肥大[成長]　secondary thickening [growth]

にじぶんげつ　二次分げつ　secondary tiller

にじぶんれつそしき　二次分裂組織　secondary meristem

にじもくぶ　二次木部　secondary xylem

にじゅうせんしょく　二重染色　double staining

にじゅうとつぜんへんい　二重突然変異　double mutation

にじゅうのりかえ　二重乗換え　double crossing-over

にじゅうれっせい　二重劣性　double recessive

にじょうオオムギ　二条オオムギ　two-rowed barley, *Hordeum vulgare* L.

にじょうやせいオオムギ　二条野生オオムギ　two-rowed wild barley, *Hordeum spontaneum* K. Koch

ニシンかす　ニシン粕　herring cake, herring meal

にちおんどかくさ　日温度較差　daily temperature range

にちか　二値化　binarization

にちかくさ　日較差　diurnal range

にちぞうたい[りょう]　日増体[量]【家畜】　daily gain (DG)

にちへいきんきおん　日平均気温　daily mean [air] temperature

にちへんか　日変化　diurnal change, diurnal variation

にちょうきょくせん　二頂曲線　bimodal curve

にっかんかんそう　日干乾燥【タバコ】　sun-curing

ニッケイ(肉桂)　Saigon cinnamon, *Cinnamomum sieboldii* Meisn. (= *C. loureirii* Nees)

にっこうくっせい　日光屈性　heliotropism

にっしゃけい　日射計　pyranometer,

solarimeter
にっしゃりょう　日射量　amount of solar radiation, amount of insolation
にっしゃりようこうりつ　日射利用効率　radiation use efficiency, 光利用効率　light use efficiency
にっしょう　日照　sunshine
にっしょうけい　日照計　sunshine recorder, heliograph
にっしょうじかん　日照時間　duration of sunshine
にっちょう　日長　daylength
にっちょうこうか　日長効果⇒光周性　photoperiodism
にっちょうしょり　日長処理　photoperiodic treatment
にっちょうちょうせつ　日長調節　photoperiodic control
にっちょうはんのう　日長反応　photoperiodic response
ニッパヤシ　nipa palm, *Nypa fruticans* Wurmb.
にとう　二糖　disaccharide
にねんさんさく　二年三作　triple cropping in two years
にねんせいさくもつ　二年生作物　biennial crop
にねんせいざっそう　二年生雑草　biennial weed
にねんせいしょくぶつ　二年生植物　biennial plant
にねんせい[の]　二年生[の]　biennial
にねんそう　二年草　biennial herb (grass), biennial
にねんりんさく　二年輪作　two-year rotation
にばいせい　二倍性　diploidy
にばいたい[しょくぶつ]　二倍体[植物]　diploid [plant], diplont
にばんがり　二番刈り　1) second cutting【作業】2) second cut【牧草】
にばんこ　二番粉【ソバ】　No.2 flour
にばんじょそう　二番除草　second weeding
にばんそう　二番草　second crop
にばんだち　二番立ち　second flush
にばんめ　二番芽　second flush
にほうぶんぷ　二峰分布　bimodal distribution
にほんがた　日本型【イネ】　japonica type
ニホンカボチャ　pumpkin, winter squash, *Cucurbita moschata* (Duch. ex Lam.) Duch. ex Poir.
にほんこうぎょうきかく　日本工業規格　Japanese Industrial Standards (JIS)
にほんさくもつがっかい　日本作物学会　The Crop Science Society of Japan
にほんさくもつがっかいきじ　日本作物学会紀事　1) Japanese Journal of Crop Science【1977年から】2) Proceedings of the Crop Science Society of Japan【1976年まで】
ニホンナシ→ナシ
にほんのうりんきかく　日本農林規格　Japanese Agricultural Standards (JAS)
ニホンハッカ(日本薄荷)→ハッカ
にめいほう　二名法, 二命名法　binomial nomenclature
にめいめいほう　二命名法, 二名法　binomial nomenclature
にもうさく　二毛作　double cropping
にゅうえき　乳液　latex
にゅうかざい　乳化剤　emulsifier, emulsifying agent
にゅうかん　乳管　laticifer, latex duct, latex tube, latex vessel
にゅうざい　乳剤　emulsifiable concentrate
にゅうさいぼう　乳細胞　laticiferous cell
にゅうさん　乳酸　lactic acid
にゅうさんはっこう　乳酸発酵　lactic acid fermentation
にゅうしゃこう　入射光　incident light

にゅうじゅくき　乳熟期　milky [ripe] stage, milk-ripe stage
にゅうしょく　入植　settlement
にゅうしょくち　入植地　settlement [site], reclaimed settlement
にゅうとうじょうとっき　乳頭状突起　papilla
にゅうばい　入梅, つゆ(梅雨)入り　beginning of bai-u
にゅうはくまい　乳白米　milky white rice kernel
にゅうばち　乳鉢　mortar
にゅうびょう　乳苗【イネ】　nursling seedling
にゅうぼう　乳棒　pestle
ニュージーランドアサ（ニュージーランド麻）　New Zealand hemp, *Phormium tenax* J.R. Forst. et G. Forst.
ニューマン・ケウルスほう　ニューマン・ケウルス法　Newman-Keuls method
ニューラルネットワーク【情報処理】　neural network
にょうそ　尿素　urea
にょうそたいちっそ　尿素態窒素　urea nitrogen
によんジクロロフェノキシさくさん　2,4-ジクロロフェノキシ酢酸　2,4-dichlorophenoxy acetic acid (2,4-D)
ニラ（韮）　Chinese chive, *Allium tuberosum* Rottler ex. Spreng.
にりゅうけいコムギ　二粒系コムギ　emmer wheat
にれつたいせい　二列対生　distichous opposite
ニワゼキショウ　blue-eyed grass, *Sisyrinchium atlanticum* Bicknell
ニワホコリ　eragrostis, *Eragrostis multicaulis* Steud.
ニワヤナギ→ミチヤナギ
ニンジン（人参）　carrot, *Daucus carota* L.
ニンニク（葫, 大蒜）　garlic, *Allium sativum* L.
ニンヒドリンはんのう　ニンヒドリン反応　ninhydrin reaction

[ぬ]

ぬか（糠）　rice bran
ぬかあぶら　ぬか(糠)油　rice oil, bran oil
ヌカボ　*Agrostis clavata* Trin. ssp. *matsumurae* (Hack) Tateoka
ぬきとり　抜取り, 標本抽出【統計】　sampling
ぬきとりけんさ　抜取り検査　sampling inspection
ヌグ→ニガーシード
ヌクレオソーム　nucleosome
ヌクレオチド　nucleotide
ヌメリグサ　*Sacciolepis indica* (L.) Chase ssp. *oryzetorum* (Makino) T. Koyama

[ね]

ね　根　root
ネガティブせんしょく　ネガティブ染色　negative staining
ねがりしたて　根刈り仕立て, 低幹仕立て【クワ】　low cut training
ネギ（葱）　Welsh onion, *Allium fistulosum* L.
ねぐされ[びょう]　根腐れ[病]　root rot
ネクロシス, え(壊)死　necrosis
ねこぶ[びょう]　根こぶ[病]　club root
ネザサ　dwarf bamboo, *Pleioblastus chino* (Franch. et Savat.) Makino var. *viridis* (Makino) S. Suzuki (= *Arundinaria pygmaea* Mitford var. *glabra* Ohwi)
ねざし　根挿し　root cutting

ネジレフサマメノキ　pete, *Parkia speciosa* Hassk. (= *Peltogyne speciosa* Hassk).
ネズミムギ　→イタリアンライグラス
ネスラーしやく　ネスラー試薬　Nessler's reagent
ねつぎ　根接ぎ　root grafting, inarching【果樹】
ねづけごえ　根付け肥　starter
ねつしゅうし　熱収支　heat balance, heat budget, energy balance
ねつショックタンパクしつ　熱ショックタンパク質　heat shock protein
ねっせんふうそくけい　熱線風速計　hot-wire anemometer
ねつそんりゅう　熱損粒　fire burnt kernel, heat-damaged kernel
ねったい　熱帯　tropical zone, tropics
ねったい[せい]さくもつ　熱帯[性]作物　tropical crop
ねったい[せい]しょくぶつ　熱帯[性]植物　tropical plant
ねったいどじょう　熱帯土壌　tropical soil
ねったいの　熱帯の　tropical
ねつでんたつ　熱伝達　heat transfer
ねつでんつい　熱電対　thermocouple
ネットさいばい　ネット栽培　net culture
ねつへんせい　熱変性　heat denaturation
ねつようりょう　熱容量　heat capacity, thermal capacity
ねつりきがく　熱力学　thermodynamics
ねつりょう　熱量　heat quantity
ねつりょうけい　熱量計　calorimeter
ネナシカズラ　Japanese dodder, *Cuscuta japonica* Choisy
ねのぶんぴつぶつ　根の分泌物　root exudate
ねばこ　根箱　root box
ねばり　根張り　root spread
ねばり　粘り　stickiness

ネピアグラス　napiergrass, elephant grass, *Pennisetum purpureum* Schumach.
ねまわし　根回し　root pruning prior to transplanting
ねゆき　根雪　continuous snow cover
ねゆききかん　根雪期間　duration of continuous snow cover
ねりどこ　練り床　kneaded nursery bed
ねわけ　根分け　division of root stock, root splitting
ねんえき　粘液　mucilage
ねんかく　粘核【果実】　clingstone
ねんじそうかん　年次相関　year-to-year correlation
ねんじつ　稔実　ripening
ねんじつしょうがい　稔実障害　impediment in ripening
ねんじつぶあい　稔実歩合　percentage [of] ripening
ねんじへんい　年次変異　year-to-year variation, interannual variation
ねんせい　粘性, 粘度　viscosity
ねんせい　稔性　fertility
ねんせいかいふくけいとう　稔性回復系統　restorer
ねんだんせい　粘弾性　viscoelasticity
ねんど　粘度, 粘性　viscosity
ねんど　粘土　clay
ねんどこうぶつ　粘土鉱物　clay mineral
ねんどしつの　粘土質の　clayey
ねんへんか　年変化　annual change, annual variation
ねんりん　年輪　annual ring

[の]

のうえんのうぎょう　農園農業　estate agriculture, 企業農業　commercial agriculture, プランテーション　plantation
のうか　農家　farm household

のうがいこよう 農外雇用 off-farm employment
のうがいしょとく 農外所得 off-farm income
のうがく 農学 agronomy, agricultural sciences
のうかんき 農閑期 farmers' off-season
のうきぐ 農機具 farming machines and implements
のうきょう 農協⇒農業協同組合 agricultural cooperative
のうぎょう 農業 agriculture
のうぎょうかいりょうふきゅういん 農業改良普及員 extension agent
のうぎょうかいりょうふきゅうじぎょう 農業改良普及事業 agricultural extension service
のうぎょうかいりょうふきゅうしょ 農業改良普及所 agricultural extension office
のうぎょうかんきょうぎじゅつけんきゅうしょ 農業環境技術研究所 National Institute of Agro-Enviromental Sciences
のうぎょうきかい[がく] 農業機械[学] agricultural machinery
のうぎょうぎじゅつ 農業技術 agricultural technology
のうぎょうきしょうがく 農業気象学 agricultural meteorology, agrometeorology
のうぎょうきほんほう 農業基本法 The Agricultural Basic Law
のうぎょうきょうどうくみあい 農業協同組合(農協) agricultural cooperative
のうぎょうけいえい 農業経営 farm management
のうぎょうけいえいひ 農業経営費 agricultural expenditure
のうぎょうけいざいがく 農業経済学 agricultural economics
のうぎょうけいしつ 農業形質 agronomic character (trait)
のうぎょうけんきゅうセンター 農業研究センター National Agriculture Research Center
のうぎょうこうがく 農業工学 agricultural engineering
のうぎょうこうがくけんきゅうしょ 農業工学研究所 National Research Institute of Agricultural Engineering
のうぎょうこうぞうかいぜんじぎょう 農業構造改善事業 Agricultural Structure Improvement Project
のうぎょうしけんじょう 農業試験場 agricultural experiment station
のうぎょうしせつ 農業施設 agricultural structures, agricultural facilities
のうぎょうしゅうぎょうじんこう 農業就業人口 population engaged in agriculture
のうぎょうじょうほう 農業情報 agricultural information
のうぎょうしょとく 農業所得 agricultural income
のうぎょうせいぶつがく 農業生物学 agricultural biology, agrobiology
のうぎょうせいぶつしげんけんきゅうしょ 農業生物資源研究所 National Institute of Agrobiological Resources
のうぎょうそうごうけんきゅうしょ 農業総合研究所 National Research Institute of Agricultural Economics
のうぎょうどぼく 農業土木 irrigation, drainage and reclamation engineering
のうぎょうはくしょ 農業白書 Annual Report on Japanese Agriculture
のうぎょうりっち 農業立地 agricultural location
のうぐ 農具 agricultural implements, farm implements
のうげいかがく 農芸化学 agricultural

chemistry
のうこう 農耕 cultivation, agriculture
のうこうしりょう 濃厚飼料 concentrate [feed]
のうさぎょう 農作業 farm practices, farm working
のうさくもつ 農作物 field crop(s), farm crop(s)
のうさんぶつ 農産物 agricultural products
のうじれき 農事暦 agricultural calendar
のうすいしょう 農水省(農林水産省) MAFF (Ministry of Agriculture, Forestry and Fisheries)
ノウゼンハレン→キンレンカ
のうそんの 農村の rural
のうち 農地 agricultural land
のうどうてききゅうしゅう 能動的吸収 ⇒積極的吸収 active absorption
のうどうてきセンサー 能動的センサー active sensor
のうどうゆそう 能動輸送 active transport
のうどきせい 濃度規制 emission concentration regulation
のうはんき 農繁期 farmers' busy season
のうほう 農法 farming system
のうやく 農薬 agricultural chemicals, pesticide
のうりょくけんてい 能力検定 performance test
のうりんすいさんぎじゅつかいぎ 農林水産技術会議 Agriculture, Forestry and Fisheries Research Council (AFFRC)
のうりんすいさんしょう 農林水産省(農水省) Ministry of Agriculture, Forestry and Fisheries (MAFF)
ノーザンブロットほう ノーザンブロット法 northern blot technique, northern blotting

のぎ⇒ぼう(芒) awn, arista
ノゲシ milk thistle, sow thistle, *Sonchus oleraceus* L.
のこぎりがま のこぎり(鋸)鎌 serrated sickle
のずみ 野積み stack, field heaping
ノチドメ *Hydrocotyle maritima* Honda
ノボロギク common groundsel, *Senecio vulgaris* L.
ノミノツヅリ thymeleaf sandwort, *Arenaria serpyllifolia* L.
ノミノフスマ slender starwort, *Stellaria alsine* Grimm var. *undulata* (Thunb.) Ohwi
ノリアサ(糊麻) *Abelmoschus glutino-textilis* Kagawa
のりかえ 乗換え, 交さ(又) crossing-over
のりかえか 乗換え価, 交さ(又)価 crossing-over value
ノンパラメトリックほう ノンパラメトリック法 nonparametric method

[は]

は 葉 leaf
はあみ 葉あみ【タバコ】 sewing, stringing
はい 胚 embryo
はいいしょく 胚移植 embryo transplanting
はいいろしんりんど 灰色森林土 Gray Forest soil
はいいろていちど 灰色低地土 Gray Lowland soil
ばいう 梅雨 bai-u, rainy spell in early summer
バイオアッセイ, 生物検定 bioassay
バイオセンサー biosensor
バイオタイプ, 同遺伝子型個体群 biotype
バイオテクノロジー biotechnology
バイオトロン biotron

バイオニクス, 生体工学　bionics
バイオマス, 生物[体]量, 現存量　biomass, standing crop
バイオリアクター　bioreactor
バイオリズム, 生物リズム　biorhythm
バイオレメディエーション, 生物的浄化　bioremediation
ばいかいこんちゅう　媒介昆虫　1) insect vector【病原体】 2) insect pollinator【花粉】
ばいかいせいぶつ　媒介生物, ベクター　vector
ばいかはんすうたい　倍加半数体　doubled haploid
はいがまい　胚芽米　milled rice with embryo
ハイキビ　torpedo grass, *Panicum repens* L.
はいきぶつ　廃棄物　waste
はいぐうし　配偶子　gamete
はいぐうしちし　配偶子致死　gametic lethal
はいぐうしひ　配偶子比　gametic ratio
はいぐうたい　配偶体　gametophyte
ハイグロマイシン　hygromycin
はいけいせい　胚形成, 胚発生　embryogenesis
はいごうしりょう　配合飼料　formula feed
はいごうど　配合土　commercial compost, soil mixes
はいごうひりょう　配合肥料　blended fertilizer, mixed fertilizer
ハイコヌカガサ→クリーピングベントグラス
はいじく　胚軸　hypocotyl
はいじく[がわ]の　背軸[側]の【植物】abaxial
はいしゅ　胚珠　ovule
はいしゅつ　排出　excretion
はいしゅばいよう　胚珠培養　ovule culture
はいすい　排水　1) drainage【耕地】 2) guttation【植物】
はいすいこうぞう　排水構造　hydathode
はいすいさいぼう　排水細胞　hydathodal cell
はいすいふりょうでん　排水不良田⇒湿田　ill-drained paddy field
はいすいもう　排水毛　hydathodal hair
ばいすうせい　倍数性　polyploidy
ばいすうたい　倍数体　polyploid
はいせいちょういんし　胚成長因子　embryo factor
はいせつぶつ　排泄物　excreta
ばいせんざい　媒染剤　mordant
はいち　配置, 割付け【試験区】　layout
ばいち　培地　culture medium (*pl.* media)
ばいど　培土, 土寄せ　molding, earthing up, ridging
はいとうたい　配糖体　glucoside
パイナップル　pineapple, *Ananas comosus* (L.) Merr.
バイナリーベクター　binary vector
はいにゅう　胚乳　albumen, endosperm【慣用】
はいのう　胚のう(嚢)　embryosac
はいのうぼさいぼう　胚のう(嚢)母細胞　embryosac mother cell (EMC)
はいばいよう　胚培養　embryo culture
はいはっせい　胚発生, 胚形成　embryogenesis
はいばん　胚盤　scutellum (*pl.* scutella)
はいふくせい　背腹性　dorsiventrality
ハイブリダイゼーション, 雑種形成　hybridization
ハイブリッドライス　hybrid rice
はいへい　胚柄　suspensor
バイメタルおんどけい　バイメタル温度計　bimetallic thermometer
はいめん　背面　dorsal side
バイモ　*Fritillaria verticillata* Willd. var. *thunbergii* (Miq.) Baker (= *F. thumbergii* Miq.)

ばいよう　培養　culture
ばいようえき　培養液　culture solution
ばいようえきはくまくすいこうほう　培養液薄膜水耕法　nutrient film technique (NFT)
ばいようき　培養基　culture medium
ばいようさいぼう　培養細胞　cultured cell
ばいようど　培養土⇒配合土　commercial compost, soil mixes
バイラス⇒ウイルス　virus
ばいりつ　倍率【光学】　magnification
はいれつひょうしきぶい　配列標識部位　sequence-tagged sites (STS)
バインダ，結束機　binder, reaper and binder
パウローサ　pau rosa, *Aniba rosiodora* Ducke
バオバブ　baobab, monkey-bread tree, *Adansonia digitata* L.
はかき　葉か(搔)き【タバコ】　picking, priming
バガス，甘蔗搾粕　bagasse
ばかなえ　馬鹿苗　"bakanae", rice seedling infected with *Gibberella fujikuroi*
はがれ　葉枯れ　leaf dying
ハキダメギク　hairy galinsoga, *Galinsoga quadriradiata* Ruiz et Pav.
パキテンき　パキテン期，太糸期【減数分裂】　pachytene stage
はくあしつまい　白亜質米　chalky rice
はくか　白化⇒クロロシス　chlorosis
ばくが　麦芽　malt
ばくがとう　麦芽糖，マルトース　maltose
ばくがひんしつ　麦芽品質　malting quality
ばくかんちょくはん　麦間直播　direct sowing of rice between the rows of winter cereals
ハクサイ(白菜)　Chinese cabbage, *Brassica rapa* L. var. *amplexicaulis* Tanaka et Ono (= *B. pekinensis* Rupr.)
ばくさく(ムギさく)　麦作　1) barley cropping (culture)【オオムギ】2) wheat cropping (culture)【コムギ】
はくしょくたい　白色体　leucoplast
はくそうクロマトグラフィー　薄層クロマトグラフィー　thin layer chromatography (TLC)
バクテリオファージ　bacteriophage, ファージ　phage
バクテロイド【根粒菌】　bacteroid
はくど　白度【コムギ】　whiteness
はくひ　剥皮　peeling
はくまい　白米　milled rice, polished rice
はくまくすいこう　薄膜水耕⇒培養液薄膜水耕法　nutrient film technique (NFT)
はくりきこ　薄力粉【コムギ】　soft flour
バークローバ　bur clover, *Medicago polymorpha* L. (= *M. hispida* Gaertn.)
パクロブトラゾル　paclobutrazol
はこいくびょう　箱育苗　seedling-raising in box
はこう　は(耙)耕⇒砕土　harrowing
ばこう　馬耕　horse plowing
ハコベ　common chickweed, *Stellaria media* (L.) Villars
はさ(はざ)(稲架)　[paddy] sheaf rack
はざかいき　端境期　period of short supply
はさき　葉先　leaf tip
はざし　葉挿し　leaf cutting
はざぼし　はざ(稲架)干し　rack drying
はさみづみ　はさみ摘み【チャ】　shear plucking
はしつぎ　橋接ぎ　bridge grafting
バシームクローバ→エジプシャンクローバ
はしゅ　播種　sowing, seeding
はしゅき　播種機　seeder, seeding

machine
はしゅき　播種期　sowing (seeding) time
はしゅしんど　播種深度　sowing (seeding) depth
はしゅどこ　播種床　sowing (seeding) bed
はしゅほう　播種法　sowing (seeding) method
はしゅみつど　播種密度　sowing (seeding) density, sowing (seeding) rate
はしゅりょう　播種量　sowing (seeding) rate
はしりほ　走り穂　precocious ear
バジル　[sweet] basil, *Ocimum basilicum* L.
ハス (蓮)　Indian lotus, sacred lotus, *Nelumbo nucifera* Gaertn.
ハスイモ (蓮芋)　*Colocasia gigantea* Hook. f.
パスかいせき　パス解析, 経路分析　path analysis
パスカルぶんぷ　パスカル分布⇒負の二項分布　negative binominal distribution
バーズフット・トレフォイル　birdsfoot trefoil, *Lotus corniculatus* L. var. *corniculatus*
はずれち　外れ値　outlier
はずれちのけんしゅつ　外れ値の検出　detection of outlier
はせいけいとう　派生系統　derived line
はせいさいぼうかんげき　破生細胞間隙　lysigenous intercellular space
はせいつうきそしき　破生通気組織　lysigenous aerenchyma
ハゼノキ (櫨)　Japanese wax tree, *Rhus succedanea* L.
パセリ　parsley, *Petroselinum crispum* (Mill.) Nym. ex A.W. Hill.
パーソナルコンピュータ (パソコン)　personal computer

パターンにんしき　パターン認識　pattern recognition
ハダカエンバク (裸燕麦)　naked oat, *Avena nuda* L.
はだかせい　はだか (裸) 性【オオムギ】　naked grain type
ハダカムギ (裸麦)　naked barley, *Hordeum vulgare* L.
はたけ　畑　field, upland field
はたさく　畑作　upland farming, field crop cultivation
はた [さく] すいとう　畑 [作] 水稲　upland-cultured paddy rice
はたさくもつ　畑作物　field crop
ハタササゲ　catjang, *Vigna unguiculata* (L.) Walp. var. *catjang* (Burm. f.) H. Ohashi
はだずれまい　肌ずれ米　skin-abrased rice
はたちかんがい　畑地灌漑　upland irrigation
はたどじょう　畑土壌　upland soil
はたなえ　畑苗　upland [rice] seedling
はたなわしろ　畑苗代　upland rice-nursery
はタバコ　葉タバコ　leaf tobacco
はち　鉢, 植木鉢, ポット　pot
はちあげ　鉢上げ　potting
はちうえ　鉢植え　potting
はちかえ　鉢替え　repotting
ハチク (淡竹)　black bamboo, *Phyllostachys nigra* (Lodd. ex Loudon) Munro f. *henonis* (Mitord) Stapf ex Rendle
はちさいばい　鉢栽培　pot culture
ハチジョウススキ (八丈薄)　Hachijo plume grass, *Miscanthus sinensis* Andersson var. *condensatus* (Hack.) Makino
ハチジョウナ　*Sonchus brachyotus* DC.
パーチメントコーヒー　parchment coffee
はちもの　鉢物　pot plant, potted plant

バーチャルリアリティー，仮想現実 virtual reality
パチョリ patchouli, *Pogostemon cablin* (Blanco) Benth. (= *P. patchouli* Pell.)
はついく 発育 development
はついくし 発育枝 vegetative shoot, vegetative branch
はついくせいり 発育生理 developmental physiology
はついくそう 発育相 developmental phase
はついくだんかい 発育段階 developmental stage
ハッカ（薄荷）Japanese mint, *Mentha arvensis* L. var. *piperascens* Malinv. ex Holmes
はつが 発芽 germination
はつがおんど 発芽温度 germination temperature
はつがしけん 発芽試験 germination test
はつがしけんき 発芽試験器 germinator [chamber]
はつがしょう 発芽床 germination bed
はつがぜい 発芽勢 germination rate
はつがそくしんぶっしつ 発芽促進物質 germination stimulator
はつがぶあい 発芽歩合，発芽率 percentage of germination, germination percentage
はつがよくせいぶっしつ 発芽抑制物質 germination inhibitor
はつがりつ 発芽率，発芽歩合 percentage of germination, germination percentage
はつがりゅう 発芽粒 germinated grain
はつがりょく 発芽力 germination ability, viability of seed, germinability
ばっかん 麦稈 1) barley straw【オオムギ】2) wheat straw【コムギ】
はっこう 発酵 fermentation

はっこうぶんこうぶんせき 発光分光分析 emission spectrochemical analysis
はっこうまい 発酵米 fermented rice
はっこん 発根 rooting
はっこんりょく 発根力 rooting ability
はつしも 初霜 first frost
ハッショウマメ（八升豆）Yokohama [velvet] bean, *Mucuna pruriens* (L.) DC. var. *utilis* (Wight) Burck (= *Stizolobium hassjoo* Piper et Tracy)
パッションフルーツ passion fruit, *Passiflora edulis* Sims
はっせい 発生 development
はっせいがく 発生学 embryology
はっせいさいせいき 発生最盛期 peak of occurrence
はっせいしょうちょう 発生消長 seasonal prevalence
はっせいよさつ 発生予察 forecasting of occurrence
はつどばんプラウ はつ（撥）土板プラウ moldboard plow
バッファローグラス buffalograss, *Buchloe dactyloides* (Nutt.) Engelm.
はつらいき 発蕾期【タバコ】button stage
パーティクルガン，遺伝子銃 particle gun
ハーディンググラス hardinggrass, *Phalaris tuberosa* L.
ハードウェア hardware
ハードニング 硬化，順化【培養】hardening
ハトムギ（薏苡）Job's-tears, *Coix lacryma-jobi* L. var. *ma-yuen* (Roman.) Stapf (= *C. lacryma-jobi* L. var. *frumentacea* Makino)
はな 花 flower
バーナー burner
ハナイバナ *Bothriospermum tenellum*

(Hornem.) Fisch. et Mey.
ハナスゲ *Anemarrhena asphodeloides* Bunge
バナナ banana, *Musa* × *paradisiaca* L.
ハナハッカ oregano, common marjoram, *Origanum vulgare* L.
はなぶるい 花振るい【ブドウ】 shatter
パナマゴム Panama rubber, Central American rubber, *Castilla elastica* Cerv.
パナマソウ Panama hat palm (plant), *Carludovica palmata* Ruiz et Pav.
ハナマメ (花豆) → ベニバナインゲン
はなみず 花水【イネ】 irrigation at flowering stage
はなめ (かが) 花芽 flower bud
はなめ (かが) けいせい 花芽形成 flower bud formation
はなめ (かが) ぶんか 花芽分化 flower bud initiation (differentiation), floral differentiation
バーナリゼーション, 春化 [処理] vernalization
バニラ vanilla, *Vanilla planifolia* Andr.
パネリスト, パネル構成員 panelist, taster
パネル panel
パネルこうせいいん パネル構成員, パネリスト panelist, taster
はのし 葉のし【タバコ】 flattening
はのみずポテンシャル 葉の水ポテンシャル leaf water potential
パパイア papaya, *Carica papaya* L.
ハハコグサ cudweed, *Gnaphalium affine* D. Don
ババスヤシ (ババス椰子) babassu, *Orbignya speciosa* (Mart.) B. Rodr. (= *O. martiana* B. Rodr.)
バヒアグラス bahiagrass, *Paspalum notatum* Flugge
バビロフコムギ *Triticum vavilovii* Jakubz.
ハブソウ negro coffee, *Senna occidentalis* Link. (= *Cassia torosa* Cav., *C. occidentalis* L.)
ハプテン hapten
ハーベスタ, 収穫機 harvester
はへんせんしょくたい 破片染色体, 断片染色体 fragment chromosome
パーボイルドライス parboiled rice
ハマスゲ nut grass, *Cyperus rotundus* L.
ハマナス rugosa rose, *Rosa rugosa* Thunb.
ハマヒエガエリ rabbit's-foot polypogon, annual beard grass, *Polypogon mouspeliensis* (L.) Desf.
ハマボウフウ (浜防風) *Glehnia littoralis* F. Schmidt ex Miq. (= *Philopterus littoralis* Benth.)
バーミキュライト vermiculite
バーミューダグラス Bermuda grass, *Cynodon dactylon* Flugge (= *C. dactylon* (L.) Pers.)
パームかくゆ パーム核油【アブラヤシ】 palm kernel oil
パームゆ パーム油【アブラヤシ】 palm oil
はめ (ようが) 葉芽 leaf bud, foliar bud
はめ (ようが) ざし 葉芽挿し leaf-bud cutting
はもぎ 葉もぎ【タバコ】 leaf stripping
はやうえ 早植え early planting
はやうえさいばい 早植栽培 early-planting culture
はやがり 早刈り early harvesting
はやけ 葉焼け leaf burn
はやざき 早咲き early flowering
はやじも 早霜 early frost
ハヤトウリ chayote, *Sechium edule* (Jacq.) Swartz
はやどり 早取り early harvesting
はやばまい 早場米 early [season] delivery rice

はやほり　早掘り　early harvesting
はやまき　早播き　early sowing (seeding)
はらがわの　腹側の　ventral
パラゴム　Para rubber, *Hevea brasiliensis* Muell.-Arg.
はらじろまい　腹白米　white-belly rice
バラタ　balata, *Manilkara bidentata* (A.DC.) A.Chev.
パラチオン　parathion
はらつぎ　腹接ぎ【接木】　side-grafting
パラフィンほう　パラフィン法　paraffin method
ばらまき　ばら播き ⇒ 散播(さんぱ)　broadcast sowing (seeding), broadcast
パラミツ(波羅蜜)　jackfruit, *Artocarpus heterophyllus* Lam.
パラメータ　parameter
ハリイ　*Eleocharis congesta* D. Don ssp. *japonica* (Miq.) T. Koyama
ハリイモ→トゲドコロ
ハリビユ　spiny amaranth, *Amaranthus spinosus* L.
バリン　valine (Val)
はるうえ　春植え　spring planting
はるえだ　春枝　spring shoot
ハルガヤ→スイートバーナルグラス
はるぎり　春切【クワ】　spring pruning
バルクかんそう　バルク乾燥【タバコ】　bulk curing
はるごえ　春肥　spring dressing
はるこくもつ　春穀物　spring cereals
はるこむぎ　春コムギ　spring wheat
はるさく　春作　spring cropping
はるさく[もつ]　春作[物]　spring crop
バルサムモミ　balsam fir, *Abies balsamea* (L.) Mill.
ハルジオン(春紫苑)　Philadelphia fleabane, *Erigeron philadelphicus* L.
パルス-チェイスじっけん　パルス-チェイス実験　pulse-chase experiment
ハルタデ　lady's-thumb, *Persicaria vulgaris* Webb. et Moq. (= *Polygonum persicaria* L.)
はるまき　春播き　spring sowing (seeding)
はるまきがた　春播き型　spring type
はるまきせい　春播き性　spring habit
はるまきせいていど　春播き性程度　degree of spring habit
はるまきひんしゅ　春播き品種　spring variety
パルマローザ　palmarosa, *Cymbopogon martini* (Roxb.) Wats.
パルミチンさん　パルミチン酸　palmitic acid
パルミラヤシ→オウギヤシ
パールミレット　pearl millet, *Pennisetum americanum* (L.) Leeke (= *P. typhoideum* Rich.)
バレイショ(馬鈴薯)→ジャガイモ
バーレーしゅ　バーレー種【タバコ】　Burley [tobacco]
パレスチナコムギ　wild emmer wheat, *Triticum dicoccoides* (Körn.) Schwein.
ハロー　harrow, clod crusher, pulverizer
ハロイサイト　halloysite
はわけ　葉分け【タバコ】　classification of leaves [on stalk position]
パワー・スペクトル　power spectrum
はんい　範囲【統計量】　range
ばんかさいばい　晩化栽培, 晩期栽培　late-season culture
はんかんせいゆ　半乾性油　semidrying oil
はんかんそうち　半乾燥地　semiarid land
ばんきさいばい　晩期栽培, 晩化栽培　late-season culture
ぱんきじ　パン生地　dough
はんけい　半径　radius
はんげんき　半減期　half-life
パンコムギ　bread wheat, common

wheat, *Triticum aestivum* L.
パンゴラグラス　pangola grass, *Digitaria eriantha* Steud.
はんさいばいダイズ　半栽培ダイズ　semi-cultured soybean, *Glycine gracilis* Skov.
ばんさいぼう　伴細胞　companion cell
はんしゃ　反射　reflection
はんしゃきょう　反射鏡【光学】　mirror
はんしゃけいすう　反射係数　reflection coefficient
はんしゃスペクトル　反射スペクトル　reflectance spectrum
はんしゃりつ　反射率　reflectivity
ばんじゅく　晩熟　late maturation
ばんじゅくせい　晩熟性　late maturity
はんじゅん　半旬【気象】　pentad
ばんじょうたい　盤状体 ⇒ 胚盤 (はいばん)　scutellum (*pl.* scutella)
はんしょく　繁殖　propagation, multiplication, reproduction, breeding
ばんしょく　晩植　late planting
ばんしょくさいばい　晩植栽培　late-planting culture
はんしょくようしき　繁殖様式　reproductive system, mode of reproduction, breeding system
パンジロウ→グワバ
ハンスイヒユ (繁穂ヒユ)　grain amaranth, *Amaranthus hypochondriacus* L.
はんすうせい　半数性　haploidy
はんすうたい　半数体, 単相 [体]　haploid, haplont
はんすうたいいくしゅ　半数体育種　haploid breeding
はんすうたいしょくぶつ　半数体植物　haploid [plant]
はんすうたいばいかけいとう　半数体倍加系統　doubled haploid
はん [すう] ちしやくりょう　半 [数] 致死薬量　lethal dose 50 % (LD_{50}), median lethal dose

はんすうどうぶつ　反すう (芻) 動物　ruminant
ばんせいいでん　伴性遺伝　sex-linked inheritance
ばんせいの　晩生の　late maturing
ばんせい (おくて) ひんしゅ　晩生品種　late [maturing] variety
ばん [そう]　盤 [層]　pan
ばんそう　晩霜, 遅霜 (おそじも)　late frost
ばんそうがい　晩霜害　late frost damage (injury)
はんそくさいぼう　反足細胞　antipodal cell, antipode
ばんちゃ　番茶　coarse green tea
はんつきまい　半つ (搗) き米　half-milled rice, half-polished rice
はんていきじゅん　判定基準【統計】　criterion (*pl.* criteria)
はんてんこう　反転耕　upside down plowing
はんてんびょう　斑点病　1) brown spot【トウモロコシ】 2) spot blotch【コムギ】 3) leaf blight, leaf spot【サツマイモ】
ばんとう　晩稲　late variety of rice
はんとうせい　半透性　semipermeability
はんとうまく　半透膜　semipermeable membrane
ハンドトラクタ, 歩行用トラクタ　walking tractor
ばんねり　盤練り　subsoil puddling
はんのう　反応, 応答　reaction, response
はんのうきょくせん　反応曲線, 応答曲線　response curve
はんのうそくど　反応速度　reaction rate
はんのうちゅうしん　反応中心　reaction center
パンノキ　bread-fruit【種なし】, bread-nut【種あり】,

Artocarpus communis Forst.
はんばいひりょう　販売肥料, 金肥, 購入肥料　commercial fertilizer
バンバラマメ　bambara bean, bambara groundnut, *Vigna subterranea* (L.) Verdc. (= *Voandzeia subterranea* (L.) Thouars)
はんぷく　反復　replication
はんぷくおや　反復親　recurrent parent
はんぷくしけんく　反復試験区　replicated plot
はんぷくせんばつ　反復選抜　repeated selection
はんぷくはいれつ　反復配列　repetitive sequence
はんべつかんすう　判別関数　discriminant function
はんべつしゅ　判別種, 識別種　differential species
はんべつひんしゅ　判別品種, 識別品種　differential variety
はんべつぶんせき　判別分析　discriminant analysis
はんようびょう　斑葉病【オオムギ】　stripe
バンレイシ　sugar apple, sweet sop, custard apple, *Annona squamosa* L.
はんわいせいひんしゅ　半わい(矮)性品種　semidwarf variety

[ひ]

ひいでんてきへんい　非遺伝的変異　nonheritable variation
ひいれ　火入れ　burning, firing
ひいんじゅ　ひ(庇)陰樹　shade tree
ヒエ　(稗, 穇)　Japanese millet, barnyard millet, *Echinochloa utilis* Ohwi et Yabuno
ピーエイチしじやく　pH指示薬　pH indicator
ピーエイチスタット　pHスタット　pH-stat
ピーエイチメーター　pHメーター　pH-meter
ひえだ　冷田　paddy field irrigated with cold water
ピーエフち　pF値　pF value
ビーカー　beaker
ひがいりゅう　被害粒　damaged grain (kernel)
ひかぎゃくはんのう　非可逆反応　irreversible reaction
ひかくひんしゅ　比較品種, 対照品種　check variety
ひがしインドアローロート　東インドアローロート　East Indian arrowroot, *Curcuma angustifolia* Roxb.
ひがしインドレモングラス　東インドレモングラス　East Indian lemongrass, Malabar grass, *Cymbopogon flexuosus* (Nees ex Steud.) Wats.
ひかっせい　比活性　specific activity
ひかりエネルギーてんかんこうりつ　光エネルギー転換効率　efficiency of light energy conversion
ひかりエネルギーりようこうりつ　光エネルギー利用効率　efficiency of light energy utilization
ひかりきょうど　光強度　light intensity
ひかりくっせい　光屈性　phototropism
ひかりけいたいけいせい　光形態形成　photomorphogenesis
ひかりごうせい⇒こうごうせい　光合成　photosynthesis
ひかり(こう)こきゅう　光呼吸　photorespiration
ひかりしょうがい　光障害, 光阻害, 強光阻害　photoinhibition
ひかり(こう)せいちょうはんのう　光成長反応　light growth reaction
ひかりそがい　光阻害, 強光阻害, 光障害　photoinhibition
ひかり(こう)ちゅうだん　光中断　light interruption, light break
ひかりとうかりつ　光透過率　light

transmittance
ひかり(こう)はつが　光発芽　light germination
ひかり(こう)はつがしゅし　光発芽種子　light germinater, photoblastic seed
ひかり(こう)ぶんかい　光分解　photolysis, photodecomposition
ひかり(こう)ほうわ　光飽和　light saturation
ひかりほしょうてん　光補償点　light compensation point
ひかりりようこうりつ　光利用効率　light use efficiency, 日射利用効率 radiation use efficiency
ひかんげんとう　非還元糖　nonreducing sugar
ひかんこうせいひんしゅ　非感光性品種　non-photosensitive variety
ひかんじゅせい　非感受性　insensitivity
ヒガンバナ(彼岸花)　*Lycoris radiata* (L'Hér.) Herb.
ヒキオコシ(引起)　*Rabdosia japonica* (Burm. f.) H. Hara (= *Isodon japonicus* Hara)
ひきぐわ　引きぐわ(鍬)　pulling hoe
びきこう　微気候　microclimate
びきしょう[がく]　微気象[学]　micrometeorology
ひきちゃ　挽茶⇒抹茶(まっちゃ)　powdered tea
ひきっこうそがい　非拮抗阻害　noncompetitive inhibition
ひきわり　ひき割り　1) grinding【工程】 2) ground barley
ひくうね　低うね(畝), 浅うね(畝)　low ridge, low bed
ひくがり　低刈り　low-level cutting
ひげね　ひげ根　fibrous root
ひげねがたこんけい　ひげ根型根系　fibrous root system
ひこう　肥効　fertilizer effect
ひこう　肥厚　thickening

びこうさくもつ　備荒作物, 救荒作物　emergency crop
ひこうぞうせいたんすいかぶつ　非構造性炭水化物　nonstructural carbohydrate
ひこうちょうせつがたひりょう　肥効調節型肥料　controlled availability fertilizer
ひこばえ　1)(蘖)　ratoon【イネ, サトウキビ, パイナップルなど】 2) sucker【園芸】
ひこんちょう　比根長, 根長/根重比　specific root length
びさ(びしゃ)　微砂　very fine sand
びさいこうぞう　微細構造　ultrastructure
ヒシ(菱)　water chestnut, *Trapa bispinosa* Roxb. var. *iinumai* Nakano
ひししょくぶつ　被子植物　angiosperm
ひしつ　比湿　specific humidity
びしゃ(びさ)　微砂　very fine sand
ひじゅう　比重　specific gravity
ひじゅうせん　比重選　seed selection by specific gravity
ひじゅうせんべつき　比重選別機　gravity separator
びしょうかん　微小管　microtubule
ひしょくけい　比色計　colorimeter
ひしょくそう　被食草【草地】　consumed herbage
ひしょくていりょう　比色定量　colorimetry
ひしょくほう　比色法　colorimetric method
ひしょく[りょう]　被食[量]【草地】　herbage consumption
ひしんちょうけいぶ　非伸長茎部　unelongated stem part
ピスタチオ　pistachio, *Pistacia vera* L.
ヒスチジン　histidine (His)
ヒステリシス, 履歴現象　hysteresis
ヒストグラム　histogram
ヒストン　histone

びせいぶつ　微生物　microorganism, microbe
びせいぶつてきぼうじょ　微生物的防除　microbial control
びせいぶつのうやく　微生物農薬　microbial pesticide
びせいぶつぶんかい　微生物分解　microbial breakdown
ひせんたくせいじょそうざい　非選択性除草剤　nonselective herbicide
ひそう　皮層　cortex
ひそうかてき　非相加的　nonadditive
ひそうか[てき]いでんこうか　非相加[的]遺伝効果　nonadditive genetic effect
ひたいしょう　非対称　asymmetry
ひだい[せいちょう]　肥大[成長]　thickening [growth]
ひだいりゅう　肥大粒　plump kernel
ビタミン　vitamin
ひタンパク[たい]ちっそ　非タンパク[態]窒素　nonprotein nitrogen
びちくしゅし　備蓄種子　reserved seed
ビッグトレフォイル　big trefoil, *Lotus uliginosus* Schkuhr.
ビッグブルーステム　big bluestem, *Andropogon gerardii* Vitman var. *gerardii*
ひっすアミノさん　必須アミノ酸　essential amino acid
ひっすげんそ　必須元素　essential elements
ひでりあおだち　ひでり(旱)青立　straighthead due to drought
ヒデリコ　globe fringe-rush, *Fimbristylis miliacea* (L.) Vahl
ひど　被度　cover, coverage, cover degree
ビート→テンサイ
びどういでんし　微働遺伝子　minor gene, ポリジーン　polygene
ひとかぶく　一株区　hill plot
ひとざとしょくぶつ　人里植物　ruderal plant
ヒートバランスほう　ヒートバランス法　heat balance method
ヒートパルスほう　ヒートパルス法　heat pulse method
ビートパルプ　beet pulp
ひとほ(いっすい)じゅう　一穂重　weight of a head
ひとほ(いっすい)りゅうすう　一穂粒数　number of grains per head
ひとめざし　一芽挿し　eye cutting
ヒナガヤツリ　*Cyperus flaccidus* R. Br.
ヒナタイノコズチ　*Achyranthes bidentata* Blume. var. *tomentosa* (Honda) H. Hara (= *A. fauriei* Lév. et Van.)
ひなたぎり　日向切り　earthing up on the north side of row
ピーナッツ　peanut→ラッカセイ
ビニルしょうじ　ビニル障子　vinyl sash
ビニルハウス　vinyl house
ビニルフィルム　vinyl film
ひねつ　比熱　specific heat
ひばいかんがい　肥培灌漑　manuring irrigation
ひばいかんり　肥培管理　manuring practice
ひはんぷくおや　非反復親, 一回親　nonrecurrent parent
ひふくさいばい　被覆栽培　1) mulch culture【土を覆う】 2) shade culture【作物を覆う】
ひふくさくもつ　被覆作物　cover crop
ひふくしざい　被覆資材　covering material
ひふくしゅし　被覆種子　coated seed, pelleted seed
ひふくひりょう　被覆肥料　coated fertilizer
ヒマ(蓖麻)　castor bean, castor, *Ricinus communis* L.
ヒマしゆ　ひまし油　castor oil

ヒマワリ (向日葵) sunflower, *Helianthus annuus* L.
ヒマワリゆ ヒマワリ油 sunflower oil
ヒメイヌビエ barnyard grass, *Echinochloa crus-galli* (L.) P. Beauv. var. *praticola* Ohwi
ヒメウイキョウ, キャラウェー caraway, *Carum carvi* L.
ヒメオドリコソウ purple deadnettle, *Lamium purpureum* L.
ヒメガヤツリ *Cyperus tenuispica* Steud.
ヒメクグ *Kyllinga brevifolia* Rottb. (= *Cyperus brevifolius* (Rottb.) Hassk. var. *leiolepis* (Franch. et Savat.) T. Koyama)
ヒメコバンソウ little quaking grass, *Briza minor* L.
ヒメコーラ cola, kola, *Cola acuminata* (P. Beauv.) Schott et Endl.
ヒメジョオン (姫女苑) annual fleabane, *Erigeron annuus* (L.) Pers. (= *Stenactis annuus* (L.) Cass.)
ヒメスイバ red sorrel, *Rumex acetosella* L.
ヒメタイヌビエ barnyard grass, *Echinochloa crus-galli* (L.) P. Beauv. var. *formosensis* Ohwi
ヒメホタルイ *Scirpus lineolatus* Franch. et Savat.
ヒメミソハギ *Ammannia multiflora* Roxb.
ヒメムカシヨモギ horseweed, *Erigeron canadensis* L.
ひメンデルしきいでん 非メンデル式遺伝 non-Mendelian inheritance
ひもく 皮目 lenticel
ヒモゲイトウ→センニンコク
ビャクダン (白檀) sandalwood, white sandalwood, *Santalum album* L.
ひゃくようばこ 百葉箱 instrument shelter, instrument screen
ひやけ 日焼け sunburn, sunscald

ピーユーシーけいプラスミド pUC系プラスミド pUC plasmid
ビュレット buret
ひょうがい ひょう(雹)害 hail damage
びょうがい 病害 disease injury
びょうがいちゅうぼうじょ 病害虫防除 pest control
びょうがいていこうせい 病害抵抗性, 耐病性 disease resistance
びょうがいはっせいよさつ 病害発生予察 disease forecasting, forecasting of disease outbreak
びょうき 病気 disease
ひょうげんがた 表現型 phenotype
ひょうげんがた 病原型, レース race
ひょうげんがたそうかん 表現型相関 phenotypic correlation
ひょうげんがたぶんさん 表現型分散 phenotypic variance
ひょうげんがたもしゃ 表現型模写 phenocopy
びょうげんきん 病原菌 pathogenic fungus
びょうげんさいきん 病原細菌 pathogenic bacterium
びょうげんたい 病原体 pathogen
ひょうしきいでんし 標識遺伝子 marker gene
ひょうしきかごうぶつ 標識化合物 labelled compound
ひようじゅう 比葉重 specific leaf weight (SLW)
ひょうじゅんごさ 標準誤差 standard error
ひょうじゅん[しけん]く 標準[試験]区, 対照[試験]区 control plot
ひょうじゅんせいきぶんぷ 標準正規分布 standard normal distribution
ひょうじゅんせひりょう 標準施肥量 standard dosage of fertilizer
ひょうじゅんひんしゅ 標準品種 standard variety

ひょうじゅんへんさ　標準偏差　standard deviation
びょうじょう　苗条⇒茎葉部　shoot
ひょうそうせひ　表層施肥　surface application of fertilizer
ひょう[そう]ど　表[層]土　surface soil, top soil
びょうちゅうがい　病虫害　disease and insect damage
びょうちょう　病徴　symptom
ひょうてん　氷点　freezing point
ひょうてん　評点，スコア　score
ひょうどこう　表土耕　surface tillage
びょうはん　病斑　lesion
ひょうひ　表皮　epidermis
ひょうひけい　表皮系　epidermal system, dermal system
びょうほ　苗圃　nursery garden, nursery field
ひょうほん　標本　sample, specimen
ひょうほんたんい　標本単位　sample unit
ひょうほんちゅうしゅつ　標本抽出，抜取り【統計】　sampling
ひょうほんちゅうしゅつごさ　標本抽出誤差　sampling error
ひょうほんちゅうしゅつりつ　標本抽出率　sampling ratio
ひょうほんちょうさ[ほう]　標本調査[法]　sampling survey
ひょうほんのおおきさ　標本の大きさ　sample size
ひょうほんびん　標本びん(瓶)　specimen bottle
ひょうほんぶんさん　標本分散　sample variance
ひょうほんへいきん　標本平均　sample mean
ひょうめんかっせいざい　表面活性剤⇒界面活性剤　surface-active agent, surfactant
ひょうめんせき　表面積　surface area
ひようめんせき　比葉面積　specific leaf area (SLA)
ひょうめんりゅうしゅつ　表面流出　surface runoff
びょうれい　苗齢　seedling age
ひよくち　肥沃地　fertile land
ひよくど　肥沃度⇒土壌肥沃度　soil fertility
ひよくど　肥沃土　fertile soil
ひよけ　日除け　shade, shading
ヒヨコマメ　chick pea, common gram, *Cicer arietinum* L.
ヒヨス　henbane, black henbane, *Hyoscyamus niger* L.
ひらうえ　平植え　level planting
ひらうち　平耕　flat break
ひらうね　平うね(畝)　level row
ひらぐわ　平ぐわ(鍬)，風呂ぐわ(鍬)　wooden base hoe
ひらさく　平作　level culture
ひらどこ　平床　level seedbed, flat seedbed
ひらばち　平鉢　pan
ひらまき　平播き　level sowing (seeding)
ヒラマメ(扁豆)　lentil, *Lens culinaris* Medik. (= *L. esculenta* Moench)
びりゅうざい　微粒剤　fine granule
ひりょう　肥料　fertilizer, manure
びりょうげんそ　微量元素　microelement, minor element, trace element
ひりょうさんぷき　肥料散布機　fertilizer distributor
ひりょうさんようそ　肥料三要素　three major nutrients, NPK elements
ひりょうしけん　肥料試験　fertilizer test
ひりょうはんのう　肥料反応　fertilizer response
びりょうぶんせき　微量分析　microanalysis
びりょうようそ　微量要素　micronutrient

びりょうようそひりょう　微量要素肥料　micronutrient fertilizer
ヒルガオ　*Calystegia japonica* Choisy
ヒルはんのう　ヒル反応　Hill reaction
ピルビンさん　ピルビン酸　pyruvic acid
ビルマウルシ　Burmese varnish tree, *Melanorrhoea usitata* Wall.
ビールムギ(ビール麦)　beer brewing barley, malting barley
ヒルムシロ　*Potamogeton distinctus* A. W. Benn.
ひれいちゅうしゅつ[ほう]　比例抽出[法]【統計】proportional sampling
ひれいひょうほんちゅうしゅつ[ほう]　比例標本抽出[法]【統計】proportionate sampling
ヒレタゴボウ　winged waterprimrose, *Ludwigia decurrens* Walt.
ひろうね　広うね(畝)　broad ridge
ヒロハイヌノヒゲ　*Eriocaulon robustius* (Maxim.) Makino
ひろはざっそう　広葉雑草　broadleaved weed
ヒロハセネガ　*Polygala senega* L. var. *latifolia* Torr. et A. Gray
ヒロハノウシノケグサ→メドーフェスク
ひろはばまき　広幅播き　broad sowing (seeding)
ヒロハラベンダー→スパイクラベンダー
ピロリンさん　ピロリン酸　pyrophosphate (PPi)
ビワ(枇杷)　loquat, Japanese medlar, *Eriobotrya japonica* (Thunb.) Lindl.
ひんえいようか　貧栄養化　oligotrophication
ヒンジガヤツリ　*Lipocarpha microcephala* (R.Br.) Kunth
ひんしつ　品質　quality
ひんしつけんさ　品質検査　quality inspection
ひんしつけんてい　品質検定　quality test
ひんしゅ　品種　variety, cultivar (cv.)
ひんしゅかんこうざつ　品種間交雑　varietal cross
ひんしゅかんさい　品種間差異　varietal difference
ひんしゅせいたいがく　品種生態学　genecology
ひんしゅたいか　品種退化　degeneration of variety
ひんしゅちゃえん　品種茶園　cultivar tea field
ひんしゅとうろく　品種登録　variety registration
ひんしゅとくせい　品種特性　varietal characteristics
ひんしゅひかくしけん　品種比較試験　variety test
ひんしゅぶんか　品種分化　varietal differentiation
ひんしゅほぞん　品種保存　variety preservation
ピンセット　forceps
ピンチコック　pinchcock
ひんど　頻度, 度数　frequency
ひんどぶんぷ　頻度分布, 度数分布　frequency distribution
ビンロウジュ(檳榔樹)　betel palm, *Areca catechu* L.

[ふ]

ふ(麩)　bran
ふ(稃)　glume, hull, chaff
ぶあい　歩合　percentage
ファイトトロン　phytotron
ファイトプラズマ　phytoplasma
ファイトマー　phytomer
ファイル【コンピュータ】　file
ファージ　phage, バクテリオファージ bacteriophage
ファージミド　phagemid, ファスミド phasmid
ファストグリーン【染色】　fast green

ファスミド　phasmid, ファージミド　phagemid
ファゼインさん　ファゼイン酸　phaseic acid
ブアンえき　ブアン液【固定液】　Bouin's fluid
ブイエーきんこん　VA菌根　vesicular-arbuscular mycorrhiza (VAM)
フィトクロム　phytochrome
フィードバック　feedback
フィードバックそがい　フィードバック阻害　feedback inhibition
ふいり　斑入り　variegation, mottle
ふうがい　風害　wind damage (injury)
ふうか[さよう]　風化[作用]　weathering
ふうかん　風乾　air drying
ふうかんじゅう　風乾重　air-dry weight
ふうかんど　風乾土　air-dry soil
ふうこうけい　風向計　anemoscope, wind vane
ふうこうふうそくけい　風向風速計　wind vane and anemometer
ふうしょく　風食　wind erosion
ふうすいがい　風水害　wind and flood damage
ふうせん　風選　winnowing, wind selection
ふうせんき　風選機　seed blower
ふうそく　風速　wind speed, wind velocity
ふうそくけい　風速計　anemometer
ふうにゅうざい　封入剤　mounting agent
ふうばい　風媒　anemophily, wind pollination
ふうばいでんせん　風媒伝染　wind-borne infection
ふうみ　風味　flavor and taste
フェイス　FACE (Free-Air CO_2 Enrichment)

ふえいようか　富栄養化　eutrophication
フェニルアラニン　phenylalanine (Phe)
フェノール　phenol
フェノールフタレイン　phenolphthalein
フェーリングはんのう　フェーリング反応　Fehling's reaction
フェーン　foehn
フォイルゲンはんのう　フォイルゲン反応　Feulgen's reaction
フォニオ　fonio, fundi, hungry rice, *Digitaria exilis* (Kippist) Stapf
フォーレージハーベスタ　forage harvester
ふか　負荷　load
ふかうえ　深植え　deep planting
ふかがり　深刈り【チャ】　deep trimming of canopy
ふかきゅうたいようぶん　不可給態養分　unavailable nutrient
ふかぎり　深き(剪)り　heavy pruning
ふかつざい　ふ(賦)活剤　activator
ふかっせいか　不活性化　inactivation
ふかっせいせいぶん　不活性成分　inert ingredient
ふかまき　深播き　deep sowing (seeding)
ふかみず　深水　deep flooding
ふかみずいね　深水稲　deepwater rice
ふかみずかんがい　深水灌漑　deep-flood irrigation
ふかみぞまき　深溝播き　deep-furrow sowing (seeding)
ぶかんせいちょう　部間成長, 介在成長　intercalary growth
ふかんせいゆ　不乾性油　nondrying oil
ふかんぜんか　不完全花　imperfect flower
ふかんぜんまい　不完全米　imperfect rice kernel
ふかんぜんゆうせい　不完全優性　incomplete dominance
ふかんぜんよう　不完全葉　incomplete

leaf
ふかんび[がた]ブロックせっけい 不完備[型]ブロック設計 incomplete block design
ぶかんぶんれつそしき 部間分裂組織, 介在分裂組織 intercalary meristem
フキ (蕗) Japanese butterbur, *Petasites japonicus* (Sieb. et Zucc.) Maxim.
ふきゅうみこみめんせき 普及見込面積【品種】 expected cover area
ふきんいつせい 不均一性 heterogeneity
ふくが 副芽 accessory bud
ふくがせいの 伏臥性の trailing
ふくこう 腹溝, 縦溝【ムギ類】 crease, furrow
ふくごうおせん 複合汚染 combined pollution
ふくこうざつ 複交雑, 複交配 double cross
ふくごうとつぜんへんい 複合突然変異 complex mutation
ふくこうはい 複交配, 複交雑 double cross
ふくごうひりょう 複合肥料 compound fertilizer, mixed fertilizer
ふくごえい 副護穎 rudimentary glume
ふくさいぼう 副細胞【気孔】 subsidiary cell
ふくさく 複作 polyculture
ふくさくもつ 副作物 minor crop, side crop
ふくさんぶつ 副産物 by-product
ふくし ふく(匐)枝, ほふく(匍匐)枝, ストロン runner, stolon
ふくしき 複糸期, ディプロテン期【減数分裂】 diplotene stage
ふくじひょうほん 副次標本 subsample
ふくしゃ ふく(輻)射⇒放射 radiation
ふくじょうまき 複条播き paired row sowing (seeding)

フクシン【染色】 fuchsine
ふくせい 複製【核酸】 replication
ふくそう[たい] 複相[体], 二倍体 diploid
ふくたいりついでんし 複対立遺伝子 multiple alleles
ふくちせいの 伏地性の procumbent
ふくど 覆土 covering with soil
ふくにばいせい 複二倍性 amphidiploidy
ふくにばいたい 複二倍体 amphidiploid
ふくへいりついかんそく 複並立維管束 bicollateral vascular bundle
ふくめん 腹面 ventral side
ふくよう 複葉 compound leaf
ふくりゅう 複粒 compound grain (granule)
ふくろかけ 袋掛け【果実】 bagging
ふけっかせい 不結果性 unfruitfulness
ふけつじつ 不結実 unfruitfulness
ふけまい ふけ米 "Fuke" rice, absidia diseased rice
ふこうきまき 不耕起播き 1) nontillage sowing (seeding) 2) sod seeding【牧草】
ふごうけんてい 符号検定 sign test
ふさく 不作 poor harvest, poor crop, bad crop
ふじかいか 不時開花 unseasonable flowering
フシコクシン fusicoccin
ふじさいばい 不時栽培 off-season culture, out-of-season culture
ふじしゅっすい 不時出穂 premature heading, unseasonable heading
ふじちゅうだい 不時抽だい(苔) unseasonable bolting
ふしつ ふ(麩)質⇒グルテン gluten
フジマメ (鵲豆) lablab, hyacinth bean, *Lablab purpureus* (L.) Sweet (= *Dolichos lablab* L.)
ふじゅせい 不受精 unfertilization

ふしょく　腐植　humus
ふしょくしつどじょう　腐植質土壌　humic soil
ふしょくふ　不織布　non-woven fabric
ふすいざっそう　浮水雑草　floating weed
ふすま(麬)　wheat bran
ふせいちゅうしんちゅう　不整(斉)中心柱　atactostele
ふせこみ　伏込み【サツマイモ】　laying-in
ふせつごう　不接合　asynapsis
ブタクサ　common ragweed, *Ambrosia artemisiifolia* L. var. *elatior* (L.) Desc.
ブタノール　buthanol ⇒ブチルアルコール　butyl alcohol
フタバムグラ　*Hedyotis diffusa* Willd.
ふたまたぶんし　二又分枝, 二叉(にさ)分枝　dichotomous branching, dichotomy
ふちうえ　縁植え, 周縁植物　border plant
ふちぐされ　縁腐れ【レタス, ハクサイ等】　tip burn
ふちゃくすい　付着水　adhering water
ふちゃくせんしょくたい　付着染色体　attached chromosome
ブチルアルコール　butyl alcohol
ふつうがたコンバイン　普通型コンバイン　conventional combine
ふつうきさいばい　普通期栽培　normal-season culture
ふつうけいコムギ　普通系コムギ　dinkel wheat
ふつうさくもつ　普通作物　field crop
ふつうなえどこ　普通苗床　ordinary nursery
ふつうよう　普通葉, 本葉(ほんよう)　foliage leaf
ふっきとつぜんへんい　復帰突然変異, 逆突然変異　back mutation, reverse mutation

ふっきゅうかく　復旧核　restitution nucleus
ぶっしつじゅんかん　物質循環　material cycle, cycle of matter
ぶっしつせいさん　物質生産, 乾物生産　dry-matter production
フッソ　フッ素　fluorine (F)
フットプリントほう　フットプリント法　footprinting
ぶつりてきいでんしちず　物理的遺伝子地図　physical genetic map
ぶつりてきふうじこめ　物理的封じ込め　physical containment
ふていが　不定芽　adventitious bud
ふていこん　不定根　adventitious root
フトイ(太藺, 莞)　black rush, *Schoenoplectus lacustris* (L.) Palla ssp. *validus* (Vahl) T. Koyama (= *Scirpus lacustris* L.)
ふといとき　太糸期, パキテン期【減数分裂】　pachytene stage
ブドウ(葡萄)　grape, *Vitis vinifera* L.
ふとうせい　不透性　impermeability
ブドウとう　ブドウ糖, グルコース　glucose
ふとうぶんれつ　不等分裂　unequal [cell] division
ふとうめいの　不透明の　opaque
ふとうようせい　不等葉性　anisophylly
ふとね　太根【チャ】　thick root
ぶどまり　歩留り　yielding percentage, finishing ratio
プトレッシン　putrescine, ptorescine
ふなぞこうえ　船底植え, 船底挿し【サツマイモ】　concave planting
ふなぞこざし　船底挿し, 船底植え【サツマイモ】　concave planting
ふねん　不稔　sterility
ふねんか　不稔花　sterile flower
ふねんしゅし　不稔種子　sterile seed
ふねんせい　不稔性　sterility
ふねんの　不稔の　sterile
ふねんぶあい　不稔歩合　percentage

sterility
ふのにこうぶんぷ　負の二項分布　negative binomial distribution, 逆二項分布　inverse binomial distribution
ふはい　腐敗　putrefaction
ぶぶんがり　部分刈り　quadrat sampling
ぶぶんこうかん　部分交換【染色体】segmental interchange
ぶぶんふねん[せい]　部分不稔[性]　partial sterility
ぶぶんゆうせい　部分優性　partial dominance
ふほうわしぼうさん　不飽和脂肪酸　unsaturated fatty acid
ふみきりみぞ　踏切溝　foot furrow of rice nursery
ふみこみ　踏込み　stamping
ふみすき　踏すき(鋤)　stamping spade
ふもうち　不毛地, やせ(痩)地　infertile land
ふゆいちねんそう　冬一年草, 越年草　winter annual[s], biennial[s]
ふゆうばいよう　浮遊培養, 懸濁培養　suspension culture
ふゆこくもつ　冬穀物　winter cereals
ふゆコムギ　冬コムギ　winter wheat
ふゆさく　冬作　winter cropping
ふゆさく[もつ]　冬作[物]　winter crop
ふゆざっそう　冬雑草　winter weed
ふゆまき　冬播き　winter sowing (seeding)
ふゆめ　冬芽　winter bud
ふようざっそう　浮葉雑草　floating leaved weed
ふようど　腐葉土　leaf mold
プライマー　primer
プラウ, 洋梨　plow
プラウこう　プラウ耕, り(犁)耕　plowing
プラーク　plaque
フラクション, 画分, 分画　fraction

フラクションコレクター　fraction collector
フラクタル　fractal
フラクタルじげん　フラクタル次元　fractal dimension
フラサバソウ　ivy-leaf speedwell, *Veronica hederifolia* L.
ブラシノステロイド　brassinosteroid
ブラシノライド　brassinolide
ブラジルスオウ→ブラジルボク
ブラジルボアドローズ　Brazilian bois de rose→パウロ―サ
ブラジルボク　brazilwood, *Caesalpinia echinata* Lam.
ブラジルロウヤシ→カルナウバヤシ
プラスチックフィルム　plastic film
プラスチド, 色素体　plastid
プラストキノン　plastoquinone
プラストクロン, 葉間期　plastochron[e]
プラスミド　plasmid
フラックス, 流束　flux
ブラックマッペ, ケツルアズキ(毛蔓小豆)　black matpe, black gram, urd, *Vigna mungo* (L.) Hepper (= *Phaseolus mungo* L.)
ブラックメディック　black medic, *Medicago lupulina* L.
ブラベンダーたんい　ブラベンダー単位　brabender unit
プランテーション　plantation, 農園農業　estate agriculture, 企業農業　commercial agriculture
プラントオパール, 植物タンパク石　plant opal
フーリエきゅうすう　フーリエ級数　Fourier series
フーリエへんかん　フーリエ変換　Fourier transform
フリーズエッチング, 凍結エッチング【電顕】　freeze-etching
フリーズフラクチャ, 凍結割断【電顕】　freeze-fracture
ふりょうどじょう　不良土壌　poor soil

フリントコーン　flint corn, *Zea mays* L. var. *indurata* Bailey
ブル【イネの生態型】　bulu
ふるいつち　ふるい(篩)土　sieved soil
ふるいわけ　ふるい(篩)分け　sieving, screening
フルクタン　fructan, フルクトサン fructosan
フルクトース, 果糖　fructose
フルクトース -1,6- ビスホスファターゼ fructose-1,6-bisphosphatase
フルクトース -1,6- ビスリンさん　フルクトース -1,6- ビスリン酸 fructose-1,6-bisphosphate
フルクトサン　fructosan, フルクタン fructan
ブルーグラマ　blue grama, *Bouteloua gracilis* (Kunth) Lag. ex Steud.
ふるだね　古種子　aged seed
ブルーパニックグラス　blue panicgrass, *Panicum antidotale* Retz.
ブレークダウン【デンプン特性】 breakdown
プレッシャーチャンバーほう　プレッシャーチャンバー法, 圧ボンベ法 pressure chamber method
プレッシャープローブ【膨圧測定】 pressure probe
プレパラート　preparation
プレリーグラス→レスキュグラス
ふれんぞくぶんぷ　不連続分布 discontinuous distribution
ふれんぞくへんい　不連続変異 discontinuous variation
プログラム　program
ふろぐわ　風呂ぐわ(鍬), 平ぐわ(鍬) wooden base hoe
ふろすき　風呂すき(鋤)　wooden base spade
プロセスモデル, 過程模型 process-based model
プロセッサ　processor
ブロック　block

フロッピィーディスク　floppy disk
プロテアーゼ, タンパク[質]分解酵素 protease
プロトプラスト, 原形質体　protoplast
プロトンポンプ　proton pump
プロプラスチド, 原色素体　proplastid
プロモーター　promoter
ふろゆしんほう　風呂湯浸法　seed disinfection in hot bath
フロラ, 植物相　flora
プロラミン　prolamin
フロリゲン　florigen
フロリダアロールート　Florida arrowroot, coontie, *Zamia floridana* A. DC.
プロリン　proline (Pro)
ふわごうせい　不和合性 incompatibility
ふん(糞)　feces, dung
ぶんあつ　分圧　partial pressure
ふんい　粉衣　dust coating, dust dressing
ぶんえきろうと　分液ろうと(漏斗) separating funnel, separatory funnel
ぶんか　分化　differentiation
ぶんかいのう　分解能　resolving power, resolution
ぶんかく　分画, フラクション, 画分 fraction
ぶんかくえんしん　分画遠心 differential centrifugation
ぶんかぜんのうせい　分化全能性, 全能性　totipotency
ぶんかつしけんくせっけい　分割試験区設計　split-plot design
ぶんかつひょう　分割表　contingency table
ぶんげつ　分げつ(蘗) 1) tiller【形態】 2) tillering【発育現象】
ぶんげつが　分げつ芽　tiller bud
ぶんげつき　分げつ期　tillering stage
ぶんげつごえ　分げつ肥　topdressing at tillering stage

ぶんげつじい　分げつ次位　order of tiller
ぶんげつすう　分げつ数　number of tiller, tiller number
ぶんげつせいき　分げつ盛期　active-tillering stage
ぶんげつせつ　分げつ節　tillering node
ぶんげつせつい　分げつ節位　tillering position on stem
ぶんこうこうどけい　分光光度計　spectrophotometer
ぶんこうはんしゃりつ　分光反射率　spectral reflectance
ぶんこうぶんせき　分光分析　spectrochemical analysis
ぶんこうほうしゃけい　分光放射計　spectro-radiometer
ふんさい　粉砕　milling, trituration
ふんざい　粉剤【農薬】　dust
ぶんさん　分散【統計】　variance
ぶんさんかんがい　分散灌漑　dispersed irrigation
ぶんさんきょうぶんさんぎょうれつ　分散共分散行列　variance-covariance matrix
ぶんさんざい　分散剤　dispersing agent
ぶんさんせいぶん　分散成分　variance component, component of variance
ぶんさんひ　分散比　variance ratio
ぶんさんぶんせき　分散分析　analysis of variance (ANOVA)
ぶんし　分施【肥料】　split application, split dressing
ぶんし　分枝　1) branch【形態】 2) branching【発育現象】
ぶんしいくしゅ　分子育種　molecular breeding
ぶんしかくさん　分子拡散　molecular diffusion
ぶんしがた　分枝型　branching form, branching type
ぶんしこうぞう　分子構造　molecular structure
ぶんしこん　分枝根　branch root, 側根　lateral root
ぶんしさいぼうせいぶつがく　分子細胞生物学　molecular cell biology
ぶんししき　分子式　molecular formula
ぶんしせいぶつがく　分子生物学　molecular biology
ふんしつの　粉質の　mealy
ぶんしふるい　分子ふるい(篩)　molecular sieve
ぶんしマーカー　分子マーカー　molecular marker
ふんじょうしつりゅう　粉状質粒【ムギ類】　chalky kernel, mealy kernel
ぶんしりょう　分子量　molecular weight
ぶんぱいクロマトグラフィー　分配クロマトグラフィー　partition chromatography
ぶんぴ[つ]　分泌　secretion
ぶんぴ[つ]そしき　分泌組織　secretory tissue
ぶんぴ[つ]ぶつ　分泌物　secretion, exudate
ぶんぷ　分布　distribution
ぶんぷかんすう　分布関数　distribution function, 累積分布関数　cumulative distribution function
ぶんぷパラメータ　分布パラメータ　distribution parameter
ぶんべつ　分別　fractionation
ぶんみつとう　分蜜糖　centrifuged sugar
ふんむ　噴霧　spraying
ふんむき　噴霧機　sprayer
ふんむこう　噴霧耕　mist culture, aeroponics
ぶんり　分離【遺伝】　segregation
ぶんりいくしゅ[ほう]　分離育種[法]　breeding by separation
ぶんりせだい　分離世代　segregating generation
ぶんるい　分類　classification

ぶんるいがく 分類学 systematics, taxonomy
ぶんれつき 分裂期 mitotic phase, M期 M phase
ぶんれつそうち 分裂装置 mitotic apparatus
ぶんれつそしき 分裂組織 meristem

[ヘ]

ヘアリベッチ hairy vetch, wooly vetch, *Vicia villosa* Roth
べいか 米価 price of rice, rice price
へいかじゅせい 閉花受精 cleistogamy, close fertilization
べいかしんぎかい 米価審議会(米審) Rice Price Council
へいかつか 平滑化 smoothing
へいきん 平均 mean, average
へいきんきおん 平均気温 mean air temperature
へいきんち 平均値 mean [value]
へいきんにじょうごさ 平均二乗誤差 mean-square error
へいきんへいほう 平均平方 mean square
へいこうげんしょう 平行現象 parallelism
へいこうさいぼう 平衡細胞 statocyst, statocyte
へいこうじょうたい 平衡状態 equilibrium
へいこうしんか 平行進化 parallel evolution
へいこうせき 平衡石 statolith
へいこうみゃく 平行脈 parallel venation, parallel vein
べいこくけんさ 米穀検査 rice inspection
べいこくねんど 米穀年度 rice year
へいさか 閉鎖花 cleistogamous flower
べいさく 米作⇒稲作 rice cropping, rice cultivation
へいさけい 閉鎖系 closed system
べいしつ 米質 rice quality
へいしんがた 閉心型 closed-center type
ベイズのていり ベイズの定理 Bayes' theorem
ベイズりゅうすいてい ベイズ流推定 Bayesian estimation
ベイズりゅうすいろん ベイズ流推論 Bayesian inference
べいせんき 米選機 rice sorter, grain sorter
へいそうぶんれつ 並層分裂 periclinal division
へいたんちちゃえん 平坦地茶園 flat level tea field
へいねんさく 平年作 normal crop, average crop
へいねんち 平年値 normal value, normals
へいばんばいよう 平板培養 plate culture
へいひせっぺん 並皮切片 paradermal section
へいふくけい 平伏茎 procumbent stem
べいふん 米粉 rice flour
へいほうわ 平方和 sum of squares
へいほうわせきわぎょうれつ 平方和積和行列 matrix of sum of squares and products
へいめんこう 平面耕 flat breaking
へいりついかんそく 並立維管束 collateral vascular bundle
べいりゅう 米粒 rice grain, rice kernel
ヘイレージ haylage
へきあつ 壁圧 wall pressure
へきこう 壁孔 pit
ヘキソース, 六単糖 hexose
ヘクソカズラ *Paederia scandens* (Lour.) Merr.

ベクター，媒介生物　vector
ペクチナーゼ　pectinase
ペクチン　pectin
ベクトルデータ　vector data
ヘシアンクロース，黄麻布（こうまふ）Hessian cloth
へそ（臍）　hilum, navel
へた　calyx
ベータグルクロニダーゼ　β-グルクロニダーゼ　β-glucuronidase (GUS)
ベタシアニン　betacyanin
ベチベル　vetiver, khus khus, *Vetiveria zizanioides* Stapf (= *V. zizanioides* (L.) Nash ex Small)
ヘチマ（糸瓜）　sponge gourd, *Luffa cylindrica* M. Roem.
ヘテロシス　heterosis, 雑種強勢 hybrid vigor
ヘテロせつごうたい　ヘテロ接合体⇒異型接合体　heterozygote
べとびょう　べと病　downy mildew
ペトリざら　ペトリ皿　petri dish, シャーレ　culture dish
ヘドロ　sludge deposit, muddy sediment
ペドロジー　pedology, 土壌学　soil science
ベニノキ（紅木）　annatto tree, *Bixa orellana* L.
ベニバナ（紅花）　safflower, *Carthamus tinctorius* L.
ベニバナインゲン（紅花隠元）　scarlet runner bean, flower bean, *Phaseolus coccineus* L.
ベニバナツメクサ→クリムソンクローバ
ベニバナボロギク　*Crassocephalum crepidioides* (Benth.) S. Moore
ベニバナゆ　ベニバナ（紅花）油 safflower oil
ペニーロイヤルミント　pennyroyal mint, *Mentha pulegium* L.
ヘネケン　henequen, *Agave fourcroydes* Lem.

ペーパークロマトグラフィー　paper chromatography
ペーパーポット　paper pot
ペパーミント　peppermint, *Mentha* × *piperita* L.
ヘビイチゴ　Indian strawberry, *Duchesnea chrysantha* (Zoll. et Mor.) Miq. (= *D. indica* var. *japonica* Kitam.)
ヘビウリ　snake gourd, *Trichosanthes cucumerina* Buch.-Ham. ex Wall. (= *T. anguina* L.)
ペポカボチャ　summer squash, pumpkin, *Cucurbita pepo* L.
ヘマトキシリン【染色】　hematoxylin
ヘミセルロース　hemicellulose
へら，スパチュラ　spatula
ヘラオオバコ　buck-horn plantain, *Plantago lanceolata* L.
ヘラオモダカ　water plantain, *Alisma canaliculatum* A. Br. et Bouché ex Sam.
ベラドンナ　belladonna, deadly nightshade, *Atropa belladonna* L.
ヘリウム　helium (He)
ヘリオトロープ　common heliotrope, *Heliotropium arborescens* L. (= *H. perviamum* L.)
ペルオキシソーム　peroxisome, グリオキシソーム　glyoxysome, ミクロボディ　microbody
ペルオキシダーゼ，過酸化酵素 peroxidase
ベルガモット　bergamot, bergamot orange, *Citrus bergamia* Risso et Poit.
ペルシアコムギ　Persian wheat, *Triticum carthlicum* Nevski
ペルシャンクローバ　Persian clover, *Trifolium resupinatum* L. (= *T. suaveolens* Willd.)
ベルヌーイぶんぷ　ベルヌーイ分布 Bernoulli distribution
ベルベットグラス　common

velvetgrass, Yorkshire fog, *Holcus lanatus* L.
ペレット, 固形飼料　pellet
ペレニアルライグラス　perennial ryegrass, *Lolium perenne* L.
へんい　変異　variation
へんいかでん　変異荷電　variable charge
へんいけいすう　変異係数⇒変動係数　coefficient of variation (CV)
へんいせい　変異性　variability
へんい[の]はば　変異[の]幅　range of variation
へんかいき[けいすう]　偏回帰[係数]　partial regression [coefficient]
へんこう　偏光　polarization, polarized light
へんこういでんし　変更遺伝子　modifying gene, modifier
へんこうけんびきょう　偏光顕微鏡　polarizing microscope
へんさ　偏差　deviation
へんさへいほうわ　偏差平方和⇒平方和　sum of squares [of deviation]
へんしゅ　変種　variety
へんしょく　変色　discoloration
ベンジルアデニン　benzyl adenine (BA)
へんすう　変数　variable
へんせい　変性　denaturation
へんそうかん[けいすう]　偏相関[係数]　partial correlation [coefficient]
へんたい　変態　metamorphosis
へんどうけいすう　変動係数　coefficient of variation (CV)
ペントース, 五単糖　pentose
ベントナイト　bentonite
へんぱ　偏波　polarization
へんりょう　変量　variate, 確率変数　random variable
へんりょう[こうか]もけい　変量[効果]模型　random [effects] model
ヘンルーダ　rue, common rue, *Ruta graveolens* L.

[ほ]

ほ　穂　ear, head, panicle, spike
ポアソンぶんぷ　ポアソン分布　Poisson distribution
ボアドローズ　bois de rose, *Aniba* spp.
ボイルゆ　ボイル油　boiled oil
ほう　苞, 苞葉　bract, bract leaf
ぼう(芒)　awn, arista
ほうあつ　膨圧　turgor pressure
ほういいかんそく　包囲維管束　concentric vascular bundle
ほうが　萌芽　sprout[ing]
ほうがき　萌芽期【チャ】　sprouting time, time of bud opening
ホウキギク　saltmarsh aster, *Aster subulatus* Michx.
ホウキモロコシ(箒蜀黍)　broom corn, *Sorghum bicolor* (L.) Moench var. *hoki* Ohwi
ほうきょうこうしょうしけん　豊凶考照試験　crop-weather relationship experiment
ほうけいく　方形区, わく(枠), コドラート　quadrat
ほうげんさいぼう　胞原細胞　archesporial cell
ほうこうせい　膨こう性【タバコ】　filling power, filling capacity
ほうごうせん　縫合線　suture
ほうこうへんい　彷徨変異　fluctuation
ほうこうまい　芳香米⇒香り米　aromatic rice, scented rice
ほうさ　飽差　saturation deficit
ほうさく　豊作　good harvest, bumper crop
ほうし　胞子　spore
ほうしたい　胞子体, 造胞体　sporophyte
ほうしゃ　放射　radiation
ほうしゃいかんそく　放射維管束　actinostele

ほうしゃせいどういげんそ　放射性同位元素, 放射性同位体　radioisotope (RI), radioactive isotope (RI)
ほうしゃせいめんえきけんていほう　放射性免疫検定法, ラジオイムノアッセイ　radioimmunoassay (RIA)
ほうしゃせんいくしゅ　放射線育種　radiation breeding
ほうしゃせんいでんがく　放射線遺伝学　radiation genetics
ほうしゃせんせいぶつがく　放射線生物学　radiation biology
ほうしゃせんとつぜんへんいせいせい　放射線突然変異生成　radiation mutagenesis
ほうしゃそしき　放射組織　ray
ほうしゃのう　放射能　radioactivity
ほうしゃぶんれつ　放射分裂　radial division
ほうしゃへき　放射壁【細胞】　radial wall
ほうしゃれいきゃく　放射冷却　radiative cooling, radiation cooling
ほうしゅうぜんげんのほうそく　報酬漸減の法則, 収量漸減の法則　law of diminishing returns
ほうじゅん　膨潤　swelling
ほうじょ　防除　control
ほうじょほう　防除法　control method
ほうすいし　紡錘糸　spindle fiber
ほうすいたい　紡錘体　spindle body
ほうせきせんい[りょう]さくもつ　紡績繊維[料]作物　textile fiber crop
ホウそ　ホウ素　boron (B)
ほうそう　防霜, 霜除け　frost protection
ほうそうファン　防霜ファン, ウインドマシン【チャ】　wind machine
ほうちょうもう　防鳥網　bird-proof net
ほうなんど　膨軟土　mellow soil
ほうなんりゅう　豊軟粒　mellow kernel
ほうなんりゅう　膨軟粒　mellow kernel
ほうにんさいしゅ　放任採種　seed production by open pollination
ほうにんじゅふん　放任受粉　open pollination, natural pollination
ほうひ　包皮　hull, husk
ほうふう　防風　wind protection
ほうふうがき　防風垣　shelter hedge
ほうふうもう　防風網　windbreak net
ほうふうりん　防風林　windbreak forest
ほうふうりんたい　防風林帯　shelter belt
ほうぶつせん　放物線　parabola
ほうぼく　放牧　grazing, pasturing
ほうぼくち　放牧地　pasture, grazing land
ボウマ (萵麻)　China jute, Indian mallow, *Abutilon avicennae* Gaertn.
ほうまい　包埋　embedding
ほうまいざい　包埋剤　embedding agent
ほうよう　苞葉, 苞　bract, bract leaf
ホウレンソウ　spinach, *Spinacia oleracea* L.
ほうわしぼうさん　飽和脂肪酸　saturated fatty acid
ほうわすいじょうきあつ　飽和水蒸気圧　saturation vapor pressure
ほうわすいぶんりょう　飽和水分量, 最大容水量　maximum water holding capacity
ほうわとうすいけいすう　飽和透水係数【土壌】　saturated hydraulic conductivity
ほうわようえき　飽和溶液　saturated solution
ほおんせっちゅうなわしろ　保温折衷苗代　protected semiirrigated rice-nursery
ほおんなわしろ　保温苗代, 保護苗代　protected rice-nursery
ほおんはたなわしろ　保温畑苗代

protected upland rice-nursery
ぼかいきけいすう　母回帰係数⇒回帰パラメータ　regression parameter
ほがた　穂型　ear type, head type, panicle type
ぼかぶ　母株　stock plant
ほがり　穂刈り　ear plucking, ear reaping
ぼがん　母岩　parent rock
ほぎ　穂木, 接ぎ穂　scion
ほぎれ　穂切れ　ear breaking
ぼく　牧区　paddock
ぼくじょう　牧場　livestock farm, ranch
ぼくそう　牧草　pasture plant
ぼくそうかんそうき　牧草乾燥機　hay dryer
ぼくそうち　牧草地　pasture
ほくび　穂首　neck of panicle (spike), panicle base
ほくびせつ　穂首節　neck node of panicle (spike)
ほくびせっかん　穂首節間　neck internode of panicle (spike)
ほくび[せつ]ぶんかき　穂首[節]分化期　panicle (spike) neck node differentiation stage
ぼくめつ　撲滅　eradication
ぼくや　牧野　range
ホーグランドえき　ホーグランド液　Hoagland's solution
ほくりくのうぎょうしけんじょう　北陸農業試験場　Hokuriku National Agricultural Experiment Station
ぼけいせんばつ[ほう]　母系選抜[法]　maternal line selection
ほけいとう　穂系統　head row
ほこう　補光　supplementary illumination (lighting)
ほこういくびょう　補光育苗　seedling raising by illuminated nursery
ほこうそ　補酵素　coenzyme
ほこうようトラクタ　歩行用トラクタ, ハンドトラクタ　walking tractor
ほごえ　穂肥　topdressing at panicle (ear) formation stage
ほごさくもつ　保護作物　nurse crop
ほごなわしろ　保護苗代, 保温苗代　protected rice-nursery
ほごばいよう　保護培養, ナース培養　nurse culture
ぼざい　母材　parent material
ほさき　穂先　ear tip
ほじく⇒すいじく　穂軸　1) rachis【イネ・ムギ類】2) cob【トウモロコシ】
ホシクサ　Eriocaulon cinereum R. Br.
ぼじゅ　母樹　mother tree
ほじゅう　穂重　ear weight, panicle weight
ほじゅうがた　穂重型　ear-weight type, heavy-panicle type, panicle-weight type
ぼしゅうだん　母集団【統計】population
ぼしゅうだんパラメータ　母集団パラメータ　population parameter
ぼしゅうだんへいきん[ち]　母集団平均[値], 母平均[値]　population mean
ほじょう　圃場　field
ほじょうけんてい　圃場検定　field test
ほしょうさよう　補償作用　compensatory effect
ほじょうしけん　圃場試験　field experiment, field trial
ほしょうしゅし　保証種子　certified seed
ほじょうせいび　圃場整備　farmland consolidation
ほじょうていこうせい　圃場抵抗性　field resistance
ほじょうようすいりょう　圃場容水量　field capacity
ほしょく　補植　complementary planting, supplementary planting
ほじょざい　補助剤【農薬】adjuvant
ほじょしきそ　補助色素　accessory

pigment
ほじょしりょう　補助飼料　supplement, supplementary feed
ほじょしりょうきゅうよ　補助飼料給与　supplementary feeding, supplemental feeding
ほすいせい　保水性　water retentivity
ほすう　穂数　ear number, panicle number
ぼすう　母数⇒母集団パラメータ　population parameter
ほすうがた　穂数型　ear-number type, many-tillering type, panicle-number type
ぼすうもけい　母数模型⇒固定[効果]模型　fixed [effects] model
ホースグラム　horsegram, *Macrotyloma uniflorum* (Lam.) Verdc. (= *Dolichos uniflorus* Lam.)
ポストハーベスト　postharvest
ホスファターゼ　phosphatase
ホスホエノールピルビンさん　ホスホエノールピルビン酸　phosphoenolpyruvic acid (PEP)
ホスホエノールピルビンさんカルボキシラーゼ　ホスホエノールピルビン酸カルボキシラーゼ　phosphoenolpyruvate (PEP) carboxylase
ホスホグリセリンさん　ホスホグリセリン酸　phosphoglyceric acid (PGA)
ホスホリラーゼ, リン酸化酵素　phosphorylase
ぼせいいでん　母性遺伝　maternal inheritance
ほせいいんし　補正因子　correction factor
ほせいこう　補正項　correction term
ぼせいこうか　母性効果　maternal effect
ほぜん　保全　conservation
ほせんばつほう　穂選抜法　ear-selection method, head-selection method
ホソアオゲイトウ　spleen amaranth, *Amaranthus patulus* Bert.
ほそいとき　細糸期, レプトテン期【減数分裂】　leptotene stage
ほそう　穂相　ear characteristics
ほぞう　保蔵, 保存　storage, preservation
ほそくいでんし　補足遺伝子　complementary gene
ホソバザミア→フロリダアロールート
ホソバヒメミソハギ　purple ammannia, *Ammannia coccinea* Rottb.
ホソムギ→ペレニアルライグラス
ほぞろいき　穂揃期　full heading time
ほぞん　保存, 保蔵　storage, preservation
ほぞんえき　保存液　preservative
ホタルイ　Japanese bulrush, *Schoenoplectus juncoides* (Roxb.) Palla ssp. *hotarui* (Ohwi) Soják (= *Scirpus juncoides* Roxb. var. *hotarui* Ohwi)
ボタン (牡丹)　moutan, tree p[-a-]eony, *Paeonia suffruticosa* Andr. (= *P. moutan* Sims)
ほちょう　穂長　ear length, panicle length
ホッカイトウキ (北海当帰)　*Angelica acutiloba* Kitagawa var. *sugiyamae* Hikino
ほっかいどうのうぎょうしけんじょう　北海道農業試験場　Hokkaido National Agricultural Experiment Station
ポット, 植木鉢, 鉢　pot
ホップ (忽布)　hop, *Humulus lupulus* L.
ポップコーン　pop corn, *Zea mays* L. var. *everta* Bailey
ホテイアオイ (布袋葵)　water hyacinth, *Eichhornia crassipes* (Mart.) Solms-Laub.

ポテトハーベスタ potato harvester
ポテトプランタ potato planter
ホテリングのティーにじょう[とうけいりょう] ホテリングのT^2[統計量] Hotelling's T^2 [-statistics]
ポテンシャルしゅうりょう ポテンシャル収量, 潜在収量 potential yield
ホトケノザ henbit, *Lamium amplexicaule* L.
ポドコーン pod corn, *Zea mays* L. var. *tunicata* St. Hil.
ポドゾル Podzol
ボトムプラウ bottom plow
ポトメーター, 吸水計 potometer
ホナガイヌビユ→アオビユ
ほはつが 穂発芽 preharvest sprouting
ほばらみき 穂ばら(孕)み期 booting stage, boot stage
ほふくがた ほふく(匍匐)型 prostrate type, creeping type
ほふくし ほふく(匍匐)枝, ふく(匐)枝, ストロン runner, stolon
ほふくしょくぶつ ほふく(匍匐)植物 creeper
ほふくせいの ほふく(匍匐)性の creeping
ぼへいきん[ち] 母平均[値], 母集団平均[値] population mean
ホホバ jojoba, *Simmondsia chinensis* (Link) C. K. Schneid.
ぼほん 母本 mother plant
ホメオスタシス homeostasis
ホメオチックとつぜんへんい ホメオチック突然変異 homeotic mutation
ホメオドメイン homeodomain
ホメオボックス homeobox
ボーメひじゅうけい ボーメ比重計 Baumé's hydrometer
ホモジナイザー homogenizer
ホモせつごう ホモ接合, 同型接合 homozygosis
ホモせつごうせい ホモ接合性, 同型接合性 homozygosis

ホモせつごうたい ホモ接合体, 同型接合体 homozygote
ポーランドコムギ Polish wheat, *Triticum polonicum* L.
ポリアクリルアミドでんきえいどうほう ポリアクリルアミド電気泳動法 polyacrylamide gel electrophoresis (PAGE)
ポリアミン polyamine
ポリエチレングリコール polyethylene glycol (PEG)
ポリエチレンフィルム polyethylene film
ポリジーン polygene, 微動遺伝子 minor gene
ポリソーム polysome, ポリリボソーム polyribosome
ポリデオキシリボヌクレオチドシンターゼ polydeoxyribonucleotide synthase, DNAリガーゼ DNA ligase
ほりとりき 掘取り機 digger
ボリビアキナ *Cinchona ledgeriana* Moens ex Trimen
ポリペプチド polypeptide
ポリメラーゼれんさはんのう ポリメラーゼ連鎖反応 polymerase chain reaction (PCR)
ポリリボソーム polyribosome, ポリソーム polysome
ホールクロップサイレージ whole crop silage
ホールピペット transfer pipette
ボルドーえき ボルドー液 Bordeaux mixture
ホルマリンさくさんアルコール ホルマリン酢酸アルコール formalin acetic alcohol (FAA)
ホルモン hormone
ホルモンがたじょそうざい ホルモン型除草剤 hormonal herbicide
ホルモンへんいたい ホルモン変異体 hormone mutants

ボロ【イネの生態型】 boro
ポロメーター porometer
ホワイトクローバ→シロクローバ
ポンキン→ペポカボチャ
ほんぞうがく 本草学 herbalism
ほんでん 本田 paddy field, rice field
ホントウキ→トウキ(当帰)
ポントクタデ *Persicaria pubescens* (Blume) Hara (= *Polygonum pubescens* Blume)
ほんぱ 本葉【タバコ】 leaf
ほんばしゅし 本場種子 home seed
ほんぽ 本圃 field
ほんやく 翻訳 translation
ほんよう 本葉, 普通葉 foliage leaf

[ま]

マイクロコンピュータ(マイコン) microcomputer
マイクロは マイクロ波 microwave
マイクロメーター micrometer
マイコプラズマ mycoplasma
まいどしゅし 埋土種子 buried seed
マウンテンブロムグラス mountain bromegrass, *Bromus marginatus* Nees ex Steud.
マオラン(麻緒蘭)→ニュージーランドアサ
マガタマノキ→カシュー
マーガリン margarine
マカロニコムギ macaroni wheat, デュラムコムギ durum wheat, *Triticum durum* Desf.
まきつきしょくぶつ 巻きつき植物 twining plant, volubile plant
まきどこ 播き床 seed bed
まきはば 播き幅 sowing (seeding) width
まきひげ 巻きひげ tendril
まきみぞ 播き溝 sowing (seeding) furrow
まくこう 膜孔⇒壁孔 pit

マグネシウム magnesium (Mg)
マグネチックスターラー magnetic stirrer
まくらじ まくら(枕)地【桑園】 headland
まぐわ 馬ぐわ(鍬) drag, comb harrow
マゲー maguey, *Agave* spp.
まごいも 孫芋 secondary tuber
マコモ(真菰) *Zizania latifolia* (Griseb.) Turcz. ex Stapf
マコモダケ(菰筍, 茭白筍) *Zizania* shoot
マコンブ kelp, *Laminaria japonica* Areschoug
まさい 摩砕 grinding, trituration
まじり 混り contamination
まぜまき 混播き⇒混播(こんぱ, こんぱん) mix-sowing (-seeding)
またい 麻袋 jute bag, gunny bag
マダケ (真竹) madake bamboo, *Phyllostachys bambusoides* Sieb. et Zucc.
マチン nux vomica, strychine tree, *Strychnos nux-vomica* L.
まっきじゅふん 末期受(授)粉 end-season pollination
マッシュルーム common mushroom, *Agaricus bisporus* Sing.
マツタケ matsutake fungus, *Trichloma matsutake* Sing.
まっちゃ 抹茶 powdered tea
マットいでんし MAT遺伝子 MAT gene
マツバイ needle spikerush, slender spikerush, *Eleocharis acicularis* (L.) Roem. et Schult.
マッハコムギ macha wheat, *Triticum macha* Dek. et Men.
マツモ(松藻) hornwort, *Ceratophyllum demersum* L.
マツリカ(茉莉花) Arabian jasmine, *Jasminum sambac* Aiton

マテチャ　mate, *Ilex paraguayensis* A. St. Hil.
マトリックポテンシャル　matric potential
マニピュレーター　manipulator
マニホットゴム　manihot rubber, ceara rubber, *Manihot glaziovii* Muell.-Arg.
マニュアスプレッダ，堆肥散布機　manure spreader
マニラアサ(マニラ麻)　abaca, Manila hemp, *Musa textilis* Née
まびき　間引き　thinning
まびきせんてい　間引きせん(剪)定　thinning-out pruning
ママコノシリヌグイ　*Persicaria senticosa* (Franch. et Sav.) H. Gross (= *Polygonum senticosum* (Meisn.) Franch. et Sav.)
マメ　pulse [crop], leguminous crop
マメうち　マメ打ち　pulse threshing
マメか　マメ科　Leguminosae
マメかさくもつ　マメ科作物　leguminous crop, pulse crop
マメかしょくぶつ　マメ科植物　legume, leguminous plant
マメかぼくそう　マメ科牧草　forage legume
マメグンバイナズナ　Virginia pepperweed, peppergrass, *Lepidium virginicum* L.
マメシンクイガ　soybean pod borer
マメダオシ　Australian dodder, *Cuscuta australis* R. Br.
マメるい　マメ類　pulses, pulse crops, leguminous crops
マヨラナ　sweet majoram, *Origanum majorana* L. (= *Majorana hortensis* Moench)
マラバールグラス→東インドレモングラス
まるうね　丸うね(畝)　round ridge
マルコフかてい　マルコフ過程　Markov process
マルコフれんさ　マルコフ連鎖　Markov chain
まるぞこフラスコ　丸底フラスコ　round-bottom flask
マルチ　1) mulch【状態】 2) mulching【作業】
マルチメディア　multimedia
マルチャー　mulcher
マルトース，麦芽糖　maltose
まるぬき　丸抜き【ソバ】dehulled seed
マルバタバコ(丸葉煙草)→ルスチカタバコ
マルバヤハズソウ　Korean lespedeza, Korean bush-clover, *Kummerowia stipulacea* (Maxim.) Makino
まわりこう　回り耕　roundabout plowing
まんかい　満開　full bloom
まんかき　満花期　full bloom stage
マンガン　manganese (Mn)
マンゴー　mango, *Mangifera indica* L.
まんごく　万石　grain sorter, grain screen
マンゴスチン　mangosteen, mangis, *Garcinia mangostana* L.
マンジュシャゲ(曼珠沙華)→ヒガンバナ
マンナン　mannan
マンニット　Mannit ⇒マンニトール　mannitol
マンニトール　mannitol
マンネンロウ(迷迭香)→ローズマリー
まんのう　万能　Japanese hoe
マン-ホイットニーけんてい　マン-ホイットニー検定　Mann-Whitney test

[み]

み(箕)　winnow
み　実　fruit
ミカエリスていすう　ミカエリス定数

Michaelis constant
みかけのこうごうせい　みかけの光合成 apparent photosynthesis,　純光合成 net photosynthesis
みき　幹　trunk
みきがり　幹刈り【タバコ】　stalk cutting
ミキサー　mixer
みきぼしかんそう　幹干し乾燥【タバコ】　stalk-cut curing
ミクロオートラジオグラフィー　microautoradiography
ミクロソーム　microsome
ミクロトーム　microtome
ミクロフィブリル　microfibril
ミクロボディ　microbody,　グリオキシソーム　glyoxysome,　ペルオキシソーム　peroxisome
みこうち　未耕地　uncultivated arable land, potential arable land
みごえ　実肥　topdressing at ripening stage
みこぼれ　実こぼれ　shattering, grain shedding
みじかいさんざいはんぷくはいれつ　短い散在反復配列　short interspersed repetitive sequence (SINE)
ミシマサイコ　Bupleurum scorzonerifolium Willd. var. stenophyllum Nakai (= B. falcatum L.)
みじゅく　未熟　immaturity
みじゅくど　未熟土　Regosol
みじゅくどじょう　未熟土壌　immature soil
みじゅくりゅう　未熟粒　immature grain (kernel)
みしょう　実生, 苗, 幼植物　seedling
みしょうせんばつ　実生選抜　seedling selection
ミズアオイ　Monochoria korsakowii Regel et Maack
ミズオオバコ　Ottelia alismoides (L.) Pers. (= O. japonica Miq.)
ミズガヤツリ　Juncellus serotinus (Rottb.) C. B. Clarke (= Cyperus serotinus Rottb.)
みずかんけい　水関係, 水分生理　water relations
みずかんり　水管理　water control, water management
みずけっさ　水欠差　water saturation deficit
みずごえ　水肥 ⇒ 液肥　liquid fertilizer
ミズゴケ(水苔)　peat moss, sphagnum
みずしゅうし　水収支　water balance
みずストレス　水ストレス　water stress
みずチャンネル　水チャンネル　water channel
ミストき　ミスト機, 送風式噴霧機　mist blower
みずなえ　水苗　paddy rice-seedling
みずなわしろ　水苗代　paddy rice-nursery
ミズハコベ　water starwort, Callitriche palustris L.
ミズハナビ→ヒメガヤツリ
ミズフトモモ→ミズレンブ
みずポテンシャル　水ポテンシャル　water potential
ミズマツバ　Rotala mexicana Cham. et Schl. (= R. pusilla Tulasne)
みずゆそう　水輸送　water transport
みずりようこうりつ　水利用効率　water use efficiency (WUE)
ミズレンブ　water apple, Syzygium aqueum (Burm. f.) Alston
ミセル　micell[e]
みそ　味噌　soybean paste
みぞうえ　溝植え　furrow planting
ミゾソバ　Persicaria thunbergii (Sieb. et Zucc.) H. Gross (= Polygonum thunbergii Sieb. et Zucc.)
ミゾハコベ　Elatine triandra Schkuhr
みぞはば　溝幅　furrow width
みぞほりき　溝掘り機, トレンチャ

trencher
みぞまき 溝播き furrow sowing (seeding)
ミチヤナギ prostrate knotweed, *Polygonum aviculare* L.
みつ 蜜 nectar
みつげんさくもつ 蜜源作物 honey crop
みつじょうは 密条播 dense drilling
みっしょく 密植 dense planting, close planting
みっすい 密穂 compact ear, dense ear
みっせい 密生 thick stand
みつせん 蜜腺 nectary
みつど 密度 density
みつどかんすう 密度関数【確率分布】 density function
みつどこうか 密度効果 density effect
みつどこうばいえんしん 密度勾配遠心 density gradient centrifugation
ミツバ (三葉) Japanese hornwort, mitsuba, *Cryptotaenia japonica* Hassk.
みっぱ (みっぱん) 密播, 厚播き dense sowing (seeding), thick sowing (seeding)
ミツマタ (三椏) mitsumata, *Edgeworthia chrysantha* Lindl. (= *E. papyrifera* Sieb. et Zucc.)
ミトコンドリア mitochondrion (*pl.* mitochondria)
ミドリハッカ→スペアミント
みなくち 水口 water inlet
みなじり 水尻 water outlet
ミニマックスげんり ミニマックス原理 minimax principle
ミノゴメ→カズノコグサ
ミブヨモギ sea wormwood, *Artemisia maritima* L.
みほんえん 見本園 exhibition garden (field)
ミミナグサ *Cerastium fontanum* Baumg. ssp. *triviale* (Link) Jalas var. *angustifolium* (Franch.) H. Hara
みゃくけい 脈系, 葉脈系 venation
みゃくたん 脈端 vein ending
ミョウガ (茗荷) mioga ginger, *Zingiber mioga* (Tumb. ex Murray) Roscoe
みわれ 実割れ fruit cracking
みんかんいくしゅ 民間育種 personal breeding

[む]

むがい 霧害 fog damage
むがいざっそう 無害雑草 innoxious weed
むかく 無核 seedless
むかご (零余子) aerial tuber, bulbil
ムカゴコンニャク *Amorphophallus bulbifer* Bl.
ムギ[るい] 麦[類] wheat, barley etc., winter cereals
ムギうち 麦打ち 1) barley threshing【オオムギ】 2) wheat threshing【コムギ】
むきえいよう 無機栄養 mineral nutrition
むきか 無機化 mineralization
ムギこき 麦こ(扱)き 1) barley threshing【オオムギ】 2) wheat threshing【コムギ】
むきこきゅう 無気呼吸, 無酸素呼吸, 嫌気的呼吸 anaerobic respiration
ムギさく (ばくさく) 麦作 1) barley cropping (culture)【オオムギ】 2) wheat cropping (culture)【コムギ】
むきしつひりょう 無機質肥料 inorganic fertilizer
むきずの 無傷の, 健全な intact
むきせい 無機組成 mineral composition
ムギふみ 麦踏み treading
ムギわら 麦わら 1) straw of barley 【オオムギ】 2) straw of rye【ライム

ギ】 3) straw of wheat【コムギ】
むきんばいよう　無菌培養　sterile culture, aseptic culture
むげんかじょ　無限花序　indeterminate inflorescence
むげんしんいくがた　無限伸育型　indeterminate type
むげんぼしゅうだん　無限母集団　infinite population
むこうか　無効花　fruitless flower
むこうすい[ぶん]　無効水[分]　unavailable water (moisture)
むこうぶんげつ　無効分げつ　non-productive tiller
むさくいか　無作為化　randomization
むさくいこうはい　無作為交配　random mating
むさくいはいち　無作為配置　random arrangement
むさくいひょうほん　無作為標本　random sample
むさくい[ひょうほん]ちゅうしゅつ　無作為[標本]抽出　random sampling
むさくいブロックせっけい　無作為ブロック設計, 乱塊法　randomized block design (method)
むさんそこきゅう　無酸素呼吸, 無気呼吸, 嫌気的呼吸　anaerobic respiration
ムシゲル　mucigel
むしぶんれつ　無糸分裂　amitosis
むしんうねたて　無心うね(畝)立て　ridging on tilled land
むせいせいしょく　無性生殖　asexual reproduction
むせいはんしょく　無性繁殖　asexual propagation
むそうきかん　無霜期間　frostless period
むてきしん　無摘心　non-topping
むてきフィルム　無滴フィルム　drip-free film
むどじょうさいばい　無土壌栽培　soilless culture
むはいぐうせいしょく　無配偶生殖　apomixis
むはいしゅし　無胚種子　embryoless seed
むはいせいしょく　無胚生殖　apogamy
むはいにゅうしゅし　無胚乳種子　exalbuminous seed
むへいの　無柄の　sessile
むべんか　無弁花　apetalous flower
むぼうの　無芒の　awnless
むもうの　無毛の　glabrous
ムラサキ　*Lithospermum erythrorhizon* Sieb. et Zucc.
ムラサキアカザ　*Chenopodium purpurescens* Jacq.
むらさきイネ　紫稲　purple rice
ムラサキウマゴヤシ→アルファルファ
ムラサキカタバミ　Dr. Martius' wood-sorrel, *Oxalis corymbosa* DC.
ムラサキサギゴケ→サギゴケ
ムラサキツメクサ→アカクローバ
むりゅうさんこんひりょう　無硫酸根肥料　non-sulfate fertilizer
むろ　室　pit

[め]

め　芽　bud
め　目【ジャガイモなど】　eye
めいがら　銘柄　brand
めいがらひんしゅ　銘柄品種　branded variety, costly registered variety, marketing standard variety
めいき　明期　light period
めいきょはいすい　明きょ(渠)排水　open ditch drainage
めいはつがしゅし　明発芽種子⇒光発芽種子　light germinater, photoblastic seed
めいはんのう　明反応　light reaction
めいめいほう　命名法　nomenclature
めかき　芽か(掻)き⇒摘芽　disbudding, bud picking

メカニスティックモデル ⇒機構的モデル　mechanistic model
めかぶ　雌株　female plant
メキシコマゲー→リュウゼツラン
めしべ　雌しべ(蕊)　pistil
メス　dissecting scalpel
メース【ニクズクの仮種皮】　mace
めす[の]　雌[の]　female
メスシリンダー　graduated cylinder
メストムシース　mestom[e] sheath
メスピペット　measuring pipette
メスフラスコ　volumetric flask
メタキセニア　metaxenia
めだしまき　芽出し播き　sprout sowing (seeding)
メタノール　methanol⇒メチルアルコール　methyl alcohol
メタン　methane (CH_4)
メタンはっこう　メタン発酵　methane fermentation
メチオニン　methionine (Met)
メチルアルコール　methyl alcohol
メチルグリーン【染色】　methylgreen
めつぎ　芽接ぎ　budding, bud grafting
めっきん　滅菌　sterilization
メッシュこうち　メッシュ気候値　mesh climatic data
メッシュデータ　mesh data
メッセンジャーアールエヌエー　メッセンジャー RNA, 伝令 RNA　messenger RNA (mRNA)
メディアン, 中央値　median
メドハギ　serisea lespedeza, *Lespedeza cuneata* (Dum. Cours.) G.Don
メドーフェスク　meadow fescue, *Festuca pratensis* Huds.
メドーフォックステール　meadow foxtail, *Alopecurus pratensis* L.
メナモミ　*Siegesbeckia pubescens* Makino
めばえ　芽ばえ　seedling, budding
めばな(しか)　雌花　female flower, pistillate flower

メヒシバ　crabgrass, *Digitaria adscendens* (H.B.K.) Henr. (= *D. ciliaris* (Retz.) Koeler)
めべり　目減り　loss in weight
メボウキ(目箒)→バジル
めぼし　芽干し　drainage after sprouting
メモリ　memory
メリケンカルカヤ　broomsedge, *Andropogon virginicus* L.
メルカプトエタノール　mercaptoethanol
メロン　melon, *Cucumis melo* L.
めんえき　免疫　immunity
めんえきグロブリン　免疫グロブリン　immunoglobulin
めんえきけいこうほう　免疫蛍光法　immunofluorescence technique, 蛍光抗体法　fluorescent antibody technique
めんえきけんていほう　免疫検定法, イムノアッセイ　immunoassay
めんえきせいひんしゅ　免疫性品種　immune variety
めんえきブロットほう　免疫ブロット法　immunoblotting
めんか　綿花　1) cotton【広義】　2) lint【狭義】
めんじつ　棉実　cotton seed
めんじつゆ　棉実油　cotton seed oil
めんじょ　綿絮　floss
めんじょうしんしょく　面状侵食　sheet erosion
メンデル[がく]せつ　メンデル[学]説　Mendelism
メンデルしゅうだん　メンデル集団　Mendelian population
メンデルせいいでん　メンデル性遺伝　Mendelian inheritance
メンデルのほうそく　メンデルの法則　Mendel's law
めんもう　綿毛　cotton fiber, lint
めんるい　麺類　noodles

めんろう　棉ろう(蠟)　cotton wax

[も]

モーア, 草刈り機　mower
もうかんげんしょう　毛管現象　capillarity, capillary phenomenon
もうかんすい　毛管水　capillary water
もうかんポテンシャル　毛管ポテンシャル　capillary potential
もうじょうこん　毛状根　hairy root
もうじょうたい　毛状体　emergence
もうじょうちゅうしんちゅう　網状中心柱　dictyostele
もうじょうみゃく　網状脈　netted vein, netted venation, reticulate venation
モウソウチク(孟宗竹)　Moso bamboo, *Phyllostachys heterocycla* (Carrière) Matsum. f. *pubescens* (Mazel ex Houz.) D. C. McClint.
もうもんどうかん　網紋導管　reticulate vessel
もく　目【分類】　order
もくか　木化　lignification
もくぶ　木部　xylem
もくぶじゅうそしき　木部柔組織　xylem parenchyma
もくぶせんい　木部繊維　xylem fiber
もくほん　木本　arbor
もくほんさくもつ　木本作物　arbor crop
もくほんしょくぶつ　木本植物　woody plant
もくほん[せい]の　木本[性]の　arboreous, woody
もくめん　木棉　arboreous cotton
モグラあんきょ　モグラ暗きょ(渠), 弾丸暗きょ(渠)　mole drain
もくろう　木ろう(蠟)　vegetable wax
もけい　模型, モデル　model
モザイク　mosaic
モスビーン　moth bean, mat bean, *Vigna aconitifolia* (Jacq.) Maréchal (= *Phaseolus aconitifolius* Jacq.)
もすまい　もす米　"Mosu" rice, penicillium diseased rice
もち　餅　rice cake
もちうるちせい　もちうるち(糯粳)性　endosperm type
もちオオムギ　もち(糯)オオムギ　waxy barley, glutinous barley
もちコムギ　もち(糯)コムギ　waxy wheat, glutinous wheat
もちごめ　もち(糯)米　glutinous rice, waxy rice
もちせいの　もち(糯)性の　glutinous, waxy
モチソバ→ムラサキアカザ
モデリング　modeling
モデル, 模型　model
モデルのけんしょう　モデルの検証　model validation
モード, 最頻値　mode
もとごえ(きひ)　基肥, 元肥　basal dressing, basal application
もどしこうざつ　戻し交雑　backcross, backcrossing
もどしこうざつほう　戻し交雑法　backcross method
もとじろまい　基白米　white-based rice kernel
モニタリングシステム　monitoring system
モノリス, 土壌モノリス　soil monolith
もはんのうじょう　模範農場　pilot farm
もみ　籾　rough rice, unhulled rice, paddy
もみがら　籾殻　hull, husk, chaff
もみすり　籾す(摺)り　hulling, husking, shelling
もみすりき　籾す(摺)り機　huller, husker
もみすりぶあい　籾す(摺)り歩合　husking ratio
もみずれ　籾ずれ　skin abrasion of

rough rice
もみわらひ　もみわら(籾藁)比　grain-straw ratio
モーメント, 積率【統計】　moment
モモ(桃)　peach, *Prunus persica* (L.) Batch
モモミヤシ　peach palm, *Guilielma gasipaes* (H.B.K.) L. H. Bailey
モーリシャスアサ(モーリシャス麻)　Mauritius hemp, *Furcraea gigantea* (D. Dietr.) Vent.
もりつち(もりど)　盛土　bank, mound
モリブデン　molybdenum (Mo)
モルのうど　モル濃度　molarity
モルひ　モル比　mole ratio
モロコシ(蜀黍)　sorghum, grain sorghum, *Sorghum bicolor* (L.) Moench
もん　門【分類】　phylum
もんがれびょう　紋枯病【イネ】　sheath blight
もんだいどじょう　問題土壌　problem soil
モンテカルロ・シミュレーション　Monte Carlo simulation
モンモリロナイト　montmorillonite

[や]

ヤウテア→アメリカサトイモ
やえざき　八重咲き　double flower
ヤエナリ→リョクトウ
ヤエムグラ　catchweed, bedstraw, *Galium spurium* L. var. *echinospermon* (Wallr.) Hayek
やおん　夜温　night temperature
やがいちょうさ　野外調査　field survey
やきごめ　焼米　burnt rice
やきつち　焼土　1) heated (roasted) soil 2) soil burning【作業】
やきばたのうこう　焼畑農耕, 移動耕作　shifting cultivation, slash-and-burn agriculture
やきもみがら　焼籾殻　carbonized rice husks, carbonized chaff
やく　葯　anther
やくがい　薬害　chemical injury
やくかく　葯隔　connective, connectivum
やくざいかんそう　薬剤乾燥　chemical desiccation
やくざいさんぷ　薬剤散布　chemical spraying
やくざいじょそう　薬剤除草　chemical weeding
やくざいてきか　薬剤摘果　chemical fruit thinning
やく[さ]じ　薬[さ]じ　medicine spoon
やくしつ　葯室　anther loculus (*pl.* loculi)
やくばいよう　葯培養　anther culture
やくほうし　薬包紙　powder paper
やくようしょくぶつ　薬用植物　medicinal crop
ヤクヨウニンジン(薬用人参)　ginseng, *Panax ginseng* C. A. Mey. (= *P. schinseng* Nees)
ヤーコン　yacon, *Polymnia sonchifolia* Poepp. et Endl.
やさい　野菜　vegetable [crop]
やさいえんげい　野菜園芸　vegetable gardening, olericulture
やさい・ちゃぎょうしけんじょう　野菜・茶業試験場　National Research Institute of Vegetable, Ornamental Plants and Tea
ヤシかくゆ　ヤシ核油⇒パーム核油 palm kernel oil
ヤシゆ　ヤシ油【ココヤシ】　coconut oil, copra oil
やせいいちりゅうけいコムギ　野生一粒系コムギ　wild einkorn wheat, *Triticum boeoticum* Boiss.
やせいか　野生化　escape
やせいがた　野生型　wild type

やせいげんしゅ　野生原種　wild ancestor
やせいしゅ　野生株　wild strain
やせいしゅ　野生種　wild species
やせいチモフェービけいコムギ　野生チモフェービ系コムギ→アルメニアコムギ
やせち　やせ(痩)地，不毛地　infertile land
やそう　野草　1) wild herb【一般】 2) wild grass, natural grass【イネ科】
やそうち　野草地　native grassland (pasture), range, rangeland
やちだ　谷地田　ill-drained paddy field in inland valley
ヤックベクター　YACベクター，酵母人工染色体ベクター　YAC (yeast artificial chromosome) vector
ヤッコササゲ→ハタササゲ
やつだ　谷津田⇒谷地田　ill-drained paddy field in inland valley
ヤナギタデ　water pepper, *Persicaria hydropiper* (L.) Spach (= *Polygonum hydropiper* L.)
ヤノネグサ　*Persicaria nipponensis* (Makino) H. Gross
ヤハズエンドウ→カラスノエンドウ
ヤハズソウ　striate lespedeza, Japanese clover, *Kummerowia striata* (Thunb. ex Murray) Schindl.
ヤブカラシ　*Cayratia japonica* (Thunb.) Gagn.
ヤブジラミ　Japanese hedgeparsley, *Torilis japonica* (Houtt.) DC.
ヤブタビラコ　*Lapsana humilis* (Thunb.) Makino
ヤマイモ→ヤマノイモ
やませ[かぜ]　山背[風]　Yamase [wind]
ヤマチドメ→オオチドメ
ヤマトトウキ→トウキ
ヤマノイモ　Japanese yam, *Dioscorea japonica* Thunb. ex Murray
ヤムイモ　yam

[ゆ]

ゆういさ　有意差　significant difference
ゆういすいじゅん　有意水準　significance level, level of significance
ゆういせい　有意性　significance
ゆういせいけんてい　有意性検定　significance test, test of significance
ゆういな　有意な　significant
ゆういん　誘引【園芸】　training
ゆういんざい　誘引剤　attractant
ゆうか(おばな)　雄花　male flower, staminate flower
ゆうがいざっそう　有害雑草　noxious weed
ゆうかく　雄核　male nucleus, sperm nucleus
ゆうきえんそざい　有機塩素剤　chlorinated organic compound
ゆうきか　有機化【土壌窒素など】　immobilization
ゆうきこきゅう　有気呼吸，好気呼吸，酸素呼吸　aerobic respiration
ゆうきさいばい　有機栽培　organic culture, organic cultivation
ゆうきさん　有機酸　organic acid
ゆうきしつどじょう　有機質土壌　organic soil
ゆうきしつひりょう　有機質肥料　organic fertilizer
ゆうきすいぎんざい　有機水銀剤　organic mercury pesticide
ゆうきのうぎょう　有機農業　organic farming, organic agriculture
ゆうきリンざい　有機リン剤　organic phosphorus pesticide
ゆうげんかじょ　有限花序　determinate inflorescence
ゆうげんしんいくがた　有限伸育型

determinate type
ゆうげんぼしゅうだん 有限母集団 finite population
ゆうこういでんりつ 有効遺伝率 effective heritability
ゆうこうおんど 有効温度 effective temperature
ゆうこうか 有効花 fruitful flower
ゆうこうけい 有効茎 productive culm (stem), fruitful culm (stem)
ゆうこうけいぶあい 有効茎歩合 percentage of productive culms (stems)
ゆうこうすい[ぶん] 有効水[分] available water (moisture)
ゆうこうせい 有効性【統計】 efficiency
ゆうこうせいぶん 有効成分 active ingredient
ゆうこうたいようぶん 有効態養分⇒可給態養分 available nutrient
ゆうこうどそう 有効土層 effective soil depth
ゆうこうなしゅうだんのおおきさ 有効な集団の大きさ effective population size
ゆうこうぶんげつ 有効分げつ productive tiller, bearing tiller
ゆうこうぶんげつしゅうしき 有効分げつ終止期 last productive-tiller emergence stage
ゆうしぶんれつ 有糸分裂 mitosis (*pl.* mitoses), karyokinesis, caryokinesis
ゆうしょくたい 有色体 chromoplast, chromoplastid
ゆうしんうねたて 有心うね(畝)立て ridging with untilled core
ゆうすい 雄穂 1) staminate inflorescence 2) tassel【トウモロコシ】
ゆうずい 雄ずい(蕊), 雄しべ(蕊) stamen
ゆうすいじょきょ 雄穂除去 detasseling
ゆうずいせんじゅく 雄ずい(蕊)先熟 protandry, proterandry
ゆうずいせんじゅくか 雄ずい(蕊)先熟花 protandrous flower
ゆうせい 優性 dominance
ゆうせいこうか 優性効果 dominant effect
ゆうせいせいしょく 有性生殖 sexual reproduction
ゆうせいとつぜんへんい 優性突然変異 dominant mutation
ゆうせいはんしょく 有性繁殖 sexual propagation
ゆうせいふねん 雄性不稔 male sterility
ゆうせんざっそう 優占雑草 dominant weed
ゆうせんしゅ 優占種 dominant species
ゆうでんりつ 誘電率 dielectric constant
ゆうど 尤度 likelihood
ゆうどうけつごうこうしゅうはプラズマぶんこうぶんせき 誘導結合高周波プラズマ分光分析, ICP分光分析 inductively coupled plasma spectrometery
ゆうどかんすう 尤度関数 likelihood function
ゆうどくざっそう 有毒雑草 poisonous weed
ゆうどひけんてい 尤度比検定 likelihood ratio test
ゆうはいにゅうしゅし 有胚乳種子 albuminous seed
ゆうはついんし 誘発因子 inducer
ゆうはつとつぜんへんい 誘発突然変異 induced mutation
ゆうへいの 有柄の petiolate
ゆうぼうの 有芒の awned, aristate, bearded
ゆうようやそう 有用野草 useful wild

plant
ゆうりオーキシン　遊離オーキシン　free auxin
ゆうりき　遊離基　[free] radical
ゆうりちっそ　遊離窒素　free nitrogen
ゆうわんさくじょうそしきさいぼう　有腕柵状組織細胞　arm palisade cell
ユーカリ　eucalyptus, *Eucalyptus* spp.【*E. citriodora* Hook., *E. macarthurii* Decne. et Maiden , *E. globulus* Labill. など】
ゆきぐされびょう　雪腐病　1) snow mold　2) snow blight【ムギ類】
ゆごう　ゆ(癒)合　healing
ゆさいぼう　油細胞　oil cell
ゆし　油脂　fat and oil
ゆしょうそしき　ゆ(癒)傷組織⇒カルス　callus (*pl.* calli)
ゆしょうホルモン　ゆ(癒)傷ホルモン　wound hormone
ゆしんほう　油浸法【顕微鏡】　oil immersion method
ゆそうえんげい　輸送園芸　truck gardening
ゆそうさいぼう　輸送細胞, 転送細胞　transfer cell
ゆちゃく　ゆ(癒)着　cohesion
ゆりょうさくもつ　油料作物　oil crop
ゆりょうしゅし　油料種子　oil seed

[よ]

ようい　葉位　leaf position on stem
ようイオンこうかんようりょう　陽イオン交換容量　cation exchange capacity (CEC)
よういん　要因, 因子　factor
よういんじっけん　要因実験, 因子実験　factorial experiment
よういんせっけい　要因設計, 因子設計　factorial design
ようえき　溶液　solution
ようえき　葉腋　leaf axil
ようえきさいばい　養液栽培　hydroponics, nutriculture, solution culture
ようえん　葉縁　leaf margin
ようおん　葉温　leaf temperature
ようが　幼芽　plumule
ようが(はめ)　葉芽　leaf bud, foliar bud
ようかいど　溶解度　solubility
ようが(はめ)ざし　葉芽挿し　leaf-bud cutting
ようかんき　葉間期, プラストクロン　plastochron[e]
ようかんせつ　葉関節, 葉節　lamina joint
ようかんせつしけん　葉関節試験　lamina joint test
ようきょ　葉裾【タバコ】　leaf skirt
ようぐん　葉群　foliage
ようけい　幼形　juvenile form
ようけい　葉形　leaf shape
ようげき　葉隙　leaf gap
ようげんき　葉原基　leaf primordium
ようこん　幼根　radicle
ヨウサイ(蕹菜)　water convolwulus, water spinach, swamp cabbage, *Ipomoea aquatica* Forsk.
ようざい　溶剤　solvent
ようさいるい　葉菜類　leaf vegetables
ようさん　養蚕　sericulture
ようじ　葉耳, 小耳　auricle
ようじかんちょう　葉耳間長【イネ】　distance between auricles of flag and penultimate leaves
ようじく　葉軸【複葉】　rachis
ようしつ　溶質　solute
ようじゅ　陽樹　sun tree, intolerant tree
ヨウシュチョウセンアサガオ　thorn apple, Jimson weed, *Datura stramonium* L.
ようしゅナタネ　洋種ナタネ　rape, *Brassica napus* L.

ヨウシュヤマゴボウ　pokeweed, scoke, *Phytolacca americana* L.
ようじょ　葉序　phyllotaxis, phyllotaxy
ようしょう　葉鞘　leaf sheath
ようじょうかんすい　葉上灌水⇒頭上灌水　overhead watering, overhead irrigation
ようじょく　葉じょく(褥)⇒葉枕　pulvinus, leaf cushion
ようしょくぶつ　幼植物, 実生(みしょう), 苗　seedling
ようしょくぶつけんてい　幼植物検定　seedling test
ようしん　葉身　leaf blade, lamina
ようすい　幼穂　young panicle
ようすいき　揚水機　water lifting machinery
ようすいけいせいき　幼穂形成期　panicle formation stage
ようすいぶんかき　幼穂分化期　panicle differentiation stage, panicle initiation stage
ようすいりょう　用水量　irrigation requirement, duty of [irrigation] water
ようすいりょう　要水量　water requirement
ようすいろ　用水路　irrigation canal, irrigation ditch
ようせいしょくぶつ　陽生植物　sun plant, heliophyte
ようせいリンぴ　溶性リン肥　fused magnesium phosphate
ようせき　容積　volume
ようせき　葉積　leaf area duration (LAD)
ようせき　葉跡　leaf trace
ようせきじゅう　容積重, かさ密度　bulk density
ようせつ　葉節, 葉関節　lamina joint
ようぜつ　葉舌, 小舌　ligule
ヨウそ　ヨウ素　iodine (I)
ヨウそか　ヨウ素価　iodine value

ようそかじょう　要素過剰【土壌窒素など】　nutrient excess
ようそけつぼう　要素欠乏　mineral deficiency
ヨウそ-デンプンはんのう　ヨウ素-デンプン反応　iodo-starch reaction
ようぞんさんそ　溶存酸素　dissolved oxygen (DO)
ようだつ　溶脱　leaching, eluviation
ようちゅう　幼虫　larva (*pl.* larvae)
ようちん　葉枕　pulvinus, leaf cushion
ようどうせんべつ　揺動選別　seed selection by oscillation
ようどうせんべつき　揺動選別機　shaking table sorter
ようにく　葉肉　mesophyll
ようにくコンダクタンス　葉肉コンダクタンス, 葉肉伝導度　mesophyll conductance
ようにくていこう　葉肉抵抗　mesophyll resistance
ようにくでんどうど　葉肉伝導度, 葉肉コンダクタンス　mesophyll conductance
ようばい　溶媒　solvent
ようびょうき　幼苗期　seedling stage
ようぶん　養分, 栄養素　nutrient
ようぶんきゅうしゅう　養分吸収　nutrient absorption, nutrient uptake
ようぶんけつぼう　養分欠乏　nutrient deficiency
ようぶんけつぼうしょう　養分欠乏症　nutrient deficiency symptom
ようぶんせき　葉分析　leaf analysis
ようへい　葉柄　petiole
ようぼく　幼木　young tree
ようぼくえん　幼木園【チャ】　young tea field
ようみゃく　葉脈　vein, nerve
ようみゃくけい　葉脈系, 脈系　venation
ようめんきゅうしゅう　葉面吸収　foliar uptake, foliar intake, foliar

absorption
ようめんさんぷ　葉面散布　foliar application, foliar spray
ようめんせき　葉面積　leaf area
ようめんせきけい　葉面積計　leaf area meter
ようめんせきしすう　葉面積指数　leaf area index (LAI)
ようめんせきひ　葉面積比　leaf area ratio (LAR)
ようめんせひ　葉面施肥　foliar application of fertlilizer
ようよう　陽葉　sun leaf
ようり　洋梨, プラウ　plow
ようりょう　用量　dose, dosage
ようりょうはんのうきょくせん　用量-反応曲線　dose-response curve
ようりょくそ　葉緑素, クロロフィル　chlorophyll
ようりょくたい　葉緑体, クロロプラスト　chloroplast
ようれい　葉齢　1) plant age in leaf number【植物体】2) leaf age【葉自身】
ようれいしすう　葉齢指数　leaf number index
よかん　予乾【飼料】　wilting
よくあついでんし　抑圧遺伝子　suppressor [gene]
よくこういくが　浴光育芽, 浴光催芽　green-sprouting [under diffused light]
よくこうさいが　浴光催芽, 浴光育芽　green-sprouting [under diffused light]
ヨークシャーフォッグ→ベルベットグラス
よくせい　抑制, 阻害　inhibition
よくせいいでんし　抑制遺伝子　repressor [gene], inhibitory gene, inhibiting gene
よくせいいんし　抑制因子　inhibitory factor
よくせいさいばい　抑制栽培　late raising

よこじく　横軸　abscissa
ヨコバイ　leaf hopper
よさつとう　予察燈　light trap
ヨシ（葭）common reed, *Phragmites australis* (Cav.) Trin. ex Steud. (= *P. communis* Trin.)
よじのぼりけい　よじ登り茎　climbing stem
よせうね　寄せうね（畝）grouped row
よせつぎ　寄せ接ぎ　approach grafting, inarching
よそ　予措【種子】pretreatment
よそく　予測　prediction
よそくち　予測値　predicted value
よそくりょう　予測量　predictor
よびちょうさ　予備調査　preliminary survey
よびつぎ　呼び接ぎ⇒寄せ接ぎ approach grafting, inarching
ヨメナ　*Kalimeris yomena* (Kitam.) Kitam.
ヨモギ（蓬）Japanese mugwort, *Artemisia princeps* Pamp.
より　撚り【棉毛】twist, convolution
よれい　予冷　precooling
ヨロイグサ　*Angelica dahurica* (Fisch.) Benth. et Hook.f.
よんばいせい　四倍性　tetraploidy
よんばいたい [しょくぶつ]　四倍体 [植物]　tetraploid [plant]
よんほんぐわ　四本ぐわ（鍬）four-prong digging hook

［ら］

らいう　雷雨　thunderstorm
ライコムギ　triticale
ライシメータ　lysimeter
ライスセンター　rice processing plant, rice center
ライトグリーン【染色】light green
ライマメ　Lima bean, butter bean, *Phaseolus lunatus* L. (= *P. limensis*

ライム　lime, sour lime, *Citrus aurantifolia* (Christm.) Swingle
ライムギ　rye, *Secale cereale* L.
ラインこうさ[てん]ほう　ライン交さ(又)[点]法　line intersection method
らくさく　落さく(蒴)　boll shedding
らくさん　酪酸　butyric acid
らくすい　落水　drainage of residual water, ponding water release
らくのう　酪農　dairy farming, dairying
らくよう　落葉　leaf abscission, leaf fall, defoliation
らくようこうようじゅ　落葉広葉樹　deciduous broad-leaved tree
らくようざい　落葉剤, 枯葉剤　defoliant, defoliator
らくようせいの　落葉性の　deciduous
らくようらくし　落葉落枝, リター　litter
らくらい　落らい(蕾)　flower-bud abscission
ラジオイムノアッセイ, 放射性免疫検定法　radioimmunoassay (RIA)
らししょくぶつ　裸子植物　gymnosperm
ラジノクローバ　ladino clover, *Trifolium repens* L. var. *giganteum*
ラスターデータ【ディスプレイ】　raster data
ラズベリー　raspberry【*Rubus idaeus* L. およびその園芸品種を含む】
らせんかいだんがたたいせいようじょ　らせん階段型対生葉序　spiroscalate phyllotaxis
らせんこうぞう　らせん構造　helical structure
らせんもんどうかん　らせん紋導管　helical vessel
らせんようじょ　らせん葉序　spiral phyllotaxis
らち　裸地　bare land, bare ground

らちきゅうかん　裸地休閑　bare fallow
らっか　落果　fruit shedding, fruit drop
らっか　落花　flower shedding, flower abscission
ラッカセイ(落花生)　peanut, groundnut, *Arachis hypogaea* L.
ラッカセイゆ　ラッカセイ油　peanut oil, groundnut oil
ラッキョウ(辣韮)　rakkyo, Baker's garlic, *Allium chinense* G. Don
らっきょう　落きょう(莢)　pod shedding
ラテライト　laterite
ラテンほうかく[せっけい]　ラテン方格[設計]　Latin square [design]
ラトソル　Latosol
ラノリン　lanolin
ラフィアヤシ　raphia palm, *Raphia pedunculata* Beauv. (= *R. ruffia* Mart.)
ラフブルーグラス　rough bluegrass, rough-stalked meadow grass, *Poa trivialis* L.
ラベンダー　lavender, true lavender, *Lavandula angustifolia* Mill. (= *L. officinalis* Chaix, *L. vera* DC.)
ラマルクせつ　ラマルク説　Lamarckism
ラミー→チョマ
ラムシュいくしゅほう　ラムシュ育種法　Ramsch method of breeding
ラムダファージ　λファージ　λ (lambda) phage
ラメラ　lamella (*pl.* lamellae)
らん　卵　egg, ovum
らんかいほう　乱塊法, 無作為ブロック設計　randomized block design (method)
らんかく　卵核　egg nucleus
らんさいぼう　卵細胞　egg cell
らんざつうえ　乱雑植え　random planting
らんすう　乱数　random number

らんすうひょう　乱数表　table of random numbers
らんそうるい　藍藻類　blue green algae, シアノバクテリア cyanobacteria
ランブータン　rambutan, *Nephelium lappaceum* L.
らんりゅう　乱流【気象】　turbulence, turbulent flow

[り]

りかく　離核【果実】　free stone
リクチメン（陸地棉）　upland cotton, *Gossypium hirsutum* L.
りくとう（おかぼ）　陸稲　upland rice
りくなわしろ（おかなわしろ）　陸苗代 ⇒畑苗代　upland rice-nursery
リグニン　lignin
りこう　り(犁)耕, プラウ耕　plowing
りさんぶんぷ　離散分布　discrete distribution
りさんへんすう　離散変数　discrete variable
りさんへんりょう　離散変量　discrete variate
リシン，リジン　lysine (Lys)
リスクぶんせき　リスク分析　risk analysis
りせいさいぼうかんげき　離生細胞間隙　schizogenous intercellular space
りそう　離層　abscission layer
りそうがた　理想型　ideotype
リソソーム　lysosome
リター，落葉落枝　litter
りたいトラクタ　履帯トラクタ　crawler tractor
リチウム　lithium (Li)
りっちじょうけん　立地条件　locational conditions, condition of site
りつもう　立毛　stand
りつもうちょうさ　立毛調査, 検見(けみ)　stand observation
リードカナリーグラス　reed canarygrass, *Phalaris arundinacea* L.
リトルブルーステム　little bluestem, *Schizachyrium scoparium* (Michx.) Nash
リノールさん　リノール酸　linoleic acid
リノレンさん　リノレン酸　linolenic acid
リーパ，刈取り機　reaper
リパーゼ　lipase
りはんいでんし　離反遺伝子　oppositional gene
りびょう　り(罹)病　disease
りびょうか　り(罹)病化【病理】　breakdown
りびょうせい　り(罹)病性　susceptibility
りびょうせいひんしゅ　り(罹)病性品種　susceptible variety
リーフディスクほう　リーフディスク法　leaf disc method
リプレッサー　repressor
リブロースいちごビスリンさん　リブロース-1,5-ビスリン酸　ribulose-1,5-bisphosphate (RuBP)
リブロースいちごビスリンさんカルボキシラーゼ/オキシゲナーゼ　リブロース-1,5-ビスリン酸カルボキシラーゼ/オキシゲナーゼ　ribulose-1,5-bisphosphate carboxylase/oxygenase (Rubisco)
リベットコムギ　rivet wheat, *Triticum turgidum* L.
リベリアコーヒー[ノキ]　Liberian coffee, *Coffea liberica* W. Bull. ex Hiern
リボかくさん　リボ核酸　ribonucleic acid (RNA)
リボソーム　ribosome
リボソームアールエヌエー　リボソーム RNA　ribosomal RNA (rRNA)
リモートセンシング, 遠隔測定, 隔測

remote sensing
りゅう　粒　grain, kernel
りゅうあん　硫安(硫酸アンモニウム)　ammonium sulfate
りゅうかすいそ　硫化水素　hydrogen sulfide (H_2S)
リュウキュウイ(琉球藺)→シチトウイ
りゅうきょ　流去, 流出　runoff, outflow, efflux
りゅうけい　粒径　particle size
りゅうけい　粒形　grain shape, kernel shape
りゅうけいぶんぷ　粒径分布　particle size distribution
りゅうこう　粒厚　grain thickness
りゅうざい　粒剤　granule
りゅうさんアンモニウム　硫酸アンモニウム(硫安)　ammonium sulfate
りゅうさんカリ　硫酸カリ⇒硫酸カリウム　potassium sulfate
りゅうさんカリウム　硫酸カリウム　potassium sulfate
りゅうしつ　粒質　grain texture, kernel texture
りゅうじゅう　粒重　grain weight, kernel weight
りゅうしゅつ　流出, 流去　runoff, outflow, efflux
りゅうじょうひりょう　粒状肥料　granular fertilizer
リュウゼツサイ(竜舌菜)　Indian lettuce, *Lactuca indica* L. var. *dracoglossa* Kitam.
リュウゼツラン(竜舌蘭)　century plant, *Agave americana* L. var. *variegata* Nichols.
りゅうせんき　粒選機　grain sorter
りゅうそく　流束, フラックス　flux
りゅうだい　粒大　grain size
りゅうちゃくみつど　粒着密度　grain density
りゅうちょう　粒長　grain length, kernel length

りゅうにゅう　流入　inflow, influx
りゅうぼう　流亡【土壌肥料】　loss
りょうがわけんてい　両側検定　two-sided test, two-tailed test
りょうしか　量子化　quantization
りょうししゅうりつ　量子収率　quantum efficiency, 量子収量　quantum yield
りょうししゅうりょう　量子収量　quantum yield, 量子収率　quantum efficiency
りょうしんへいきん　両親平均, 中間親　midparent [value]
りょうせいか　両性花　hermaphrodite flower, bisexual flower
りょうせいざっしゅ　両性雑種, 二遺伝子雑種　dihybrid
りょうてきけいしつ　量的形質, 計量形質　quantitative character (trait)
りょうめんきこうよう　両面気孔葉　amphistomatous leaf
りょうめんよう　両面葉　bifacial leaf
りょくしざし　緑枝挿し　softwood cutting
りょくしょくきゅうかん　緑色休閑　green fallow
りょくそうるい　緑藻類　green algae
りょくたいしゅんか　緑体春化　green vernalization
りょくちがく　緑地学, 造園学　landscape architecture
りょくちゃ　緑茶　green tea
リョクトウ(緑豆)　mung bean, green gram, *Vigna radiata* (L.) R. Wilczek (= *Phaseolus aureus* Roxb.)
りょくひ　緑肥　green manure
りょくひさくもつ　緑肥作物　green manure crop
りょっか　緑化【育苗】　greening
りれきげんしょう　履歴現象, ヒステリシス　hysteresis
リン(燐)　phosphorus (P)
りんが　鱗芽　bulbil

りんかいおんど　臨界温度　critical temperature
りんかいてんかんそうほう　臨界点乾燥法　critical point drying
りんかいにっちょう　臨界日長⇒限界日長　critical daylength
りんかじゅふん　隣花受粉　geitonogamy, neighboring pollination
りんかん　林冠　canopy
りんかんしきのうほう　輪換式農法　ley farming
りんかん[すい]でん　輪換[水]田　paddy field under paddy-upland rotation
りんかんばた　輪換畑　upland field under paddy-upland rotation
りんけい　鱗茎　bulb
リンゴ（林檎）　apple, *Malus pumila* Mill.
リンゴさん　リンゴ酸　malic acid
りんさく　輪作　crop rotation
りんさくそうち　輪作草地　rotational grassland, ley
リンさん　リン(燐)酸　phosphoric acid
リンさんえん　リン酸塩　phosphate
リンさんかこうそ　リン酸化酵素, ホスホリラーゼ　phosphorylase
リンさんかんしょうえき　リン酸緩衝液　phosphate buffer
リンさんきゅうしゅうけいすう　リン酸吸収係数　phosphate absorption coefficient
リンさんひりょう　リン酸肥料　phosphatic fertilizer
リンさんほじようりょう　リン酸保持容量　phosphate retention capacity
りんししつ　リン脂質　phospholipid
りんせいの　輪生の　whorled
リンターさいしゅき　リンター採取機　linter
りんないそうち　林内草地, 混牧林　woodland pasture, wood-pasture, grazable forestland
リンネしゅ　リンネ種　linneon, Linnean species
りんばんかんがい　輪番灌漑　rotational irrigation
りんばんこうざつ　輪番交雑　rotational crossing
りんぴ　鱗被　lodicule
りんぺんよう　鱗片葉　scale leaf, scaly leaf

[る]

るいえん　類縁　relationship
るいじたい　類似体, アナログ　analogue
るいせききよりつ　累積寄与率　cumulative contribution
るいせきせんばつ　累積選抜　cumulative selection
るいせきどすう　累積度数, 累積頻度　cumulative frequency
るいせきひんど　累積頻度, 累積度数　cumulative frequency
るいせきぶんぷかんすう　累積分布関数　cumulative distribution function, 分布関数　distribution function
ルイセンコせつ　ルイセンコ説　Lysenko hypothesis
ルーサン→アルファルファ
ルシフェラーゼ　luciferase
ルシフェリン　luciferin
ルスチカタバコ　rustica tobacco, aztec tobacco, *Nicotiana rustica* L.
ルタバガ　rutabaga, Swedish turnip, swede, *Brassica napus* L. var. *napobrassica* (Mill.) Reichb.
ルートスキャナー　root [length] scanner
ルーペ　magnifier

[れ]

れい　齢, エイジ　age
れいおんたい　冷温帯　cool temperate zone
れいおんたいきこう　冷温帯気候　cool temperate climate
れいか　冷夏　cool summer
れいがい　冷害　cool summer damage, cool weather damage
れいきじょうたい　励起状態　excited state
れいきゃくえんしんき　冷却遠心機　refrigerated centrifuge
レイシ(茘枝)　litchi, *Litchii chinensis* Sonn.
れいしょう　冷床　cold bed, cold frame
れいすいがい　冷水害　cold water damage
れいすいでん　冷水田　paddy field irrigated with cold water
れいせんしょくたいしょくぶつ　零染色体植物　nullisomic plant
れいとうこ　冷凍庫　freezer
れいとうちょぞう　冷凍貯蔵　freezing storage
れいとうれいぞうこ　冷凍冷蔵庫, 超低温庫　deep freezer
れき(礫)　gravel
レーキ　rake
れきこう　れき(壢)溝　furrow ditch
れきこう[さいばい]　礫耕[栽培]　gravel culture
れきこうてい　れき(壢)溝底, れき(壢)底　furrow bottom
れきじつ　暦日　calendar day
れきじょう　れき(壢)条　furrow slice
れきてい　れき(壢)底, れき(壢)溝底　furrow bottom
レゴソル⇒未熟土　Regosol
レーザー　laser
レス, 黄土　loess
レース, 病原型　race
レスキュグラス　rescuegrass, prairie grass, *Bromus catharticus* Vahl (= *B. unioloides* (Willd.) H.B.K.)
レスクグラス→レスキュグラス
レセプター, 受容体, 受容器　receptor
レタス　lettuce, *Lactuca sativa* L.
れっか　劣化　deterioration
れっかい　裂開　dehiscence
れっかいか　裂開果　dehiscent fruit
れつじょうしけんく　列条試験区　row plot
れっせいとつぜんへんい　劣性突然変異　recessive mutation
れっせいの　劣性の　recessive
レッドクローバ→アカクローバ
レッドトップ　redtop, *Agrostis gigantea* Roth
レッドフェスク　red fescue, *Festuca rubra* L.
レトロトランスポゾン　retrotransposon
レープ→ナタネ
レプテンき　レプテン期, 細糸期【減数分裂】　leptotene stage
レポーターいでんし　レポーター遺伝子　reporter gene
レモン　lemon, *Citrus limon* (L.) Burm. f.
レモングラス　lemongrass, 1) *Cymbopogon flexuosus* (Nees ex Steud.) Wats. (東インドレモングラス) 2) *C. citratus* (D.C. ex Nees) Stapf (西インドレモングラス)
レールそうこうしきてきさいき　レール走行式摘採機【チャ】　rail-tracking tea plucker
れん　連【順位統計量】　run
レンゲソウ(蓮華草, 紫雲英)　genge, Chinese milkvetch, *Astragalus sinicus* L.
レンコン(蓮根)→ハス
れんさ　連鎖　linkage
れんさく　連作　continuous cropping,

sequential cropping
れんさくしょうがい　連作障害　injury by continuous cropping
れんさぐん　連鎖群　linkage group
レンジナ　Rendzina
レンズマメ→ヒラマメ
れんぞくせんばつ　連続選抜　continuous selection
れんぞくぶんぷ　連続分布　continuous distribution
れんぞくへんい　連続変異　continuous variation
れんぞくへんすう　連続変数　continuous variable
れんぞくへんりょう　連続変量　continuous variate
レンブ　wax apple, Java apple, *Syzygium samarangense* (Blume) Merr. et Perry
れんぼしかんそう　連干し乾燥【タバコ】　prime and curing
れんらくいかんそく　連絡維管束　commissural vascular bundle

[ろ]

ロイシン　leucine (Leu)
ろう(蠟)　wax
ろうか　老化　senescence
ろうかじゅふん　老花受粉　old flower pollination
ろうきゅうかすいでん　老朽化水田　degraded paddy field
ろうすい　漏水　leakage, seepage
ろうすいでん　漏水田　water-leaking paddy field, high permeable paddy field
ろうすいぼうし　漏水防止　water leak prevention
ろうと(漏斗)　filter funnel
ろうどうしゅうやくてき　労働集約的　labor-intensive
ろうどうせいさんせい　労働生産性　labor productivity
ろうどうせつやくてき　労働節約的, 省力的　labor-saving
ろうどうひ　労働費　labor expense, labor cost
ろうりょうさくもつ　ろう(蠟)料作物　wax crop
ろえき　ろ(濾)液　filtrate
ろか　ろ(濾)過　filtration
ろくじょうオオムギ　六条オオムギ　six-rowed barley, *Hordeum vulgare* L.
ろくじょうやせいオオムギ　六条野生オオムギ　six-rowed wild barley, *Hordeum agriocrithon* Åberg
ろくたんとう　六炭糖, ヘキソース　hexose
ろくばいせい　六倍性　hexaploidy
ろくばいたい　六倍体　hexaploid
ろし　ろ(濾)紙　filter paper
ろじ　露地　open field, outdoors
ロシアタンポポ→ゴムタンポポ
ロシアワイルドライ　Russian wildrye, *Psathyrostachys juncea* (Fisch.) Nevski (= *Elymus junceus* Fisch.)
ろじさいばい　露地栽培　open culture, open-field culture, outdoor culture
ロジスティックきょくせん　ロジスティック曲線　logistic curve
ろじなえどこ　露地苗床　open-field nursery
ローズグラス　Rhodes grass, *Chloris gayana* Kunth
ローズマリー　rosemary, *Rosmarinus officinalis* L.
ロゼット　rosette
ロゼットじょうの　ロゼット状の　rosulate
ロゼル　roselle, *Hibiscus sabdariffa* L.
ロゾク(蘆粟)→サトウモロコシ
ロータリこう　ロータリ耕　rotary tilling
ロータリこううんき　ロータリ耕うん

(耘)機 rotary tiller
ロッグウッド logwood, *Haematoxylon campechianum* L.
ロックウール rock wool
ローディング loading
ろてんおんど 露点温度 dew-point temperature
ろてんけい 露点計 dew-point hygrometer
ロビッチじきにっしゃけい ロビッチ自記日射計 Robitzsch bimetallic actinograph
ロビンソンふうそくけい ロビンソン風速計 Robinson anemometer
ロブスタコーヒー[ノキ] robusta coffee, Congo coffee, *Coffea robusta* Linden (= *C. canephora* Pierr. ex Froeh.)
ロボットシステム robot system
ローマカミツレ noble camomile, chamomile, *Anthemis nobilis* L.
ローラージン【ワタ】 roller gin

[わ]

わいか わい(矮)化 dwarfing, stunting
わいかざい わい(矮)化剤, 成長抑制剤, 成長抑制物質 growth retardant
わいかびょう わい(矮)化病【イネ】 waika
ワイじかん Y字管 Y-shaped tubing connector
わいせい わい(矮)性 dwarfness, dwarfism
わいせい[の] わい(矮)性[の] dwarf
わいせいだいぎ わい(矮)性台木 dwarfing rootstock
わいせいへんいたい わい(矮)性変異体 dwarf mutants
ワイブルぶんぷ ワイブル分布 Weibull distribution

わかがえり 若返り rejuvenescence, rejuvenation
わかぎ 若木 sapling, young tree
ワカメ wakame seaweed, *Undaria pinnatifida* (Harvey) Suringar
ワキシーコーン waxy corn, *Zea mays* L. var. *amylosaccharata* Sturt.
わきめ わき芽【タバコ】 sucker
わきめよくせい わき芽抑制【タバコ】 sucker control
わきめよくせいざい わき芽抑制剤【タバコ】 suckercide
わく(枠), 方形区, コドラート quadrat
わぐされびょう 輪腐病【ジャガイモ】 ring rot
わくしけん 枠試験 frame test, frame experiment
わくづみ 枠摘み【チャ】 quadrate plucking
ワグナーポット Wagner pot
わごうせい 和合性 compatibility
ワサビ(山葵) wasabi, *Eutrema japonica* (Miq.) Koidz. (= *E. wasabi* Maxim.)
ワサビダイコン(山葵大根) horseradish, *Armoracia rusticana* Gaertn., Mey. et Scherb. (= *Cochlearia armoracia* L.)
わすき(わり) 和犂 Japanese plow
ワセビエ jungle rice, *Echinochloa colonum* (L.) Link
わせひんしゅ 早生品種 early [maturing] variety
ワタ(棉) cotton, *Gossypium* spp.
わたくり 綿繰り ginning
わたくりき 綿繰り機 gin, cotton gin
わたくりぶあい 綿繰り歩合⇒繰綿歩合(そうめんぶあい) ginning percentage
わめい 和名 Japanese name
わら(藁) straw, stover
わらじゅう わら(藁)重 straw weight
わらだて わら(藁)立て straw

windbreak
わらばい　わら(藁)灰　straw ash
ワラビ(蕨)　eastern brackenfern, bracken, *Pteridium aquilinum* (L.) Kuhn var. *latiusculum* (Desv.) Underw. ex A. Heller
わり(わすき)　和犂　Japanese plow
わりこみせいちょう　割込み成長, 侵入成長　intrusive growth
わりつぎ　割り接き　cleft grafting
わりつけ　割付け, 配置【試験区】layout

ワーリングブレンダー　Waring blender
ワルナスビ　horse nettle, *Solanum carolinense* L.
ワールブルグけんあつけい　ワールブルグ検圧計　Warburg's manometer
ワールブルグこうか　ワールブルグ効果　Warburg effect
われ　割れ【ソバ】cracked groats
われきずもみ　割れ傷籾　hull-cracked paddy
われもみ　割れ籾　split-hull paddy
ワングル→カンエンガヤツリ

英和の部

English – Japanese

[A]

A (adenine) アデニン
ABA (abscisic acid) アブシジン酸
A-D (analog-digital) conversion A-D 変換
abaca, Manila hemp, *Musa textilis* Née マニラアサ (マニラ麻)
abaxial 背軸 [側] の【植物】
Abelmoschus esculentus (L.) Moench (= *Hibiscus esculentus* L.), okura, lady's fingers オクラ
Abelmoschus glutino-textilis Kagawa ノリアサ (糊麻)
Abelmoschus manihot Medik. (= *Hibiscus manihot* L.), sunset hibiscus トロロアオイ (黄蜀葵)
Abies balsamea (L.) Mill., balsam fir バルサムモミ
abnormal early ripening 枯熟れ (かれうれ)
abnormal occurrence 異常発生
abnormal value 異常値
abnormal weather, unusual weather 異常気象
abortive grain, empty grain しいな (粃, 秕)
abortive pollen 退化花粉
aboveground part, top, aerial part 地上部
abscisic acid (ABA) アブシジン酸
abscissa 横軸
abscission 器官脱離
abscission layer 離層
absidia diseased rice, "Fuke" rice ふけ米
absolute humidity 絶対湿度
absolute temperature 絶対温度
absolute type 絶対値型【ガス分析】
absorbance 吸光度
absorbed radiation 吸収日射量
absorbed water 吸着水
absorbent cotton 脱脂綿
absorbing root, sucking root, feeder root 吸収根
absorptance 吸収率【物理】
absorption, uptake 吸収
absorption spectrum 吸収スペクトル
absorptivity 吸収率【物理】
abundant sunshine 多照
Abutilon avicennae Gaertn., China jute, Indian mallow ボウマ (莔麻)
Abyssinian banana, *Ensete ventricosum* (Welw.) Cheesman アビシニアバショウ
Acacia catechu (L. f.) Willd., catechu, cutch, black catechu アセンヤクノキ (阿仙薬樹)
Acacia farnesiana (L.) Willd., cassie, sweet acacia キンゴウカン (金合歓)
Acacia senegal Willd., gum arabic, gum Senegal アラビアゴム
Acalypha australis L., threeseeded copperleaf エノキグサ
acaricide 殺ダニ剤
accelerated generation advancement 世代促進
accelerator 促進剤
acceptability, palatability 嗜好性
acceptable daily intake (ADI) 一日摂取許容量
accessory bud 副芽
accessory pigment 補助色素
acclimation 順化
acclimatization 順化
accumulated temperature, cumulative temperature 積算温度
Acer saccharum Marsh., sugar maple サトウカエデ (砂糖楓)
acethylene reduction method アセチレン還元法
acetic acid 酢酸
acetocarmine アセトカーミン
acetone アセトン
acetosyringone アセトシリンゴン

achene そう(痩)果
achira, edible canna, purple arrowroot, Queensland arrowroot, *Canna edulis* Ker-Gawl. 食用カンナ
Achras zapota L. (= *Manilkara zapota* (L.) P. Royen), sapodilla, naseberry サポジラ
Achyranthes bidentata Blume var. *bidentata* (= *A. japonica* (Miq.) Nakai) イノコズチ
Achyranthes bidentata Blume var. *tomentosa* (Honda) H. Hara (= *A. fauriei* Lév. et Van.) ヒナタイノコズチ
acid 酸
acid detergent fiber (ADF) 酸性デタージェント繊維
acid equivalent 酸当量
acid fertilizer 酸性肥料
acid growth 酸成長
acid rain 酸性雨
acid soil 酸性土壌
acid tolerance 耐酸性
acid value 酸価【油脂】
acidic amino acid 酸性アミノ酸
acidity 酸度
acoustic emissin method AE法
acropetal 求頂的
actin アクチン
actin filament アクチンフィラメント
actinostele 放射維管束
action spectrum 作用スペクトル
activated sludge 活性汚泥
activation 活性化
activation energy 活性化エネルギー
activator 活性化剤, ふ(賦)活剤
active absorption 積極的吸収
active alumina 活性アルミナ
active carbon 活性炭
active charcoal 活性炭
active ingredient 有効成分
active sensor 能動的センサー
active-tillering stage 分げつ盛期

active transport 能動輸送
actual yield 実収[量]
Adansonia digitata L., baobab, monkey-bread tree バオバブ
adaptability 適応性
adaptability for heavy manuring 耐肥性
adaptation 適応
adaptive character 適応形質
adaptive strategy 適応戦略
adaxial 向軸[側]の【植物】
addition 添加
additive 1)添加剤, 添加物 2)相加的
additive effect 相加効果
additive genetic effect 相加[的]遺伝効果
adenine (A) アデニン
adenosine diphosphate (ADP) アデノシン二リン酸
adenosine triphosphate (ATP) アデノシン三リン酸
adhering water 付着水
adjuvant 補助剤【農薬】
ADP (adenosine diphosphate) アデノシン二リン酸
adsorption 吸着
adsorption chromatography 吸着クロマトグラフィー
adult 成虫
adult form 成体形
adult tree 成木
advanced generation 後期世代
adventitious bud 不定芽
adventitious root 不定根
adverse selection, reverse selection 逆選抜
adzuki bean, azuki bean, small red bean, *Vigna angularis* (Willd.) Ohwi et Ohashi (= *Phaseolus angularis* L.) アズキ(小豆)
Aegilops speltoides Tausch. クサビコムギ
Aegilops squarrosa L. タルホコムギ
aeration 通気

aerenchyma 通気組織
aerial application 空中散布
aerial part, top, aboveground part 地上部
aerial photographs 空中写真
aerial root 気根
aerial tuber, bulbil むかご(零余子)
aerial yam, *Dioscorea burbifera* L. カシュウイモ
aerobic 好気的
aerobic condition 好気的の条件
aerobic respiration 有気呼吸, 好気的呼吸, 酸素呼吸
aeroponics, mist culture 噴霧耕
aerosol エアロゾル
Aeschynomene indica L., Indian jointvetch クサネム(草合歓)
affinity 親和性
affinity of substrate 基質親和性
African millet, finger millet, *Eleusine coracana* (L.) Gaertn. シコクビエ(龍爪稷)
African oil palm, oil palm, *Elaeis guineensis* Jacq. アブラヤシ(油椰子)
African rice, *Oryza glaberrima* Steud. グラベリマイネ
aftereffect 後作用(こうさよう)
aftermath 再生草【牧草】
afterripening 後熟(こうじゅく)
agar medium 寒天培地
Agaricus bisporus Sing., common mushroom マッシュルーム
agarose アガロース
Agave americana L., century plant アオノリュウゼツラン
Agave americana L. var. *variegata* Nichols., century plant リュウゼツラン(竜舌蘭)
Agave cantala Roxb., cantala カンタラアサ(カンタラ麻)
Agave fourcroydes Lem., henequen ヘネケン

Agave sisalana Perr. ex Engelm., sisal サイザル
Agave spp., maguey マゲー
age 齢, エイジ
age of the moon, moon's age 月齢
aged seed 加齢種子, 古種子
ag[e]ing 1) 加齢, エイジング 2) 熟成【タバコ】
Ageratum conyzoides L., tropic ageratum, white-weed カッコウアザミ
aggregate culture, substrate culture, solid medium culture 固形培地耕
aggregate fruit, multiple fruit, syncarp 集合果, 多花果
aggregated structure 団粒構造
agribusiness アグリビジネス
agricultural biology, agrobiology 農業生物学
agricultural calendar 農事暦
agricultural chemicals, pesticide 農薬
agricultural chemistry 農芸化学
agricultural cooperative 農業協同組合(農協)
agricultural economics 農業経済学
agricultural engineering 農業工学
agricultural expenditure 農業経営費
agricultural experiment station 農業試験場
agricultural extension office 農業改良普及所
agricultural extension service 農業改良普及事業
agricultural facilities, agricultural structures 農業施設
agricultural implements, farm implements 農具
agricultural income 農業所得
agricultural information 農業情報
agricultural land 農地
agricultural location 農業立地
agricultural machinery 農業機械[学]
agricultural meteorology,

agrometeorology　農業気象学
agricultural products　農産物
agricultural sciences, agronomy　農学
Agricultural Structure Improvement Project　農業構造改善事業
agricultural structures, agricultural facilities　農業施設
agricultural technology　農業技術
agriculture　農業, 農耕
Agriculture, Forestry, and Fisheries Research Council (AFFRC)　農林水産技術会議
Agrobacterium rhizogenes　アグロバクテリウム・リゾゲネス【細菌】
Agrobacterium tumefacience　アグロバクテリウム・ツメファシエンス【細菌】
agrobiology, agricultural biology　農業生物学
agroforestry　アグロフォレストリー
agrometeorology, agricultural meteorology　農業気象学
agronomic character (trait)　農業形質
agronomy　1) 農学　2) 作物栽培学
Agropyron cristatum (L.) Gaertn., crested wheatgrass　クレステッド・ホィートグラス
Agropyron intermedium Beauv., *A. trichophorum* (Link) K. Richt. および *A. pulcherrimum* Grossh., intermediate wheatgrass　インターミーディエイト・ホィートグラス
Agropyron pauciflorum Hitchc. (= *A. trachycautum* Link), slender wheatgrass　スレンダーホィートグラス
Agropyron repens (L.) P. Beauv., quackgrass　シバムギ
Agropyron trachycautum Link (= *A. pauciflorum* Hitchc.), slender wheatgrass　スレンダーホィートグラス
Agropyron tsukushiense (Honda) Ohwi var. *transiens* (Hack.) Ohwi　カモジグサ
Agrostis capillaris L. (= *A. tenuis* Sibth.), colonial bentgrass　コロニアルベントグラス
Agrostis clavata Trin. ssp. *matsumurae* (Hack) Tateoka　ヌカボ
Agrostis gigantea Roth, redtop　レッドトップ
Agrostis stolonifera L., creeping bentgrass　クリーピングベントグラス
Agrostis tenuis Sibth. (= *A. capillaris* L.), colonial bentgrass　コロニアルベントグラス
air-cured tobacco　空気乾燥種【タバコ】
air-curing　空気乾燥【タバコ】
air-dry soil　風乾土
air-dry weight　風乾重
air drying　風乾
air-drying effect on ammonification　乾土効果
air permeability　通気性
air pollutant　大気汚染物質
air pollution　大気汚染
air temperature　気温
"akiochi"　秋落ち
"akiochi" paddy field　秋落水田
Akisame, autumnal rain　秋雨
Al (aluminum)　アルミニウム
alanine (Ala)　アラニン
albedo　アルベド
albino　アルビノ
albumen　胚乳
albumin　アルブミン
albuminous seed　有胚乳種子
alcohol　アルコール
alcohol fermentation　アルコール発酵
alcohol thermometer　アルコール温度計
alcoholic crop　アルコール[料]作物
Aleurites cordata (Thunb.) R. Br. ex Steud., tung, Japanese tung-oil tree　アブラギリ(油桐)
Aleurites fordii Hemsl., tung tree　シナ

アブラギリ (支那油桐)
Aleurites moluccana (L.) Willd., candlenut　ククイノキ
Aleurites montana (Lour.) E. H. Wilson, tung　カントンアブラギリ (広東油桐)
aleurone grain　糊粉粒
aleurone layer　糊粉層
alfalfa, lucerne, *Medicago sativa* L. 【*M.* × *media* Pers. を含む】　アルファルファ
algae (*sing.* alga)　藻類
algorithm　アルゴリズム
Alisma canaliculatum A. Br. et Bouch ex Sam., water plantain　ヘラオモダカ
Alisma plantago-aquatica L. var. *orientale* Sam., oriental water plantain　サジオモダカ
alkali soil　アルカリ土壌
alkali solubility　アルカリ崩壊度【イネ】
alkaline soil　アルカリ性土壌
alkaloid　アルカロイド
alkari tolerance　耐アルカリ性
all-or-none trait　しつゆうしつむ (悉有悉無) 形質
allele　対立遺伝子
allelomorph　対立形質
allelopathy　アレロパシー, 他感作用
alley cropping　灌木間作, アレイ作付け
allied species, related species　近縁種
alligator pear, avocado, *Persea americana* Mill.　アボカド
Allium cepa L., onion　タマネギ (玉葱)
Allium chinense G. Don, rakkyo, Baker's garlic　ラッキョウ (辣韮)
Allium fistulosum L., Welsh onion　ネギ (葱)
Allium sativum L., garlic　ニンニク (葫, 大蒜)
Allium tuberosum Rottler ex Spreng., Chinese chive　ニラ (韮)
allogamous plant　他殖性植物
allogamy, outcrossing　他殖, 他家生殖
allometric coefficient　相対成長係数

allometry　アロメトリー
allopatric　異所性の
allopatry　異所性
allophane　アロフェン
alloploid　異質倍数体
allopolyploid　異質倍数体
allosteric effect　アロステリック効果
allowance　許容度
allspice, pimento, *Pimenta dioica* (L.) Merr. (= *P. officinalis* Lindl.)　オールスパイス
alluvial fan　扇状地
alluvial land　沖積地
alluvial soil　沖積土壌
Alluvial soil　沖積土
alluvium　沖積層
almond, *Prunus amygdalus* Batsch　アーモンド (扁桃)
Alocasia macrorrhiza (L.) Schott, giant taro　インドクワズイモ
Alocasia odora (Lodd.) Spach　クワズイモ
Alopecurus aequalis Sobol. var. *amurensis* (Komar.) Ohwi, water foxtail　スズメノテッポウ
Alopecurus pratensis L., meadow foxtail　メドーフォックステール
α-naphthylamine　α-ナフチルアミン
alsike clover, *Trifolium hybridum* L.　アルサイククローバ
Alternanthera sessilis (L.) R. Br. ex Roem. et Schult., sessile joy-weed　ツルノゲイトウ
alternate　互生の
alternate husbandry, convertible husbandry, ley farming　穀草式農法
alternate [year] bearing, biennial bearing　隔年結果
alternating cropping　交互作
alternation of generation　世代交代, 世代交番
alternation of nuclear phases　核相交代

alternative hypothesis 対立仮説
altitude, elevation 高度
aluminum (Al) アルミニウム
aman アマン【イネの生態型】
Amaranthus caudatus L. (= *A. edulis* Spegazzini), grain amaranth センニンコク
Amaranthus hypochondriacus L., grain amaranth ハンスイヒユ (繁穂ヒユ)
Amaranthus lividus L., livid amaranth イヌビユ
Amaranthus patulus Bert., spleen amaranth ホソアオゲイトウ
Amaranthus spinosus L., spiny amaranth ハリビユ
Amaranthus viridis L., slender amaranth, green amaranth アオビユ
ambari hemp, kenaf, *Hibiscus cannabinus* L. ケナフ
Ambrosia artemisiifolia L. var. *elatior* (L.) Desc., common ragweed ブタクサ
Ambrosia trifida L., giant ragweed オオブタクサ, クワモドキ
AMeDAS (Automated Meteorological Data Acquisition System) アメダス, 地域気象観測システム
American burnweed, *Erechtites hieracifolia* (L.) Raf. ex DC. ダンドボロギク
American sloughgrass, *Beckmannia syzigachne* (Steud.) Fernald カズノコグサ
amino acid アミノ酸
1-aminocyclopropane-1-carboxylic acid (ACC) 1-アミノシクロプロパン-1-カルボン酸
amitosis 無糸分裂
Ammannia coccinea Rottb., purple ammannia ホソバヒメミソハギ
Ammannia multiflora Roxb. ヒメミソハギ
Ammi visnaga (L.) Lam., tooth pick アンミ
ammonia plant アンモニア植物
ammonification アンモニア化成[作用]
ammonium chloride 塩化アンモニウム (塩安)
ammonium nitrate 硝酸アンモニウム (硝安)
ammonium nitrogen アンモニア態窒素
ammonium sulfate 硫酸アンモニウム (硫安)
Amorphophallus bulbifer Bl. ムカゴコンニャク
Amorphophallus konjac K. Koch, konjak, elephant foot コンニャク (蒟蒻)
Amorphophallus oncophyllus Prain ex Hook. f. ジャワムカゴコンニャク
Amorphophallus paeoniifolius (Dennst.) Nicolson (= *A. campanulatus* Bl.), elephant yam ゾウコンニャク
Amorphophallus variabilis Bl. イロガワリコンニャク
amount of fertilizer application 施肥量
amount of information 情報量
amount of insolation 日射量
amount of precipitation, precipitation 降水量
amount of rainfall 雨量
amount of solar radiation 日射量
amphidiploid 複二倍体
amphidiploidy 複二倍性
amphistomatous leaf 両面気孔葉
Amur cork-tree, *Phellodendron amurense* Rupr. キハダ
amylase アミラーゼ
amylogram アミログラム
amylographic characteristics アミログラム特性
amylopectin アミロペクチン
amylophyll, starch leaf デンプン葉
amyloplast アミロプラスト
amylose アミロース
anabolism, assimilation 同化[作用]

Anacardium occidentale L., cashew, cashew-nut tree　カシュー
anaerobic condition, anoxia　嫌気的条件
anaerobic respiration　嫌気的呼吸, 無気呼吸, 無酸素呼吸
anaerobic seed　嫌気性種子
analog　アナログ【情報】
analog-digital (A-D) conversion　A-D変換
analog-to-digital converter　A-D変換器
analogous organ　相似器官
analogue　アナローグ, 類似体
analogy　相似
analysis of covariance　共分散分析
analysis of variance (ANOVA)　分散分析
Ananas comosus Merr., pineapple　パイナップル
anaphase　後期【細胞分裂】
anatomy　解剖学
ancestral form　祖先型
ancestral species　祖先種
ancymidol　アンシミドール
Andosol　黒ぼく土
Andropogon gerardii Vitman var. *gerardii*, big bluestem　ビッグブルーステム
Andropogon virginicus L., broomsedge　メリケンカルカヤ
Aneilema keisak Hassk. (= *Murdannia keisak* (Hassk.) Hand.-Mazz.), marsh dayflower　イボクサ
Anemarrhena asphodeloides Bunge　ハナスゲ
anemometer　風速計
anemophily, wind pollination　風媒
anemoscope, wind vane　風向計
Anethum graveolens L., dill　イノンド, ディル
aneuploidy, heteroploidy　異数性
Angelica acutiloba (Sieb. et Zucc.) Kitagawa　トウキ(当帰)
Angelica acutiloba Kitagawa var. *sugiyamae* Hikino　ホッカイトウキ(北海当帰)
Angelica dahurica (Fisch.) Benth. et Hook. f.　ヨロイグサ
angiosperm　被子植物
Aniba rosiodora Ducke, pau rosa　パウローサ
Aniba spp., bois de rose　ボアドローズ
aniline blue　アニリンブルー
animal husbandry, livestock industry　畜産
animal industry, livestock industry　畜産
animal production　家畜生産, 動物生産
anion　陰イオン
anion exchange capacity (AEC)　陰イオン交換容量
anise, *Pimpinella anisum* L.　アニス
anisophylly　不等葉性
annatto tree, *Bixa orellana* L.　ベニノキ(紅木)
Annona cherimola Mill., cherimoya　チェリモヤ
Annona muricata L., sour-sop, sour apple　トゲバンレイシ
Annona squamosa L., sugar apple, sweet sop, custard apple　バンレイシ
annual　1)一年生[の] 2)一年草
annual beard grass, rabbit's-foot polypogon, *Polypogon mouspeliensis* (L.) Desf.　ハマヒエガエリ
annual bluegrass, *Poa annua* L.　スズメノカタビラ
annual change, annual variation　年変化
annual crop　一年生作物
annual fleabane, *Erigeron annuus* (L.) Pers. (= *Stenactis annuus* (L.) Cass.)　ヒメジョオン(姫女苑)
Annual Report on Japanese Agriculture　農業白書
annual ring　年輪
annual sedge, *Cyperus compressus* L.　クグガヤツリ

annual variation, annual change 年変化
annual weed 一年生雑草
ANOVA (analysis of variance) 分散分析
anoxia, anaerobic condition 嫌気的条件
antagonism 拮抗作用
antenna pigment アンテナ色素
Anthemis cotula L., mayweed chamomile カミツレモドキ
Anthemis nobilis L., noble camomile, chamomile ローマカミツレ
anther 葯
anther culture 葯培養
anther dehiscence 開葯
anther loculus (*pl.* loculi) 葯室
anthesis, flowering 開花
anthocyan アントシアン
Anthoxanthum odoratum L., sweet vernalgrass スイートバーナルグラス
anthracnose 炭そ(疽)病
antiauxin アンチオーキシン, 抗オーキシン
antibiotics 抗生物質
antibody 抗体
anticlinal division 垂層分裂
antigen 抗原
antipodal cell 反足細胞
antipode 反足細胞
antisense DNA アンチセンスDNA
antisense RNA アンチセンスRNA
antiserum 抗血清
antitranspirant 蒸散抑制剤
anu, jicamas, *Tropaeolum tuberosum* Ruiz. et Pav. アヌウ
aperture 開度【気孔】
apetalous flower 無弁花
apex, tip 頂部
aphid アブラムシ, アリマキ
apical bud, terminal bud 頂芽
apical cell 頂端細胞
apical dominance 頂芽優性, 頂部優性
apical growth 頂端成長
apical meristem 頂端分裂組織

Apios americana Medik., potato bean, groundnut アメリカホドイモ
Apocynum sibiricum Jacq., kendyr, turka ケンディル, ツルカ
apogamy 無胚生殖
apomixis 無配偶生殖, アポミクシス
apoplast アポプラスト
apoptosis アポトーシス
apparent photosynthesis みかけの光合成
appearance 外観
apple, *Malus pumila* Mill. リンゴ(林檎)
application, spray 散布
approach grafting, inarching 寄せ接ぎ
apricot, *Prunus armeniaca* L. アンズ(杏)
aquatic plant, hydrophyte 水生植物
aquatic weed 水生雑草
Arabian coffee, *Coffea arabica* L. アラビアコーヒー[ノキ]
Arabian jasmine, *Jasminum sambac* Aiton マツリカ(茉莉花)
arable land, cultivated land 耕地
arable land rate 耕地率
arable land weed 耕地雑草
arable soil, cultivated soil, topsoil 耕地土壌, 耕土
aracacha, *Arracacia xanthorrhiza* Bancr. (= *A. esculenta* DC.) イモゼリ
Arachis hypogaea L., groundnut, peanut ラッカセイ(落花生)
Aralia cordata Thunb., udo ウド(独活)
aramina, *Urena lobata* L. var. *lobata* オオバボンテンカ
arbor 木本
arbor crop 木本作物
arbor tree, tree 高木
arboreous, woody 木本[性]の
arboreous cotton 木棉(もくめん)
arbuscular mycorrhiza アーバスキュラー菌根
arc-shaped bush formation 弧状仕立て【チャ】

archesporial cell　胞原細胞
arcsine transformation　逆正弦変換
arctic　寒帯の
Arctium lappa L., edible burdock　ゴボウ(牛蒡)
Areca catechu L., betel palm　ビンロウジュ(檳榔樹)
Arenaria serpyllifolia L., thymeleaf sandwort　ノミノツヅリ
Arenga pinnata (Kuntze) Merr., gomuti palm　サトウヤシ(砂糖椰子)
arginine (Arg)　アルギニン
arid climate, dry climate　乾燥気候
arid land　乾燥地
arid region　乾燥地帯
arid zone　乾燥地帯
arista, awn　ぼう(芒)
aristate, awned, bearded　有芒の
arithmetic mean, average　算術平均
arm　鏡柱【顕微鏡】
arm palisade cell　有腕柵状組織細胞
Armoracia rusticana Gaertn., Mey. et Scherb. (= *Cochlearia armoracia* L.), horseradish　ワサビダイコン(山葵大根)
aroma, scent, fragrance　香り, 香気
aromatic crop　香料作物
aromatic rice, scented rice　香り米
Arracacia xanthorrhiza Bancr. (= *A. esculenta* DC.), aracacha　イモゼリ
Arrhenius plot　アレニウスプロット
arrowhead　1) *Sagittaria trifolia* L.　オモダカ　2) *S. trifolia* L. var. *edulis* (Sieb.) Ohwi　クワイ
arrowroot, West Indian arrowroot, *Maranta arundinacea* L.　アロールート
Arrhenatherum elatius (L.) K. Presl, tall oatgrass　トールオートグラス
Artemisia dracunculus L., tarragon, estragon　タラゴン
Artemisia japonica Thunb., western mugwort　オトコヨモギ
Artemisia kurramensis Quazilbash, kurram santonica　クラムヨモギ
Artemisia maritima L., sea wormwood　ミブヨモギ
Artemisia princeps Pamp., Japanese mugwort　ヨモギ(蓬)
Arthraxon hispidus (Thunb.) Makino, jointhead arthraxon　コブナグサ
artificial bed soil, commercial bed soil　人工床土
artificial climate room, climatron　人工気象室
artificial crossing　人為交配, 人工交配
artificial drying　人工乾燥
artificial [farmyard] manure　速成堆肥
artificial grassland, sown grassland, tame pasture　人工草地
artificial illumination　人工照明
artificial inoculation　人工接種
artificial intelligence (AI)　人工知能
artificial light　人工光
artificial light source　人工光源
artificial lighting　人工照明
artificial mutation　人為突然変異
artificial pasture, sown pasture, tame pasture　人工草地
artificial pollination, hand pollination　人工授粉
artificial selection　人為選択, 人為選抜
Artocarpus communis Forst., bread-fruit【種なし】, bread-nut【種あり】　パンノキ
Artocarpus heterophyllus Lam., jackfruit　パラミツ(波羅蜜)
arubovirus　アルボウイルス
Arundinaria pygmaea Mitford var. *glabra* Ohwi (= *Pleioblastus chino* (Franch. et Savat.) Makino var. *viridis* (Makino) S. Suzuki), dwarf bamboo　ネザサ
ascending habit　頂上性
ascorbic acid (AsA)　アスコルビン酸

aseptic culture, sterile culture　無菌培養
asexual propagation　無性繁殖
asexual reproduction　無性生殖
ash　灰分 (かいぶん)
ashing method　灰化法 (かいかほう)
Asian Development Bank (ADB)　アジア開発銀行
Asian Institute of Technology (AIT)　アジア工科大学
Asian Productivity Organization (APO)　アジア生産性機構
Asian Vegetable Research and Development Center (AVRDC)　アジア蔬菜研究開発センター
Asiatic cotton, *Gossypium arboreum* L. および *G. herbaceum* L.　アジアメン (アジア棉)
Asiatic dayflower, dayflower, *Commelina communis* L.　ツユクサ (露草)
Asiatic hawk's-beard, *Youngia japonica* (L.) DC.　オニタビラコ
Asiatic plantain, *Plantago asiatica* L. (= *P. major* L. var. *asiatica* (L.) Dec.)　オオバコ
asparagine (Asn)　アスパラギン
asparagus bean, *Vigna unguiculata* (L.) Walp. var. *sesquipedalis* (L.) H. Ohashi　ジュウロクササゲ (十六豆豆)
asparagus pea, goa bean, winged bean, four-angled bean, *Psophocarpus tetragonolobus* (L.) DC.　シカクマメ
aspartic acid (Asp)　アスパラギン酸
Assam rubber, Indian rubber, *Ficus elastica* Roxb.　インドゴム
assay plant, test plant　検定植物
assimilate　同化産物
assimilation, anabolism　同化 [作用]
assimilation organ　同化器官
assimilation product　同化産物
assimilation starch　同化デンプン
assimilation tissue　同化組織
association, community　群集

assumption　前提条件
aster　星状体
Aster subulatus Michx., saltmarsh aster　ホウキギク
Astragalus gummifer Labill., tragacanth milkvetch　トラガカントゴムノキ
Astragalus sinicus L., genge, Chinese milkvetch　レンゲソウ (蓮華草, 紫雲英)
astringency　渋み
asymmetry　非対称
asynapsis　不接合
atactostele　不整 (斉) 中心柱
atavism, reversion　先祖返り
atmometer, evaporimeter, evaporation pan　蒸発計
atmosphere　大気
atmospheric correction　大気補正
atmospheric pressure　気圧
atomic absorption analysis　原子吸光分析
atomic number　原子番号
atomic weight　原子量
ATP (adenosine triphosphate)　アデノシン三リン酸
Atractylodes lancea (Thunb.) DC.　オケラ
Atropa belladonna L., belladonna, deadly nightshade　ベラドンナ
attached chromosome　付着染色体
attainable yield　到達可能収量
attractant　誘引剤
attribute　属性
attribute information　属性情報
auricle　葉耳 (ようじ), 小耳 (しょうじ)
aus　アウス【イネの生態型】
Australian dodder, *Cuscuta australis* R. Br.　マメダオシ
autecology, species ecology　種生態学
autoclave　オートクレーブ, 高圧滅菌器
autocorrelation　自己相関
autoecology　個生態学
autogamous plant, self-fertilizing plant

自殖性植物, 自家受精植物
autogamy　自家生殖
Automated Meteorological Data Acquisition System (AMeDAS)　地域気象観測システム, アメダス
automatic irrigation　自動灌水
automatic seeding, sowing by seeder　機械播き
automatic thresher, mechanical feeding thresher, self-feeding thresher　自動脱穀機
automatic watering　自動灌水
autopolyploidy　同質倍数性
autoradiography　オートラジオグラフィー
autoregressive process　自己回帰過程
autosome　常染色体
autosynapsis　同親接合, 同親対合【染色体】
autosyndesis　同親接合, 同親対合【染色体】
autotrophism　独立栄養
autotrophy　独立栄養
autumn (fall) crop　秋作[物]
autumn (fall) cropping　秋作
autumn (fall) planting　秋植え
autumn buckwheat, late-summer buckwheat　秋ソバ
autumn crocus, colchicum, meadow saffron, *Colchicum autumnale* L.　イヌサフラン
autumn ecotype, late-summer ecotype　秋型【ソバ】
autumn manuring, fall dressing　秋肥(あきごえ)
autumn plowing　秋耕
autumn shoot　秋芽【チャ】
autumn skiffing　秋整枝【チャ】
autumn sowing (seeding), fall sowing (seeding)　秋播き
autumn soybean, late-summer soybean　秋ダイズ
autumnal rain, Akisame　秋雨
auxin　オーキシン

auxin-binding protein　オーキシン結合タンパク質
available moisture　有効水[分]
available nutrient　可給態養分
available water　有効水[分]
Avena curvature test　アベナ屈曲試験法
Avena fatua L., wild oat　カラスムギ【≠エンバク】
Avena nuda L., naked oat　ハダカエンバク(裸燕麦)
Avena sativa L., oats【通常 *pl.*】エンバク(燕麦)【≠カラスムギ】
Avena straight-growth test　アベナ伸長試験法
average　1) 平均　2) 算術平均
average crop, normal crop　平年作
Averrhoa carambola L., carambola, star fruit　ゴレンシ(五斂子)
avocado, alligator pear, *Persea americana* Mill.　アボカド
AVRDC (Asian Vegetable Research and Development Center)　アジア蔬菜研究・開発センター
awn, arista　ぼう(芒)
awned, aristate, bearded　有芒の
awning　脱芒
awnless　無芒の
axillary bud　腋芽(えきが)
axis [of co-ordinates]　座標軸
Axonopus compressus (SW) P. Beauv., carpetgrass　カーペットグラス
Azolla japonica Franch. et Savat.　オオアカウキクサ
Azolla pinnata R. Br. ssp. *asiatica* R. M. K. Saunders et K. Fowler　アカウキクサ
azotobacter　アゾトバクター
aztec tobacco, rustica tobacco, *Nicotiana rustica* L.　ルスチカタバコ
azuki bean, adzuki bean, small red bean, *Vigna angularis* (Willd.) Ohwi et Ohashi (= *Phaseolus angularis* L.)　アズキ(小豆)

[B]

B (boron)　ホウ素
BA (benzyl adenine)　ベンジルアデニン
babassu, *Orbignya speciosa* (Mart.) B. Rodr. (= *O. martiana* B. Rodr.)　ババスヤシ (ババス椰子)
back mutation, reverse mutation　復帰突然変異, 逆突然変異
backcross method　戻し交雑法
backcross[ing]　戻し交雑
Bacopa rotundifolia (Michx.) Wettst., disc waterhyssop　ウキアゼナ
bacterial leaf blight　白葉枯病【イネ】
bacterial wilt　青枯病
bactericide　殺菌剤【細菌】
bacteroid　バクテロイド【根粒菌】
bacteriophage　バクテリオファージ
bad crop, poor harvest, poor crop　不作
bagasse　甘蔗搾粕 (カンショしぼりかす), バガス
bagging　袋掛け【果実】
bahiagrass, *Paspalum notatum* Flugge　バヒアグラス
Baical skullcap, *Scutellaria baicalensis* Georgi　コガネバナ
bai-u　梅雨
bakanae　馬鹿苗
Baker's garlic, rakkyo, *Allium chinense* G. Don　ラッキョウ (辣韮)
baking quality　製パン適性
balata, *Manilkara bidentata* (A. DC.) A. Chev.　バラタ
bale　俵
ball fertilizer　団子肥料, 固形肥料
ballast weed　脚荷雑草 (あしにざっそう)
balloonflower, Japanese bellflower, Chinese bellflower, *Platycodon grandiflorum* (Jacq.) A. DC.　キキョウ (桔梗)
balsam fir, *Abies balsamea* (L.) Mill.　バルサムモミ
balsam pear, bitter gourd, *Momordica charantia* L.　ニガウリ (苦瓜)
bambara bean, *Vigna subterranea* (L.) Verdc. (= *Voandzeia subterranea* (L.) Thouars)　バンバラマメ
bambara groundnut, *Vigna subterranea* (L.) Verdc. (= *Voandzeia subterranea* (L.) Thouars)　バンバラマメ
bamboo shoot　タケノコ
banana, *Musa* × *paradisiaca* L.　バナナ
banjhi shoot　出開き芽【チャ】
bank, mound　盛土
baobab, monkey-bread tree, *Adansonia digitata* L.　バオバブ
bare fallow　裸地休閑
bare ground　裸地
bare land　裸地
bareet grass, southern cutgrass, tiger's-tongue grass, *Leersia hexandra* Sw.　タイワンアシカキ
bark　1) 樹皮　2) 粗皮 (そひ)【アサ】
barley, *Hordeum vulgare* L.　オオムギ (大麦)
barley cropping (culture)　麦作 (むぎさく, ばくさく)【オオムギ】
barley straw　麦稈 (ばっかん)【オオムギ】
barley threshing　麦こ (扱) き, 麦打ち【オオムギ】
barn, stable　畜舎
barning　収納【干草, 穀物など】
barnyard grass　1) *Echinochloa crus-galli* (L.) Beauv. var *oryzicola* (Vasing.) Ohwi　タイヌビエ　2) *E. crus-galli* (L.) Beauv. var. *crus-galli*　イヌビエ　3) *E. crus-galli* (L.) P. Beauv. var. *praticola* Ohwi　ヒメイヌビエ　4) *E. crus-galli* (L.) P. Beauv. var. *formosensis* Ohwi　ヒメタイヌビエ　5) *E. crus-galli* (L.) Beauv., cockspur grass　ケイヌビエ
barnyard manure, stable manure,

farmyard manure　きゅう(厩)肥
barnyard millet, Japanese millet, *Echinochloa utilis* Ohwi et Yabuno　ヒエ(稗, 穆)
barren land, wasteland　荒廃地, 不毛地
barrens, wasteland　荒廃地, 不毛地
basal bristle　底刺(ていし)【オオムギ】
basal cover[age]　基底被度
basal application　基肥(もとごえ, きひ), 元肥(もとごえ)
basal dressing　基肥, 元肥
basal medium　基本培地
base　1) 塩基　2) 鏡脚【顕微鏡】
base saturation　塩基飽和度
base sequence　塩基配列【遺伝子】
basic dye　塩基性色素
basic fertilizer　塩基性肥料
basic number　基本数【染色体】
basic vegetative growth　基本栄養成長性
basikaryotype　基本核型
basil, sweet basil, *Ocimum basilicum* L.　バジル
basipetal　求基的
bast　じん(靱)皮
bast fiber　じん(靱)皮繊維
Baum's hydrometer　ボーメ比重計
bay laurel, laurel, *Laurus nobilis* L.　ゲッケイジュ(月桂樹)
Bayes' theorem　ベイズの定理
Bayesian estimation　ベイズ流推定
Bayesian inference　ベイズ流推論
beaker　ビーカー
bean　子実【マメ類】
bean curd, tofu　豆腐
bearded, awned, aristate　有芒の
bearing branch (shoot), fruit-bearing branch (shoot)　結果枝(けっかし)
bearing habit, fruiting habit　結果習性
bearing tiller, productive tiller　有効分げつ
Beckmannia syzigachne (Steud.) Fernald, American sloughgrass　カズノコグサ
bed　床
bed soil, nursery bed soil　床土
bedding, litter　敷料【畜産】
bedstraw, catchweed, *Galium spurium* L. var. *echinospermon* (Wallr.) Hayek　ヤエムグラ
beer brewing barley, malting barley　ビール麦
beet pulp　ビートパルプ
beet sugar　テンサイ(甜菜)糖
belladonna, deadly nightshade, *Atropa belladonna* L.　ベラドンナ
bench-grafting, indoor-grafting　揚げ接ぎ
bending, curvature　屈曲
bentonite　ベントナイト
benzyl adenine (BA)　ベンジルアデニン
bergamot [orange], *Citrus bergamia* Risso et Poit.　ベルガモット
Bermuda grass, *Cynodon dactylon* Flugge (= *C. dactylon* (L.) Pers.)　バーミューダグラス
Bernoulli distribution　ベルヌーイ分布
berry　液果, 多肉果, しょう(奬)果
Berseem clover, Egyptian clover, *Trifolium alexandrinum* L.　エジプシャンクローバ
β-glucuronidase (GUS)　β-グルクロニダーゼ
Beta vulgaris crassa L. var. *alba* DC., fodder beet, mangold, field beet　飼料用ビート
Beta vulgaris L. var. *rapa* Dumort., sugar beet　テンサイ(甜菜)
betacyanin　ベタシアニン
betel palm, *Areca catechu* L.　ビンロウジュ(檳榔樹)
between-group variance　群間分散
beverage　飲料
beverage crop　喫飲料作物
bicollateral vascular bundle　複並立維管束

Bidens biternata (Lour.) Merr. et Sherff. センダングサ

Bidens frondosa L., devil's beggarticks アメリカセンダングサ

Bidens tripartita L., bur beggarticks タウコギ

biennial 1) 二年生[の] 2) 二年草

biennial bearing, alternate [year] bearing 隔年結果

biennial crop 二年生作物

biennial plant 二年生植物

biennial weed 二年生雑草

bifacial leaf 両面葉

big bluestem, *Andropogon gerardii* Vitman var. *gerardii* ビッグブルーステム

big trefoil, *Lotus uliginosus* Schkuhr. ビッグトレフォイル

billion dollar grass, Indian barnyard millet, *Echinochloa frumentacea* (Roxb.) Link インドビエ

bimetallic thermometer バイメタル温度計

bimodal curve 二頂曲線

bimodal distribution 二峰分布

binarization 二値化

binary vector バイナリーベクター

binder 結合機, バインダ

binding protein 結合タンパク質

binding site 結合部位

binding, bundling 結束

binomial coefficient 二項係数

binomial distribution 二項分布

binomial nomenclature 二[命]名法

binomial theorem 二項定理

bioassay 生物検定, バイオアッセイ

biochemical oxygen demand (BOD) 生化学的酸素要求量

biochemistry 生化学

biocide, biopesticide, biotic pesticide 生物農薬

bioclimatology 生気候学

biodiversity 生物多様性

bioengineering 生物工学

biological clock 生物時計

biological containment 生物的封じ込め

biological diversity 生物多様性

biological engineering 生物工学

biological membrane, biomembrane 生体膜

biological [pest] control 生物的防除

biological weed control 生物的除草

biological yield 生物学的収量

biomass, standing crop バイオマス, 生物[体]量, 現存量

biomechanics 生物機械学

biomembrane, biological membrane 生体膜

biometeorology 生気象学

biometrics 計量生物学, 生物測定学

biometry 計量生物学, 生物測定学

bionics バイオニクス, 生体工学

biopesticide, biocide, biotic pesticide 生物農薬

biophysics 生物物理[学]

bioreactor バイオリアクター

bioremediation 生物的浄化, バイオレメディエーション

biorhythm 生物リズム, バイオリズム

biosensor バイオセンサー

biosphere 生物圏

biostatistics 生物統計学

biosynthesis 生合成

biotechnology バイオテクノロジー

biotic pesticide, biocide, biopesticide 生物農薬

biotron バイオトロン

biotype 同遺伝子型個体群, バイオタイプ

bird damage 鳥害

bird-proof net 防鳥網

birdsfoot trefoil, *Lotus corniculatus* L. var. *corniculatus* バーズフット・トレフォイル

bisexual flower, hermaphrodite flower 両性花

bitter gourd, balsam pear, *Momordica charantia* L.　ニガウリ (苦瓜)

bitterness　苦み

bivalent [chromosome]　二価染色体

Bixa orellana L., annatto tree　ベニノキ (紅木)

black bamboo, *Phyllostachys nigra* (Lodd. ex Loudon) Munro f. *henonis* (Mitord) Stapf ex Rendle　ハチク (淡竹)

black catechu, catechu, cutch, *Acacia catechu* (L. f.) Willd.　アセンヤクノキ (阿仙薬樹)

black gram, urd, black matpe, *Vigna mungo* (L.) Hepper (= *Phaseolus mungo* L.)　ケツルアズキ (毛蔓小豆), ブラックマッペ

black henbane, henbane, *Hyoscyamus niger* L.　ヒヨス

black matpe, black gram, urd, *Vigna mungo* (L.) Hepper (= *Phaseolus mungo* L.)　ケツルアズキ (毛蔓小豆), ブラックマッペ

black medic, *Medicago lupulina* L.　ブラックメディック

black mustard, *Brassica nigra* (L.) Koch　クロガラシ (黒芥子)

black nightshade, *Solanum nigrum* L.　イヌホオズキ

black rice　黒稲 (こくとう), 烏稲 (うとう), 黒米 (くろまい)

black rot　黒斑病【サツマイモ】

black rush, *Schoenoplectus lacustris* (L.) Palla ssp. *validus* (Vahl) T. Koyama (= *Scirpus lacustris* L.)　フトイ (太藺, 莞)

Black soil　黒色土

black tea　紅茶

blanching　軟白

blanching culture　軟白栽培

blast, neck rot　いもち (稲熱) 病【イネ】

bleached hemp stem　洗い麻 (あらいそ)

bleached kernel　退色粒【ムギ類】

bleached wheat　退色粒【ムギ類】

bleeding, exudation　出液【現象】

bleeding sap, exudate　出液【物質】

blended fertilizer, mixed fertilizer　配合肥料

blending　系統間交雑

blight　疫病【ソバ】

bloom, waxy bloom　果粉

blooming season, flowering time, flowering stage　開花期

blossom-end rot　尻腐れ

blowing of boll, opening of boll　開絮 (かいじょ)

blue-eyed grass, *Sisyrinchium atlanticum* Bicknell　ニワゼキショウ

blue grama, *Bouteloua gracilis* (Kunth) Lag. ex Steud.　ブルーグラマ

blue green algae　藍藻類, シアノバクテリア

blue lupine, *Lupinus angustifolius* L.　アオバナルーピン (青花ルーピン)

blue panicgrass, *Panicum antidotale* Retz.　ブルーパニックグラス

Blyxa ceratosperma Maxim.　スブタ

BOD (biochemical oxygen demand)　生化学的酸素要求量

body weight (BW)　体重

body weight gain, liveweight gain (LWG)　増体 [量]

Boehmeria nivea (L.) Gaud., ramie, China grass　チョマ (苧麻)

bog, moor　泥炭地

Bog soil, Peat soil　泥炭土

boiled oil　ボイル油

boiling characteristics of rice　炊飯特性

bois de rose, *Aniba* spp.　ボアドローズ

Bolboschoenus fluviatilis (Torr.) T. Koyama ssp. *yagara* (ohwi) T. Koyama (= *Scirpus yagara* Ohwi), river bulrush　ウキヤガラ

boll, capsule　さく (蒴)

boll shedding　落さく (蒴)

bolling　結さく (蒴)

bolting, flower stalk development　抽だい(苔)
Bombyx mori L., silkworm　カイコ(蚕), 家蚕(かさん)
bone meal　骨粉【肥料】
boot[ing] stage　穂ばら(孕)み期
Borassus flabellifer L., palmyra palm　オウギヤシ(扇椰子)
Bordeaux mixture　ボルドー液
border crop　周縁作物
border effect　周縁効果
border plant　縁植え(ふちうえ), 周縁植物
borer　せん(穿)孔性害虫
boro　ボロ【イネの生態型】
boron (B)　ホウ素
botany　植物学
Bothriospermum tenellum (Hornem.) Fisch. et Mey.　ハナイバナ
bottom grass, undergrowth　下繁草, 下草
bottom plow　ボトムプラウ
Bouin's fluid　ブアン液【固定液】
bound auxin　結合型オーキシン
bound type　結合型
bound water, combined water　結合水
boundary condition　境界条件
boundary layer　境界層
boundary layer resistance　境界層抵抗
Bouteloua curtipendula (Michx.) Torr., side-oats grama　サイドオートグラマ
Bouteloua gracilis (Kunth) Lag. ex Steud., blue grama　ブルーグラマ
bovine serum albumin (BSA)　牛血清アルブミン
bowstring hemp, sansevieria, *Sansevieria nilotica* Baker 他数種　サンセベリア
brabender unit　ブラベンダー単位
brace root, prop root　支根, 支持根, 支柱根
bracken, eastern brackenfern, *Pteridium aquilinum* (L.) Kuhn var. *latiusculum* (Desv.) Underw. ex A.Heller　ワラビ(蕨)
brackish water　汽水
bract　苞(ほう)
bract leaf　苞葉
bracteole　小苞[葉]
bractlet　小苞[葉]
bran　ふ(麩)
bran oil, rice oil　糠油(ぬかゆ)
branch　枝, 分枝【形態】
branch root　分枝根, lateral root　側根
branch spread　枝張り
branching　分枝【発育現象】
branching form　分枝型
branching type　分枝型
brand　銘柄
branded variety, costly registered variety, marketing standard variety　銘柄品種
Brassica alba (L.) Boiss. (= *Sinapis alba* L.), white mustard　シロガラシ(白芥子)
Brassica campestris L. (= *B. rapa* L. var. *campestris* (L.) Clapham, rape　在来種ナタネ
Brassica campestris L. および *B. napus* L., rape　ナタネ(菜種)
Brassica juncea (L.) Czern. et Coss., leaf mustard　カラシナ, brown mustard　セイヨウカラシナ
Brassica napus L., rape　洋種ナタネ
Brassica napus L. var. *napobrassica* (Mill.) Reichb., rutabaga, Swedish turnip, swede　ルタバガ
Brassica napus L. および *B. campestris* L., rape　ナタネ(菜種)
Brassica nigra (L.) Koch, black mustard　クロガラシ(黒芥子)
Brassica oleracea L. var. *capitata* L., cabbage　キャベツ
Brassica rapa L., turnip　カブ, 飼料カブ(蕪, 蕪菁)
Brassica rapa L. var. *amplexicaulis*

Tanaka et Ono (= *B. pekinensis* Rupr.), Chinese cabbage　ハクサイ(白菜)

Brassica rapa L. var. *campestris* (L.) Clapham (= *B. campestris* L.), rape　在来種ナタネ

brassinolide　ブラシノライド

brassinosteroid　ブラシノステロイド

Brazilian bois de rose　ブラジルボアドローズ→パウローサ

Brazilian elodea, *Egeria densa* Planch.　オオカナダモ

Brazilian wax palm, carnauba wax palm, *Copernicia prunifera* H. E. Moore (= *C. cerifera* Mart.)　カルナウバヤシ(カルナウバ椰子)

brazilwood, *Caesalpinia echinata* Lam.　ブラジルボク

bread-fruit, *Artocarpus communis* Forst.　パンノキ【種なし】

bread-nut, *Artocarpus communis* Forst.　パンノキ【種あり】

bread wheat, common wheat, *Triticum aestivum* L.　パンコムギ

breakage, fragmentation　切断【染色体】

breakdown　1) ブレークダウン【デンプン特性】　2) り(罹)病化【病理】

breaking resistance, resistance to breaking　挫折抵抗

Breea setosa (Bieb.) Kitam., creeping thistle　エゾノキツネアザミ

breeder　育種家

breeder['s] seed　育種家種子

breeder's stock, foundation seed　原原種

breeder's stock farm, foundation seed farm　原原種圃

breeding　1) 育種　2) 繁殖

breeding by separation　分離育種[法]

breeding field　育種圃[場]

breeding material　育種素材, 育種材料

breeding method　育種[方]法

breeding nursery　育種圃[場]

breeding objective　育種目標

breeding process　育種経過

breeding program　育種計画

breeding science, thremmatology　育種学

breeding station　育種試験地

breeding system, reproductive system, mode of reproduction　繁殖様式

brewers' rice, rice for sake brewery　酒米

brewing　醸造

brick tea　たん(磚)茶

bridge grafting　橋接ぎ

bristle　剛毛

Brix　糖度

Briza minor L., little quaking grass　ヒメコバンソウ

broad bean, *Vicia faba* L.　ソラマメ(蚕豆)

broad-leaved tree, broadleaf tree　広葉樹

broad-leaved weed, broadleaf weed　広葉雑草

broad ridge　広うね(畝)

broad sowing (seeding)　広幅播き

broad-spectrum resistance　広範囲抵抗性

broadcast　散播(さんぱ)

broadcast sowing (seeding)　散播(さんぱ)

broadcaster　散播機

broadcasting　全面散布

broadleaf dock, *Rumex obtusifolius* L.　エゾノギシギシ

broadleaf tree, broad-leaved tree　広葉樹

broadleaf weed, broad-leaved weed　広葉雑草

broadleaved lavender, spike lavender, *Lavandula latifolia* Medik. (= *L. spica* DC.)　スパイクラベンダー

broken kernel　砕粒

broken rice　砕け米, 砕米

Bromus catharticus Vahl (= *B. unioloides* (Willd.) H. B. K.),

rescuegrass, prairie grass　レスキュグラス
Bromus inermis Leyss., smooth bromegrass　スムーズブロムグラス
Bromus japonicus Thunb. ex Murray, Japanese brome　スズメノチャヒキ
Bromus marginatus Nees ex Steud., mountain bromegrass　マウンテンブロムグラス
broom corn, *Sorghum bicolor* (L.) Moench var. *hoki* Ohwi　ホウキモロコシ(箒蜀黍)
broomsedge, *Andropogon virginicus* L.　メリケンカルカヤ
Broussonetia kazinoki Sieb., paper mulberry　コウゾ(楮)
Broussonetia papyrifera (L.) L'Hér. ex Vent., paper mulberry　カジノキ(梶の木)
Brown Earth　褐色土
Brown soil　褐色土
Brown Forest soil　褐色森林土
brown mustard, *Brassica juncea* (L.) Czern. et Coss.　カラシナ
brown rice, hulled rice, husked rice　玄米
brown rust, leaf rust　赤さび(銹)病【コムギ】
brown spot　1) 斑点病【トウモロコシ】 2) ごま(胡麻)葉枯病【イネ】
brown sugar　黒砂糖
browning　褐変
browning stage　褐変期【タバコ】
browsing　採食【木本の小枝, 葉, 芽など】
Buchloe dactyloides (Nutt.) Engelm., buffalograss　バッファローグラス
buck-horn plantain, *Plantago lanceolata* L.　ヘラオオバコ
buckwheat, *Fagopyrum esculentum* Moench　ソバ
buckwheat flour　ソバ粉
buckwheat husk　ソバ殻

buckwheat noodle, soba　そば(蕎麦), そば切り
bud　芽
bud grafting, budding　芽接ぎ
bud mutation　芽条[突然]変異, 枝変わり
bud picking, disbudding　摘芽
bud pollination　つぼみ(蕾)受粉
bud-scale　芽鱗(がりん)
bud sport　芽条[突然]変異, 枝変わり
bud variation　芽条[突然]変異, 枝変わり
budding　1) 芽接ぎ 2) 芽ばえ
buffalograss, *Buchloe dactyloides* (Nutt.) Engelm.　バッファローグラス
buffer solution　緩衝液
bulb　球根, 鱗茎
bulb formation　結球【タマネギ等】
bulbil　鱗芽(りんが), むかご(零余子)
bulbing　結球【タマネギ等】
bulblet　子球(しきゅう)【ユリ等】
bulk curing　バルク乾燥【タバコ】
bulk density　かさ密度, 容積重
bulk emasculation, mass emasculation　集団除雄
bulk method [of breeding]　1) 混合育種法 2) 集団育種法
bulking, tuber bulking　塊茎肥大
bulliform cell, motor cell　機動細胞
bulrush, *Schoenoplectus triqueter* (L.) Palla (= *Scirpus triqueter* L.)　タイコウイ(太甲藺)
bulu　ブル【イネの生態型】
bumper crop, good harvest　豊作
bunch, fruit cluster　果房
bunch-type grass　そう(叢)生草
bundle sheath extention　維管束鞘延長部
bundling, binding　結束
bunt　なまぐさ黒穂病【コムギ】
Bupleurum falcatum L. (= *B. scorzonerifolium* Willd. var. *stenophyllum* Nakai)　ミシマサイコ
bur beggarticks, *Bidens tripartita* L.　タウコギ
bur clover, *Medicago polymorpha* L.

(= *M. hispida* Gaertn.) バークローバ
buret ビュレット
buried seed 埋土種子
Burley [tobacco] バーレー種【タバコ】
Burmese varnish tree, *Melanorrhoea usitata* Wall. ビルマウルシ
burner バーナー
burning 1) 火入れ 2) 肥焼け(こえやけ) 3) 枯上がり【タバコ】
burnt rice 焼米(やきごめ)
bush, shrub 低木
bush bean, dwarf bean ツルナシインゲン(蔓無隠元)
bush fallow そう(叢)林休閑
bush training 株仕立て
butter bean, Lima bean, *Phaseolus lunatus* L. (= *P. limensis* Macf.) ライマメ
buttercup, *Ranunculus silerifolius* L v. (= *R. quelpaertensis* (L v.) Nakai) キツネノボタン
buttercup, crowfoot, *Ranunculus sceleratus* L. タガラシ
button stage 発蕾期(はつらいき)【タバコ】
butyl alcohol ブチルアルコール
butyric acid 酪酸
Butyrospermum parkii Don Kotschy (= *Vitellaria paradoxa* (A. DC.) C. F. Gaertn.), shea [butter] tree シアバターノキ(シアバターの木)
by-product 副産物

[C]

C (carbon) 炭素
C_3 plant C_3 植物
C_4 pathway C_4 経路
C_4 plant C_4 植物
C_4-dicarboxylic acid cycle C_4 ジカルボン酸回路
C-N ratio, carbon-nitrogen ratio C-N 比
Ca (calcium) カルシウム

cabbage, *Brassica oleracea* L. var. *capitata* L. キャベツ
cacao, cocoa, *Theobroma cacao* L. カカオ
cacao bean カカオ豆
cacao butter カカオバター, カカオ脂
cadmium (Cd) カドミウム
Caesalpinia echinata Lam., brazilwood ブラジルボク
Caesalpinia sappan L., sappanwood スオウ(蘇芳)
caffeine カフェイン
cajan pea, pigeon pea, *Cajanus cajan* (L.) Millsp. キマメ
Cajanus cajan (L.) Millsp., pigeon pea, cajan pea キマメ
Calamus caesius Blume 他多種, rattan [palm], rotan, cane palm トウ(籐)
Calathea allouia (Aubl.) Lindl, topee-tambu トラフヒメバショウ
calcareous soil 石灰質土壌
calcicole plant 好石灰植物, カルシウム植物
calcifuge plant 嫌石灰植物
calcination of soil 焼土(しょうど)
calciphilous plant 好石灰植物, カルシウム植物
calciphobous plant 嫌石灰植物
calcium (Ca) カルシウム
calcium carbonate 炭酸カルシウム
calcium cyanamide 石灰窒素
calcium hydroxide 水酸化カルシウム
calcium nitrate 硝酸カルシウム, 硝酸石灰
calcium oxide, calx, quicklime 生石灰
calcium silicate ケイ酸カルシウム(珪カル)
calendar day 暦日
calibration キャリブレーション, 較正
Callitriche palustris L., water starwort ミズハコベ
callose カロース
callus (*pl.* calli) カルス

calmodulin　カルモジュリン
calorie　カロリー
calorimeter　熱量計
Calvin-Benson cycle　カルビン‐ベンソン回路
Calvin cycle　カルビン回路
calx, calcium oxide, quicklime　生石灰
Calystegia hederacea Wall., Japanese bindweed　コヒルガオ
Calystegia japonica Choisy　ヒルガオ
calyx　がく(萼), へた
CAM (crasslacean acid metabolism) plant　CAM植物
cambium (*pl*. cambia)　形成層
Camellia sinensis (L.) O. Kuntze, tea　チャ(茶)
camera lucida, drawing prism　カメラルシダ
camphor tree, *Cinnamomum camphora* (L.) Presl　クス(樟)
Canada balsam　カナダバルサム
Canada bluegrass, *Poa compressa* L.　カナダブルーグラス
Canada wild rye, *Elymus canadensis* L.　カナダワイルドライ
canal, creek　水路
Canangium odoratum Hook. f. et Thoms. (= *Cananga odorata* Baill.), ylang-ylang, ilang-ilang　イランイラン
Canavalia ensiformis (L.) DC., Jack bean　タチナタマメ(立刀豆)
Canavalia gladiata (Jacq.) DC., sword bean　ナタマメ(刀豆)
candlenut, *Aleurites moluccana* (L.) Willd.　ククイノキ
cane palm, rattan [palm], rotan, *Calamus caesius* Blume 他多種　トウ(籐)
cane sugar　甘蔗糖(カンショとう)
canihua, kaniwa, *Chenopodium pallidicaule* Aellen　カニウア
Canna edulis Ker-Gawl., edible canna, purple arrowroot, Queensland arrowroot, achira　食用カンナ
Cannabis sativa L., hemp　タイマ(大麻)
canola　カノーラ【ナタネの一系統】
canonical correlation　正準相関
canonical discriminant analysis　正準判別分析
canopy　草冠, 林冠
canopy architecture　草冠構造, 個体群構造
canopy extinction coefficient　群落吸光係数, 個体群吸光係数
canopy photosynthesis　個体群光合成, 群落光合成
canopy resistance　群落抵抗
canopy structure　草冠構造, 個体群構造
cantala, *Agave cantala* Roxb.　カンタラアサ(カンタラ麻)
capacitance humidity sensor　静電容量型湿度計
cape jasmine, *Gardenia jasminoides* Ellis　クチナシ
caper, *Capparis spinosa* L.　ケーパー
capillarity　毛管現象
capillary phenomenon　毛管現象
capillary potential　毛管ポテンシャル
capillary water　毛管水
capitulum, caput　頭状花序
Capparis spinosa L., caper　ケーパー
Capsella bursa-pastoris (L.) Medik., shepherd's-purse　ナズナ
Capsicum annuum L. (= *C. frutescens* L.), red pepper, chili, capsicum　トウガラシ(唐芥子)
capsicum, red pepper, chili, *Capsicum annuum* L. (= *C. frutescens* L.)　トウガラシ(唐芥子)
capsule　さく(蒴)果, さく(蒴)
caput, capitulum　頭状花序
carambola, star fruit, *Averrhoa carambola* L.　ゴレンシ(五斂子)
caraway, *Carum carvi* L.　キャラウェー, ヒメウイキョウ

carbohydrate　炭水化物
carbohydrate metabolism　炭水化物代謝
carbon (C)　炭素
carbon assimilation　炭素同化
carbon balance　炭素収支
carbon dioxide (CO_2)　二酸化炭素
carbon dioxide assimilation　炭酸同化
carbon dioxide enrichment　炭酸ガス施肥
carbon dioxide fertilization　炭酸ガス施肥
carbon dioxide fixation　炭酸固定
carbon dioxide absorption　二酸化炭素吸収
carbon dioxide compensation point　二酸化炭素補償点
carbon dioxide exchange rate (CER)　二酸化炭素交換速度
carbon dioxide uptake　二酸化炭素吸収
carbon isotope discrimination　炭素同位体分別
carbon-nitrogen ratio, C-N ratio　C-N比
carbon source　炭素源
carbonic anhydrase　炭酸脱水酵素，カーボニックアンヒドラーゼ
carbonized chaff　焼籾殻
carbonized rice grain　炭化米
carbonized rice husks　焼籾殻
carboxylase　カルボキシラーゼ
carboxylic acid　カルボン酸
Cardamine flexuosa With., flexuous bittercress　タネツケバナ
cardamon, *Elettaria cardamomum* (L.) Maton　ショウズク(小豆蔲), カルダモン
carding　梳綿(そめん)
Carica papaya L., papaya　パパイア
Carludovica palmata Ruiz et Pav., Panama hat palm (plant)　パナマソウ
carnauba wax palm, Brazilian wax palm, *Copernicia prunifera* H. E. Moore (= *C. cerifera* Mart.)　カルナウバヤシ(カルナウバ椰子)

carob, locust bean, *Ceratonia siliqua* L.　イナゴマメ
carotene　カロテン
carotenoid　カロテノイド
carpel　心皮(しんぴ)
carpetgrass, *Axonopus compressus* (SW) P. Beauv.　カーペットグラス
carrier　担体, キャリヤー
carrot, *Daucus carota* L.　ニンジン(人参)
carrying capacity, environmental capacity　環境収容力, 環境容量
Carthamus tinctorius L., safflower　ベニバナ(紅花)
Carum carvi L., caraway　キャラウェー, ヒメウイキョウ
caruncle, strophiole　種枕(しゅちん)
caryokinesis, mitosis (*pl*. mitoses), karyokinesis　有糸分裂
caryology, karyology　核学
caryopsis　穎果(えいか)
Caryota urens Jacq., toddy palm, wine palm　クジャクヤシ(孔雀椰子)
cash crop　換金作物
cashew, cashew-nut tree, *Anacardium occidentale* L.　カシュー
Casparian strip　カスパリー線
cassava, manioc, tapioca plant, *Manihot esculenta* Crantz (= *M. utilissima* Pohl)　キャッサバ
cassava starch　キャッサバデンプン
cassia, Chinese cinnamon, *Cinnamomum cassia* J. Presl (= *C. cassia* Blume)　カシア
Cassia obtusifolia L. (= *Senna obtusifolia* (L.) H.S. Irwin et Barneby), oriental senna　エビスグサ
Cassia senna L. (= *Senna angustifolia* Batka および *S. alexandrina* Mill.), senna　センナ
Cassia torosa Cav. (= *C. occidentalis* L., *Senna occidentalis* Link.), negro coffee　ハブソウ

cassie, sweet acacia, *Acacia farnesiana* (L.) Willd.　キンゴウカン(金合歓)
Castanea crenata Sieb. et Zucc., Japanese chestnut　クリ(栗)
Castanea sativa Mill., European chestnut　セイヨウグリ
Castilla elastica Cerv., Panama rubber, Central American rubber　パナマゴム
casting plowing　外返し耕
castor [bean], *Ricinus communis* L.　ヒマ(蓖麻)
castor oil　ひまし油
castration, emasculation　除雄
catabolism, dissimilation　異化[作用]
catch crop, intercrop　間作物
catch cropping, intercropping　間作
catchweed, bedstraw, *Galium spurium* L. var. *echinospermon* (Wallr.) Hayek　ヤエムグラ
catechu, cutch, black catechu, *Acacia catechu* (L.f.) Willd.　アセンヤクノキ(阿仙薬樹)
cation exchange capacity (CEC)　陽イオン交換容量
catjang, *Vigna unguiculata* (L.) Walp. var. *catjang* (Burm. f.) H. Ohashi　ハタササゲ
cattail, *Typha latifolia* L.　ガマ(蒲)
cattle manure　牛糞きゅう(厩)肥
cauliflory　幹成花性
cavitation　空洞現象
Cayenne bois de rose, *Aniba rosaeodora* Ducke　カイエンボアドローズ
Cayratia japonica (Thunb.) Gagn.　ヤブカラシ
Cd (cadmium)　カドミウム
cDNA (complementary DNA)　相補的DNA
ceara rubber, manihot rubber, *Manihot glaziovii* Meull. -Arg.　マニホットゴム
CEC (cation exchange capacity)　陽イオン交換容量
cedar oil　シダー油

Ceiba pentandra (L.) Gaertn., kapok　カポック
Celite　セライト
cell　細胞
cell biology　細胞生物学
cell contents, cellular contents　細胞内容物
cell culture　細胞培養
cell cycle　細胞周期
cell differentiation　細胞分化
cell division　細胞分裂
cell elongation　細胞伸長
cell engineering, cell technology　細胞工学
cell expansion　細胞拡大
cell fractionation　細胞分画法
cell fusion　細胞融合
cell lineage　細胞系譜
cell membrane　細胞膜
cell plate　細胞板
cell sap　細胞液
cell technology, cell engineering　細胞工学
cell wall　細胞壁
cell wall constituents　細胞壁[構成]物質, 細胞壁成分
cell wall extensibility　細胞壁伸展性
celloidin method　セロイジン法
cellular contents, cell contents　細胞内容物
cellulase　セルラーゼ
cellulose　セルロース
Centella asiatica (L.) Urban, Indian pennywort　ツボクサ
Center for Coarse Grains, Pulses, Roots, and Tuber Crops (CGPRT Center)　湿潤熱帯地域粗粒穀物・豆類・地下作物研究開発地域調整センター
Center for International Forestry Research (CIFOR)　国際林業研究センター
center of origin　起源中心
Centipeda minima (L.) A. Br. et

Aschers. トキンソウ (吐金草)
Central American rubber, Panama rubber, *Castilla elastica* Cerv. パナマゴム
central cylinder, stele 中心柱
central processing unit (CPU) 中央処理装置
centrifugal 遠心的
centrifugation 遠心分離
centrifuge 遠心機
centrifuge tube 遠沈管
centrifuged sugar 分蜜糖
centriole 中心子, 中心小体
centripetal 求心的
centro, *Centrosema pubescens* Benth. セントロ
Centro Internacional de Agricultura Tropical (CIAT), International Center for Tropical Agriculture 国際熱帯農業研究センター
Centro Internacional de Mejoramiento de Maiz y Trigo (CIMMYT), International Maize and Wheat Improvement Center 国際トウモロコシ・コムギ改良センター
Centro Internacional de Papa (CIP), International Potato Center 国際バレイショセンター
Centrosema pubescens Benth., centro セントロ
centrosome 中心体
century plant 1) *Agave americana* L. アオノリュウゼツラン 2) *A. americana* L. var. *variegata* Nichols. リュウゼツラン (竜舌蘭)
CEPA (2-chloroethyl phosphonic acid) 2-クロロエチルホスホン酸, ethephon エテフォン
cephalosporium stripe 条斑病【コムギ, オオムギ】
Cerastium fontanum Baumg. ssp. *triviale* (Link) Jalas var. *angustifolium* (Franch.) H. Hara ミミナグサ

Cerastium glomeratum Thuill., sticky chickweed オランダミミナグサ
Ceratonia siliqua L., carob, locust bean イナゴマメ
Ceratophyllum demersum L., hornwort マツモ (松藻)
cereal cropping 穀作, 穀物生産
cereal crops 穀類, 禾穀[類] (かこく[るい])
cereal production 穀物生産, 穀作
cereals 穀類, 禾穀[類]
cerospora leaf spot すじ(条)葉枯病【イネ】
certation 受精競争
certified seed 保証種子
Ceylon cinnamon, cinnamon, *Cinnamomum verum* J. Presl (= *C. zeylanicum* (Garc.) Bl.) シナモン
Ceylon citronella grass, *Cymbopogon nardus* (L.) Rendle セイロンシトロネラソウ
CGIAR (Consultative Group on International Agricultural Research) 国際農業研究協議グループ
CGR (crop growth rate) 個体群成長速度
chaff 籾殻, 殻ふ (桴)
chalaza カラザ, 合点
chalky kernel, mealy kernel 粉状質粒【ムギ類】
chalky rice 白亜質米
chamomile, noble camomile, *Anthemis nobilis* L. ローマカミツレ
chaos カオス
character 1) 形質 2) 特性【統計】
character of seedling 苗の素質
characteristic 1) 形質 2) 特性【統計】
chayote, *Sechium edule* (Jacq.) Swartz ハヤトウリ
check variety 比較品種, 対照品種
cheese cloth 寒冷しゃ (紗)
chelate キレート

chemical control 化学[的]防除, 化学的制御
chemical defoliation 人工落葉
chemical desiccation 薬剤乾燥
chemical evolution 化学進化
chemical fertilizer 化学肥料
chemical formula 化学式
chemical fruit thinning 薬剤摘果
chemical injury 薬害
chemical oxygen demand (COD) 化学的酸素要求量
chemical spraying 薬剤散布
chemical weeding 薬剤除草
chemotropism 化学屈性
Chenopodium album L. var. *album*, common lamb's-quarters シロザ
Chenopodium album L. var. *centrorubrum* Makino アカザ
Chenopodium ambrosioides L., Mexican tea ケアリタソウ
Chenopodium ambrosioides L. var. *anthelminticum* (L.) A. Gray, wormseed goosefoot アメリカアリタソウ
Chenopodium ficifolium Smith (= *C. serotinum* L.), figleaved goosefoot コアカザ
Chenopodium pallidicaule Aellen, canihua, kaniwa カニウア
Chenopodium purpurescens Jacq. ムラサキアカザ
Chenopodium quinoa Willd., quinoa キノア
Chenopodium serotinum L. (= *C. ficifolium* Smith), figleaved goosefoot コアカザ
cherimoya, *Annona cherimola* Mill. チェリモヤ
Chernozem チェルノーゼム
cherry, *Prunus avium* L. 他数種 オウトウ(桜桃)
Chestnut soil 栗色土(くりいろど)
chiasma (*pl.* chiasmata) キアズマ

chick pea, common gram, *Cicer arietinum* L. ヒヨコマメ
chicken dropping 鶏糞
chickling vetch, grass pea, *Lathyrus sativus* L. ガラスマメ
chicle チクル【サポジラ由来のゴム状物質】
chili, red pepper, capsicum, *Capsicum annuum* L. (= *C. frutescens* L.) トウガラシ(唐芥子)
chilling injury, low-temperature injury 低温障害
chilly wind damage (injury), cold wind damage (injury) 寒風害
chimera キメラ
China grass, ramie, *Boehmeria nivea* (L.) Gaud. チョマ(苧麻)
China jute, Indian mallow, *Abutilon avicennae* Gaertn. ボウマ(莔麻)
Chinese bellflower, Japanese bellflower, balloonflower, *Platycodon grandiflorum* (Jacq.) A. DC. キキョウ(桔梗)
Chinese cabbage, *Brassica rapa* L. var. *amplexicaulis* Tanaka et Ono (= *B. pekinensis* Rupr.) ハクサイ(白菜)
Chinese chive, *Allium tuberosum* Rottler ex. Spreng. ニラ(韮)
Chinese cinnamon, cassia, *Cinnamomum cassia* J. Presl (= *C. cassia* Blume) カシア
Chinese indigo, *Persicaria tinctoria* (Aiton.) H. Gross (= *Polygonum tinctorium* Lour.) アイ(藍)
Chinese jujube, common jujube, *Ziziphus jujuba* Mill. ナツメ(棗)
Chinese mat grass, three-cornered grass, *Cyperus malaccensis* Lam. ssp. *brevifolius* (Boeck.) T. Koyama シチトウイ(七島藺)
Chinese mat rush, *Lepironia articulata* (Retz.) Domin (= *L. mucronata* Rich.) アンペラソウ

Chinese milkvetch, genge, *Astragalus sinicus* L.　レンゲソウ(蓮華草, 紫雲英)
Chinese paeony, Chinese peony, *Paeonia lactiflora* Pall. (= *P. albiflora* Pall.)　シャクヤク(芍薬)
Chinese pennisetum, *Pennisetum alopecuroides* (L.) Spreng.　チカラシバ
Chinese sprangletop, *Leptochloa chinensis* (L.) Nees　アゼガヤ
Chinese water chestnut, *Eleocharis dulcis* (Burm. f.) Trin. ex Hensch. var. *tuberosa* (Roxb.) T. Koyama　オオクログワイ
Chinese yam, *Dioscorea opposita* Thunb. (= *D. batatas* Decne.)　ナガイモ(薯蕷)
chi-square (χ^2)　カイ二乗
chi-square (χ^2) test　カイ二乗(χ^2)検定
chi-square (χ^2) distribution　カイ二乗(χ^2)分布
chi-square (χ^2) statistic　カイ二乗(χ^2)統計量
chlorella, *Chlorella* spp.　クロレラ
chlorinated organic compound　有機塩素剤
chlorine (Cl)　塩素
Chloris gayana Kunth, Rhodes grass　ローズグラス
chlorocholine chloride (CCC), 2-chloroethyltrimethyl-ammonium chloride, cycocel　サイコセル
2-chloroethyl phosphonic acid (CEPA) 2-クロロエチルホスホン酸, ethephon　エテフォン
2-chloroethyltrimethyl-ammonium chloride, chlorocholine chloride (CCC), cycocel　サイコセル
chloroform　クロロホルム
Chlorophora tinctoria Gaud., old fustic　オールドファスチク

chlorophyll　クロロフィル, 葉緑素
chloroplast　クロロプラスト, 葉緑体
chlorosis　クロロシス
chorogi, *Stachys sieboldii* Miq. (= *S. affinis* Fresen)　チョロギ
chromatic aberration　色収差
chromatin　クロマチン, 染色質
chromatography　クロマトグラフィー
chromatophore　クロマトフォア
chromoplast, chromoplastid　有色体
chromosomal mutation　染色体突然変異
chromosome　染色体
chromosome aberration　染色体異常
chromosome doubling, doubling of chromosome　染色体倍加
chromosome map　染色体地図
chromosome substitution　染色体置換
chromosome tetrad, tetrad　四分染色体
Chrysanthemum cinerariaefolium Visiani (= *Pyrethrum cinerariifolium* Trevir.), insectpowder plant, insect flower, Dalmatian chrysanthemum　ジョチュウギク(除虫菊)
CIAT (Centro Internacional de Agricultura Tropical), International Center for Tropical Agriculture　国際熱帯農業研究センター
Cicer arietinum L., chick pea, common gram　ヒヨコマメ
CIMMYT (Centro Internacional de Mejoramiento de Maiz y Trigo), International Maize and Wheat Improvement Center　国際トウモロコシ・コムギ改良センター
cinchona, quinine, *Cinchona* spp. 【*C. pubescens* Vahl など】　キナ
Cinchona ledgeriana Moens ex Trimen　ボリビアキナ
Cinchona spp.【*C. pubescens* Vahl など】, cinchona, quinine　キナ
Cinnamomum camphora (L.) Presl, camphor tree　クス(樟)
Cinnamomum cassia J. Presl (= *C.*

cassia Blume), cassia, Chinese cinnamon　カシア

Cinnamomum sieboldii Meisn. (= *C. loureirii* Nees), Saigon cinnamon　ニッケイ(肉桂)

Cinnamomum verum J. Presl (= *C. zeylanicum* (Garc.) Bl.), cinnamon, Ceylon cinnamon　シナモン

cinnamon, Ceylon cinnamon, *Cinnamomum verum* J. Presl (= *C. zeylanicum* (Garc.) Bl.)　シナモン

CIP (Centro Internacional de Papa), International Potato Center　国際バレイショセンター

circadian rhythm　概日リズム

circular DNA　環状 DNA

circumnutation　回旋運動

cis-element　シスエレメント

citric acid　クエン酸

citric acid cycle　クエン酸回路

citronella grass, 1) *Cymbopogon nardus* (L.) Rendle (セイロンシトロネラソウ) 2) *C. winterianus* Jowitt (ジャワシトロネラソウ)　シトロネラソウ

Citrullus lanatus (Thunb.) Matsum. et Nakai (= *C. vulgaris* Schrad.), watermelon　スイカ(西瓜)

Citrus aurantifolia (Christm.) Swingle, lime, sour lime　ライム

Citrus bergamia Risso et Poit., bergamot, bergamot orange　ベルガモット

Citrus limon (L.) Burm.f., lemon　レモン(檸檬)

Citrus natsudaidai Hayata, Watson pomelo　ナツミカン(夏蜜柑)

Citrus paradisi Macf., grapefruit　グレープフルーツ

Citrus sinensis Osbeck, sweet orange　スイートオレンジ

Citrus unshiu Marcovitch, satsuma mandarin　ウンシュウミカン(温州蜜柑)

Cl (chlorine)　塩素

clamp　クランプ, 締め具

class　1) 級【区間】 2) 綱【分類】

class frequency　級頻度

class interval　級区間

classification　分類

clay　粘土, 埴土

clay loam　埴壌土

clay mineral　粘土鉱物

clay pot, unglazed pot　素焼鉢

clay soil, clay　埴土

clayey　粘土質の

clean culture　清耕栽培

cleaning crop　清耕作物

clearing　1) 透明化 2) 開墾

cleft grafting　割り接ぎ

cleistogamous flower　閉鎖花

cleistogamy, close fertilization　閉花受精

climate　気候

climate change　気候変化

climate near the ground　接地気候

climatic cultivation limit　気候的栽培限界

climatic division　気候区分

climatic element　気候要素

climatic factor　気候因子

climatic fluctuation　気候変動

climatic index　気候指数

climatic productivity index　気候生産力示数

climatic variation　気候変動

climatic zone　気候帯

climatron, artificial climate room　人工気象室

climax　極相

climbing plant, vine　つる(蔓)植物

climbing stem　よじ登り茎

clingstone　粘核【果実】

clipping　切除, 刈取り

clipping height, cutting height, mowing height　刈取り高さ

clipping interval, cutting interval,

mowing interval 刈取り間隔
clipping method 切えい(穎)法
clod 土塊
clod crusher, harrow, pulverizer ハロー
clonal preservation, preservation of stock 株保存
clonal selection 栄養系選抜
clonal separation 栄養系分離
clonal strain 栄養系, クローン
clone 栄養系, クローン
cloning 1) クローニング, クローン化 2) 栄養繁殖
close fertilization, cleistogamy 閉花受精
closed-center type 閉心型
closed system 閉鎖系
clove, *Syzygium aromaticum* (L.) Merr. et Perry (= *Eugenia caryophyllata* Thunb., *E. caryophyllus* Spreng., *E. aromatica* Kuntze) チョウジ(丁子, 丁字)
club root 根こぶ[病]
club wheat, *Triticum compactum* Host クラブコムギ
cluster クラスター
cluster analysis クラスター分析
cluster bean, guar, *Cyamopsis tetragonoloba* (L.) Taub. クラスタマメ
Cnidium officinale Makino センキュウ(川芎)
CO_2 (carbon dioxide) 二酸化炭素
coarse bran 荒糠(あらぬか)
coarse grain 粗粒穀物
coarse green tea 番茶
coarse plowing, first plowing 荒起し
coarse puddling, first puddling 荒代(あらしろ)
coated fertilizer 被覆肥料
coated seed, pelleted seed 被覆種子
cob 穂軸(すいじく), 穂心(すいしん)【トウモロコシ】
Cochlearia armoracia L. (= *Armoracia rusticana* Gaertn., Mey. et Scherb.), horseradish ワサビダイコン(山葵大根)
cock, stack かたい(禾堆), にお
cocksfoot, orchardgrass, *Dactylis glomerata* L. オーチャードグラス
cockspur grass, barnyard grass, *Echinochloa crus-galli* (L.) Beauv. ケイヌビエ【イヌビエの変種】
cocoa, cacao, *Theobroma cacao* L. カカオ
coconut oil, copra oil ココヤシ油, ヤシ油
coconut palm, *Cocos nucifera* L. ココヤシ(ココ椰子)
Cocos nucifera L., coconut palm ココヤシ(ココ椰子)
COD (chemical oxygen demand) 化学的酸素要求量
coding region コード領域
codon コドン
coefficient of correlation, correlation coefficient 相関係数
coefficient of determination 決定係数
coefficient of inbreeding, inbreeding coefficient 近交係数
coefficient of parentage 近縁係数
coefficient of variation (CV) 変動係数
coenzyme 補酵素
coevolution 共進化
Coffea arabica L., Arabian coffee アラビアコーヒー[ノキ]
Coffea liberica W. Bull. ex Hiern, Liberian coffee リベリアコーヒー[ノキ]
Coffea robusta Linden (= *C. canephora* Pierr. ex Froeh.), robusta coffee, Congo coffee ロブスタコーヒー[ノキ]
Coffea spp., coffee コーヒー
coffee, *Coffea* spp. コーヒー
coffee bean コーヒー豆
cogongrass, *Imperata cylindrica* (L.) P. Beauv. チガヤ(茅)

cohesion　ゆ(癒)着
cohesion theory　凝集力説
coir　コイル【ココヤシ】
Coix lacryma-jobi L., Job's-tears　ジュズダマ
Coix lacryma-jobi L. var. *ma-yuen* (Roman.) Stapf (= *C. lacryma-jobi* L. var. *frumentacea* Makino), Job's-tears　ハトムギ(薏苡)
cola, kola　1) *Cola nitida* (Vent.) Schott et Endl.　コーラ　2) *C. acuminata* (P. Beauv.) Schott et Endl.　ヒメコーラ
Cola acuminata (P.Beauv.) Schott et Endl., cola, kola　ヒメコーラ
Cola nitida (Vent.) Schott et Endl., cola, kola　コーラ
colchicine　コルヒチン
colchicum, autumn crocus, meadow saffron, *Colchicum autumnale* L.　イヌサフラン
Colchicum autumnale L., colchicum, autumn crocus, meadow saffron　イヌサフラン
cold bed　冷床(れいしょう)
cold climate area (region)　寒冷地
cold damage (injury)　寒害
cold frame　冷床(れいしょう)
cold hardiness (resistance)　耐寒性
cold water damage　冷水害
cold water tolerance　耐冷水性
cold weather resistance　耐冷性
cold wind damage (injury), chilly wind damage (injury)　寒風害
coleoptile　鞘葉, 子葉鞘
coleorhiza　根鞘
Coleus parviforus Benth., hausa potato　サヤバナ
collar pruning　台切り【チャ】
collateral vascular bundle　並立維管束
collenchyma　厚角組織
collenchyma cell　厚角細胞
collenchymatous cell　厚角細胞

Colocasia esculenta (L.) Schott var. *antiquorum* Hubbard & Rehder, eddoe　サトイモ(里芋)
Colocasia esculenta (L.) Schott var. *esculenta* Hubbard & Rehder, taro, dasheen　タロイモ
Colocasia gigantea Hook. f.　ハスイモ(蓮芋)
colonial bentgrass, *Agrostis capillaris* L. (= *A. tenuis* Sibth.)　コロニアルベントグラス
colony　コロニー, 群体
color and gloss　色沢(しきたく)
color composite　カラー合成【画像処理】
colored film　着色フィルム
colored grain　着色粒
colored Guinea grass, Klein grass, *Panicum coloratum* L.　カラードギニアグラス
colorimeter　比色計
colorimetric method　比色法
colorimetry　比色定量
columella　コルメラ
column chromatography　カラムクロマトグラフィー
comb harrow, drag　馬ぐわ(鍬)(まぐわ)
comb thresher, threshing comb　千歯, 千歯こ(扱)き
combination　組合せ
combination breeding　組合せ育種法
combine harvester　コンバイン
combined pollution　複合汚染
combined water, bound water　結合水
combining ability　組合せ能力
Commelina communis L., dayflower, Asiatic dayflower　ツユクサ(露草)
commercial agriculture　企業農業
commercial bed soil, artificial bed soil　人工床土
commercial compost, soil mixes　配合土
commercial crop　商品作物
commercial fertilizer　金肥(きんぴ), 販売肥料, 購入肥料

commercial seed　市販種子
commercial variety　実用品種
commissural vascular bundle　連絡維管束
common　入会権(いりあいけん)
common carpetweed, *Mollugo verticillata* L.　クルマバザクロソウ
common chickweed, *Stellaria media* (L.) Villars　ハコベ
common cocklebur, *Xanthium strumarium* L.　オナモミ
common false pimpernel, *Lindernia procumbens* (Krock.) Philcox (= *L. pyxidaria* L.)　アゼナ
common gram, chick pea, *Cicer arietinum* L.　ヒヨコマメ
common groundsel, *Senecio vulgaris* L.　ノボロギク
common heliotrope, *Heliotropium arborescens* L. (= *H. pervianum* L.)　ヘリオトロープ
common indigo, indigo tree, *Indigofera tinctoria* L.　キアイ(木藍)
common jujube, Chinese jujube, *Ziziphus jujuba* Mill.　ナツメ(棗)
common juniper, *Juniperus communis* L.　セイヨウビャクシン(洋種杜松)
common lamb's-quarters, *Chenopodium album* L. var. *album*　シロザ
common land, commons　入会地(いりあいち)
common madder, madder, *Rubia tinctorum* L.　セイヨウアカネ(西洋茜)
common marjoram, oregano, *Origanum vulgare* L.　ハナハッカ
common millet, millet, proso millet, hog millet, *Panicum miliaceum* L.　キビ(黍)
common mushroom, *Agaricus bisporus* Sing.　マッシュルーム
common nasturtium, garden nasturtium, *Tropaeolum majus* L.　キンレンカ(金蓮花)
common purslane, *Portulaca oleracea* L.　スベリヒユ
common ragweed, *Ambrosia artemisiifolia* L. var. *elatior* (L.) Desc.　ブタクサ
common reed, *Phragmites australis* (Cav.) Trin. ex Steud. (= *P. communis* Trin.)　ヨシ(葭)
common rue, rue, *Ruta graveolens* L.　ヘンルーダ
common thyme, garden thyme, *Thymus vulgaris* L.　タイム
common velvetgrass, Yorkshire fog, *Holcus lanatus* L.　ベルベットグラス
common vetch, *Vicia sativa* L.　コモンベッチ
common wheat, bread wheat, *Triticum aestivum* L.　パンコムギ
common white jasmine, poet's jasmine, *Jasminum officinale* L. f.　ソケイ(素馨), ツルマツリ(蔓莉茉)
commons, common land　入会地(いりあいち)
community　群集, 群落
community ecology, synecology　群集生態学, 群落生態学
compact ear, dense ear　密穂(みっすい)
compaction　鎮圧【農作業】
companion cell　伴細胞
companion crop　随伴作物, 同伴作物
companion planting, mixed planting　混植
companion weed　随伴雑草
compatibility　和合性
compensatory effect　補償作用
competition　競合, 競争
competitive ability　競争力
competitive inhibition　拮抗阻害, 競合阻害
competitor　競争者
complementary DNA (cDNA)　相補的DNA

complementary gene　補足遺伝子
complementary planting, supplementary planting　補植
complete dominance　完全優性
complete flower, perfect flower　完全花
complete leaf　完全葉
completely randomized design　完全無作為設計
complex mutation　複合突然変異
component of covariance, covariance component　共分散成分
component of variance, variance component　分散成分
compost　堆肥
compost depot　堆肥舎
compound fertilizer　化成肥料，複合肥料
compound grain　複粒
compound granule　複粒
compound leaf　複葉
computer　コンピュータ，電子計算機
computer graphics　コンピュータグラフィックス
computer network　コンピュータネットワーク
computer vision　コンピュータビジョン
concave planting　船底植え，船底挿し【サツマイモ】
concentrate [feed]　濃厚飼料
concentric vascular bundle　包囲維管束
condenser　コンデンサー，集光器【顕微鏡】
condiment, spice　香辛料
conditional distribution　条件分布
conditional expectation　条件[付]期待値
conditional probability　条件[付]確率
conductance　コンダクタンス，伝導度
conductive tissue　通導組織
conduit, underdrain　暗きょ(渠)
cone　球果，毬果(きゅうか)【ホップ】
confidence interval　信頼区間
confidence limit　信頼限界
confocal laser microscope (CLM)　共焦点レーザー顕微鏡
confocal scanning microscope　共焦点走査型顕微鏡
confounding　交絡【統計】
Congo coffee, robusta coffee, *Coffea robusta* Linden (= *C. canephora* Pierr. ex Froeh.)　ロブスタコーヒー[ノキ]
connective　葯隔
connectivum　葯隔
conservation　保全
consistency　コンシステンシー【土壌】
constraint condition　制約条件
Consultative Group on International Agricultural Research (CGIAR)　国際農業研究協議グループ
consumed herbage　被食草【草地】
consumer　消費者
consumer rice price　消費者米価
contact herbicide　接触性除草剤
contact herbicide　接触剤【雑草】
contact insecticide　接触剤【害虫】
contact transmission　接触伝染
contaminated field, polluted field　汚染圃場
contamination　1) 汚染　2) 雑菌混入　3) 混り
content　含有量
contingency table　分割表
continuous cropping, sequential cropping　連作
continuous distribution　連続分布
continuous flooding　常時湛水
continuous layering, horizontal layering　しゅ(撞)木取り
continuous selection　連続選抜
continuous snow cover　根雪
continuous variable　連続変数
continuous variate　連続変量
continuous variation　連続変異
contour border irrigation　等高線灌漑
contour cropping　等高線栽培
contour farming　等高線栽培
contour furrow irrigation　等高線灌漑

contour cultivation　等高線栽培

contribution ratio, ratio of contribution　寄与率

control　1) 制御　2) 防除　3) 対照

control method　防除法

control plot　標準 [試験] 区, 対照 [試験] 区

controlled availability fertilizer　肥効調節型肥料

controlled release fertilizer, slow-release fertilizer　緩効性肥料

convection　対流

conventional　慣行の, 従来の

conventional combine　普通型コンバイン

conversion　転化

convertible husbandry, alternate husbandry, ley farming　穀草式農法

convolution, twist　撚り (より)

Conyza bonariensis (L.) Cronq. (= *Erigeron bonariensis* L.), hairy fleabane　アレチノギク

Conyza sumatrensis (Retz.) Walker (= *Erigeron sumatrensis* Retz.), tall fleabane　オオアレチノギク

cool-climate highland　高冷地

cool summer　冷夏

cool summer damage　冷害

cool summer damage due to delayed growth　遅延型冷害

cool temperate climate　冷温帯気候

cool temperate zone　冷温帯

cool weather damage　冷害

coontie, Florida arrowroot, *Zamia floridana* A. DC.　フロリダアロールート

cooperative control　共同防除

co-ordinates　座標

Copernicia prunifera H.E. Moore (= *C. cerifera* Mart.), carnauba wax palm, Brazilian wax palm　カルナウバヤシ (カルナウバ椰子)

copper (Cu)　銅

copra　コプラ

copra oil, coconut oil　ココヤシ油, ヤシ油

Coptis japonica (Thunb.) Makino　オウレン (黄連)

Corchorus capsularis L., jute, white jute　ジュート

Corchorus olitorius L., nalta jute　シマツナソ

core　果心

coriander, *Coriandrum sativum* L.　コエンドロ

Coriandrum sativum L., coriander　コエンドロ

cork　コルク

cork cambium, phellogen　コルク形成層

cork tissue, phellem　コルク組織

corm　1) いも　2) 球茎

corm of konjak　コンニャク玉

cormel　子球 (しきゅう)【グラジオラス等】

corn, maize, Indian corn, *Zea mays* L.　トウモロコシ (玉蜀黍)

corn flakes　コーンフレーク

corn oil　トウモロコシ油

corn speedwell, *Veronica arvensis* L.　タチイヌノフグリ

corolla　花冠

coronal root, crown root　冠根

Coronopus didymus (L.) J. E. Smith, swine cress　カラクサナズナ

corpus　内体

correction factor　補正因子

correction term　補正項

correlation　相関

correlation coefficient, coefficient of correlation　相関係数

correlation matrix　相関行列

correlation table　相関表

cortex　皮層

cosmid vector　コスミドベクター

cost function　コスト関数

costly registered variety, branded

variety, marketing standard variety　銘柄品種
cosuppression　共抑制
cotton　綿花【広義】
cotton, *Gossypium* spp.　ワタ(棉)
cotton fiber, lint　綿毛(めんもう)
cotton gin, gin　綿繰り機(わたくりき)
cotton seed　棉実(めんじつ)
cotton seed oil　棉実油(めんじつゆ)
cotton wax　棉ろう(蠟)
cotyledon　子葉
counter-current distribution　交流分配
country elevator　カントリーエレベータ
covariance　共分散
covariance component, component of covariance　共分散成分
covariance matrix　共分散行列
cover crop　被覆作物
cover [degree]　被度
cover glass　カバーグラス
coverage　被度
covered barley, hulled barley, *Hordeum vulgare* L.　カワムギ(皮麦)
covered grain type　皮性【オオムギ】
covering material　被覆資材
covering with soil　覆土
cow vetch, *Vicia cracca* L.　クサフジ
cowpea, southern pea, *Vigna unguiculata* (L.) Walp. (= *V. sinensis* Endl.)　ササゲ(豇豆, 大角豆)
CPU (central processing unit)　中央処理装置
crabgrass, *Digitaria adscendens* (H. B. K.) Henr. (= *D. ciliaris* (Retz.) Koeler)　メヒシバ
crack　亀裂
cracked groats　割れ【ソバ】
cracked rice kernel　胴割[れ]米
cracking of fruit　玉割れ
Crassocephalum crepidioides (Benth.) S. Moore　ベニバナボロギク
crassulacean acid metabolism (CAM) plant　CAM植物

crawler tractor　履帯(りたい)トラクタ, キャタピラートラクタ
crease, furrow　腹溝(ふくこう), 縦溝【ムギ類】
creek, canal　水路
creeper　ほふく(匍匐)植物
creeping　ほふく(匍匐)性の
creeping bentgrass, *Agrostis stolonifera* L.　クリーピングベントグラス
creeping thistle, *Breea setosa* (Bieb.) Kitam.　エゾノキツネアザミ
creeping type, prostrate type　ほふく(匍匐)型
creeping woodsorrel, *Oxalis corniculata* L.　カタバミ
cresson, water-cress, *Nasturtium officinale* R. Br. (= *Roripa nasturtium-aquaticum* (L.) Hayek.)　クレソン(和蘭芥)
crested wheatgrass, *Agropyron cristatum* (L.) Gaertn.　クレステッド・ホィートグラス
crimson clover, *Trifolium incarnatum* L.　クリムソンクローバ
cristae (*sing.* crista)　クリステ
criterion (*pl.* criteria)　判定基準【統計】
critical concentration　限界濃度
critical daylength　限界日長
critical point drying　臨界点乾燥法
critical region　棄却域
critical temperature　臨界温度, 限界温度
critical value　限界値, 棄却限界
Crocus sativus L., saffron　サフラン(泪夫藍)
crooked neck　首曲り【カーネーション, チューリップ等】
crop [plant]　作物
crop breeding　作物育種
crop cultivation　耕種
crop diagnosis　作物診断
crop for home consumption, home-consuming crop　自給作物

crop growth model　作物成長モデル
crop growth rate (CGR)　個体群成長速度
crop husbandry, agronomy　作物栽培学
crop index, crop situation index　作況指数
crop physiology　作物生理学
crop remain　作物遺体
crop residue　作物遺体
crop response test to weather　気象感応試験
crop rotation　輪作
crop science　作物学
crop situation　作況, 作柄
crop situation index, crop index　作況指数
crop-weather relationship experiment　豊凶考照試験
crop year　穀物年度
cropping　作付け
cropping directory　耕種概要
cropping manual　耕種概要
cropping pattern　作付様式
cropping season　作期, 作季
cropping sequence　作付順序
cropping system　作付体系
cropping type　作型
cros[sing]　交雑, 交配
cross breeding, hybridization breeding　交雑育種
cross combination　交雑組合せ, 交配組合せ
cross compatibility　交雑和合性, 交配和合性
cross fertilization　他家受(授)精
cross incompatibility　交雑不和合性, 交配不和合性
cross pollination　他家受(授)粉
cross section　横断切片
cross sectional area　断面積
cross sterile group　交雑不稔群, 交配不稔群

crossability　交雑能力
crossing-over　交さ(叉), 乗換え
crossing-over value　交さ(叉)価, 乗換え価
crossing rate　交雑率
Crotalaria juncea L., sun (sunn) hemp　サンヘンプ
crowfoot, buttercup, *Ranunculus sceleratus* L.　タガラシ
crowfootgrass, *Dactyloctenium aegyptium* (L.) P. Beauv.　タツノツメガヤ
crown　樹冠, 冠部
crown gall　クラウンゴール
crown grafting, whittle grafting　そぎ接ぎ
crown root, coronal root　冠根
crude ash　粗灰分(そかいぶん)
crude fat　粗脂肪
crude fiber　粗繊維
crude protein (CP)　粗タンパク質
crust　クラスト, 殻皮(かくひ)
cryptogam　陰花植物
Cryptotaenia japonica Hassk., Japanese hornwort, mitsuba　ミツバ(三葉)
crystal　結晶
crystal cell　結晶細胞
crystal violet　クリスタルバイオレット
crystalliferous cell　結晶細胞
Cu (copper)　銅
cucumber, *Cucumis sativus* L.　キュウリ
Cucumis melo L., melon　メロン
Cucumis sativus L., cucumber　キュウリ
Cucurbita maxima Duch. ex Lam., pumpkin, winter squash　セイヨウカボチャ
Cucurbita moschata (Duch. ex Lam.) Duch. ex Poir., pumpkin, winter squash　ニホンカボチャ
Cucurbita pepo L., summer squash, pumpkin　ペポカボチャ
Cucurbita spp.,　カボチャ(南瓜)
cudweed, *Gnaphalium affine* D. Don

ハハコグサ
culling level 淘汰水準
culm 桿(かん)
culm length 桿長
culm stiffness, straw stiffness 強桿性
culti-packer カルチパッカ, 鎮圧機
cultivar (cv.), cutivated variety 栽培品種
cultivar tea field 品種茶園
cultivated einkorn wheat, *Triticum monococcum* L. 栽培一粒系コムギ
cultivated land, arable land 耕地
cultivated soil, arable soil, topsoil 耕地土壌, 耕土
cultivated species 栽培種
cultivated variety, cultivar (cv.) 栽培品種
cultivation 耕作, 栽培, 農耕
cultivation experiment 栽培試験
cultivation limit 栽培限界
cultivation method 耕種法
cultivation system 栽培体系
cultivation technique 栽培技術
cultivation test 栽培試験
cultivation type 栽培型
cultivation under lightening, light culture 電照栽培
cultivator カルチベータ, 中耕除草機
cultural weed control 耕種的雑草防除
culture 培養
culture medium 培養基
culture medium (*pl.* media) 培地
culture solution 培養液
cultured cell 培養細胞
cumin, *Cuminum cyminum* L. クミン
Cuminum cyminum L., cumin クミン
cumulative contribution 累積寄与率
cumulative distribution function 累積分布関数
cumulative frequency 累積頻度, 累積度数
cumulative selection 累積選抜
cumulative temperature, accumulated temperature 積算温度
Curcuma angustifolia Roxb., East Indian arrowroot 東インドアロールート
Curcuma longa L. (= *C. domestica* Valet), turmeric ウコン(欝金)
Curcuma zedoaria (Christm.) Roscoe, zedoary ガジュツ(莪蒁)
cured leaf 乾葉(かんぱ)【タバコ】
curing 乾燥【タバコ】
curing barn 乾燥室【タバコ】
curvature, bending 屈曲
curve fitting 曲線の当てはめ
curvilinear regression 曲線回帰
Cuscuta australis R. Br., Australian dodder マメダオシ
Cuscuta japonica Choisy, Japanese dodder ネナシカズラ
custard apple, sugar apple, sweet sop, *Annona squamosa* L. バンレイシ
cutch, catechu, black catechu, *Acacia catechu* (L.f.) Willd. アセンヤクノキ(阿仙薬樹)
cuticle クチクラ
cuticular layer クチクラ層
cuticular resistance クチクラ抵抗
cuticular transpiration クチクラ蒸散
cutin クチン
cutinization クチン化
cuttage 挿木繁殖
cutters 合葉(あいは)【タバコ】
cutting 1) 挿し木 2) 挿し穂 3) 刈取り
cutting-back pruning, heading-back pruning 切返しせん(剪)定
cutting frequency 刈取り頻度, 刈取り回数
cutting height, clipping height, mowing height 刈取り高さ
cutting interval, clipping interval, mowing interval 刈取り間隔
cutting width, swath 刈り幅
CV (coefficient of variation) 変動係数

cv. (cultivar), cultivated variety　栽培品種

Cyamopsis tetragonoloba (L.) Taub., guar, cluster bean　クラスタマメ

cyanamide nitrogen　シアナミド態窒素

cyanide　シアン化物

cyanide toxicity　青酸中毒

cyanobacteria　シアノバクテリア，藍藻類

cybrid　サイブリッド

cycad, *Cycas* spp.　ソテツ(蘇鉄)【広義】

Cycas revoluta Bedd. (= *C. revoluta* Thunb.), Japanese sago palm　ソテツ(蘇鉄)【狭義】

Cycas spp., cycad　ソテツ(蘇鉄)【広義】

cycle of matter, material cycle　物質循環

cyclin　サイクリン

cycling of nitrogen, nitrogen cycle　窒素循環

cyclone, depression, low　低気圧

cycocel, chlorocholine chloride (CCC), 2-chloroethyltrimethyl-ammonium chloride　サイコセル

cylindrical root, pencil-like root　梗根(こうこん)，ごぼう根【サツマイモ】

Cymbopogon citratus (D.C. ex Nees) Stapf, West Indian lemongrass　西インドレモングラス

Cymbopogon flexuosus (Nees ex Steud.) Wats., East Indian lemongrass, Malabar grass　東インドレモングラス

Cymbopogon martini (Roxb.) Wats., palmarosa　パルマローザ

Cymbopogon martini Stapf var. *sofia*, gingergrass　ジンジャーグラス

Cymbopogon nardus (L.) Rendle, Ceylon citronella grass　セイロンシトロネラソウ

Cymbopogon winterianus Jowitt, Java citronella grass　ジャワシトロネラソウ

Cynodon dactylon Flugge (= *C. dactylon* (L.) Pers.), Bermuda grass　バーミューダグラス

Cyperaceae　カヤツリグサ科

cyperaceous weed, sedge weed　カヤツリグサ科雑草

Cyperus brevifolius (Rottb.) Hassk. var. *leiolepis* (Franch. et Savat.) T. Koyama (= *Kyllinga brevifolia* Rottb.)　ヒメクグ

Cyperus compressus L., annual sedge　クグガヤツリ

Cyperus difformis L., smallflower umbrella sedge　タマガヤツリ

Cyperus esculentus L., tiger nut　ショクヨウガヤツリ

Cyperus exaltatus Retz. ssp. *iwasakii* (Makino)　カンエンガヤツリ

Cyperus flaccidus R. Br.　ヒナガヤツリ

Cyperus globosus All. (= *Pycreus flavidus* (Retz.) T. Koyama), globe sedge　アゼガヤツリ

Cyperus haspan L.　コアゼガヤツリ

Cyperus iria L., rice flatsedge　コゴメガヤツリ

Cyperus malaccensis Lam. ssp. *brevifolius* (Boeck.) T. Koyama, Chinese mat grass, three-cornered grass　シチトウイ(七島藺)

Cyperus microiria Steud.　カヤツリグサ

Cyperus papyrus L., papyrus　カミガヤツリ

Cyperus polystachyos Rottb. (= *Pycreus polystachyos* (Rottb.) P. Beauv.)　イガガヤツリ

Cyperus rotundus L., nut grass　ハマスゲ

Cyperus sanguinolentus Vahl (= *Pycreus sanguinolentus* (Vahl) Nees)　カワラスガナ

Cyperus serotinus Rottb. (= *Juncellus serotinus* (Rottb.) C. B. Clarke)　ミズ

ガヤツリ
Cyperus tenuispica Steud. ヒメガヤツリ
Cyrtosperma chamissonis (Schott) Merr., [giant] swamp taro　スワンプタロ
cyst nematode　シストセンチュウ
cyst nematode disease　シスト線虫病
cysteine (Cys)　システイン
cystine ((Cys)$_2$)　シスチン
cytochemistry　細胞化学
cytochrome　シトクロム，チトクロム
cytogene, cytoplasmic gene, plasmagene　細胞質遺伝子
cytogenetics　細胞遺伝学
cytokinesis　細胞質分裂
cytokinin　サイトカイニン
cytology　細胞学
cytoplasm　細胞質
cytoplasmic gene, cytogene, plasmagene　細胞質遺伝子
cytoplasmic hybrid　細胞質雑種
cytoplasmic inheritance　細胞質遺伝
cytoplasmic male sterility　細胞質雄性不稔
cytoplasmic membrane　細胞膜
cytoplasmic mutation　細胞質突然変異
cytoskeleton　細胞骨格

[D]

2,4-D (2,4-dichlorophenoxy acetic acid)　2,4-ジクロロフェノキシ酢酸
D-A (digital-analog) conversion　D-A 変換
Dactylis glomerata L., orchardgrass, cocksfoot　オーチャードグラス
Dactyloctenium aegyptium (L.) P. Beauv., crowfootgrass　タツノツメガヤ
daily gain (DG)　日増体[量]【家畜】
daily mean [air] temperature　日平均気温
daily temperature range　日温度較差
dairy farming, dairying　酪農

Dalbergia cochinchinensis Pierre ex Laness. など数種，rosewood　シタン(紫檀)
dallisgrass, *Paspalum dilatatum* Poir.　ダリスグラス
Dalmatian chrysanthemum, insectpowder plant, insect flower, *Pyrethrum cinerariifolium* Trevir. (= *Chrysanthemum cinerariaefolium* Visiani)　ジョチュウギク(除虫菊)
damaged grain (kernel)　被害粒
damask rose, *Rosa damascena* Mill.　ダマスクバラ
damping-off　1) 立枯れ　2) 苗立枯[病]
dandelion, *Taraxacum officinale* Weber　セイヨウタンポポ
dark-field microscope　暗視野顕微鏡
dark flour, sanago　さなご，末粉(すえこ)【ソバ】
dark germinater, negative photoblastic seed　暗発芽種子
dark germination　暗発芽
dark period　暗期
dark reaction　暗反応
dark respiration　暗呼吸
darnel, poison ryegrass, *Lolium temulentum* L.　ドクムギ(毒麦)
Darwinism　ダーウィニズム，ダーウィン説
dasheen, taro, *Colocasia esculenta* (L.) Schott var. *esculenta* Hubbard & Rehder　タロイモ
database　データベース
date palm, *Phoenix dactylifera* L.　ナツメヤシ(棗椰子)
Datura inoxa Mill. (= *D. meteloides* Dunal.), sacred datura　アメリカチョウセンアサガオ
Datura metel L. (= *D. alba* Nees), white datura　チョウセンアサガオ
Datura stramonium L., thorn apple, Jimson weed　ヨウシュチョウセンア

サガオ
Daucus carota L., carrot ニンジン(人参)
daughter cell 娘細胞(じょうさいぼう), 嬢細胞
daughter nucleus 娘核(じょうかく), 嬢核
daughter tuber 子いも【地下茎】
daughter tuberous root 子いも【根】
day-neutral plant, neutral plant 中性植物
dayflower, Asiatic dayflower, *Commelina communis* L. ツユクサ(露草)
daylength 日長
dead cotton 死綿(しめん)
dead ripe 枯熟
dead-ripe stage 枯熟期
deadly nightshade, belladonna, *Atropa belladonna* L. ベラドンナ
death, dying, plant death 枯死
decapitation 頂部除去
decarboxylase 脱炭酸酵素
deciduous 落葉性の
deciduous broad-leaved tree 落葉広葉樹
decision support system 意思決定支援システム
decussate 十字対生
dedifferentiation 脱分化
deductive inference 演繹的推論
deep-flood irrigation 深水灌漑
deep flooding 深水
deep flow technique (DFT) 湛液水耕
deep freezer 超低温庫, 冷凍冷蔵庫
deep-furrow sowing (seeding) 深溝播き
deep planting 深植え
deep plowing 深耕
deep rooted 深根性の
deep-rooted crop 深根性作物
deep sowing (seeding) 深播き
deep tillage 深耕
deep trimming of canopy 深刈り【チャ】
deepwater rice 深水稲(ふかみずいね)

defatted soybean 脱脂大豆
deficiency 1) 欠乏 2) 欠失【染色体】
deficiency symptom, hunger sign 欠乏症[状]
definite bud 定芽
defloration, flower picking, flower thinning 摘花
defoliant, defoliator 枯葉剤, 落葉剤
defoliation 落葉, 摘葉, せん(剪)葉【草地】
deformation, terata, malformation, deformity 奇形
deformed grain 奇形粒
degeneracy 縮重
degeneration 退化
degeneration of variety 品種退化
degraded paddy field 老朽化水田
degree of freedom 自由度
degree of spring habit 春播き性程度
degree of winter habit 秋播き性程度
dehiscence 裂開
dehiscent fruit 裂開果
dehulling, hulling, husking 脱ぷ(稃)
dehumidifying 除湿
dehydration, desiccation 脱水
dehydrogenase 脱水素酵素
deionized water 脱イオン水, 脱塩水
delayed heading 出穂遅延
delayed pollination 遅延受粉
deleterious substance 劇物
deletion, deficiency 欠失【染色体】
delta 三角洲, デルタ
demand and supply of food 食糧(食料)需給
demonstration farm 展示圃[場]
demonstration plot 展示圃[場]
denaturation 変性
dendrogram 樹状図, デンドログラム
denitrification 脱窒[作用]
dense dibbling 多株穴播き
dense drilling 多条播き, 密条播
dense ear, compact ear 密穂(みっすい)
dense planting, close planting 密植

dense sowing (seeding), thick sowing (seeding)　厚播き, 密播 (みっぱ, みっぱん)
densitometer　デンシトメーター
density　密度
density effect　密度効果
density function　密度関数【確率分布】
density gradient centrifugation　密度勾配遠心
dent corn, *Zea mays* L. var. *indentata* Bailey　デントコーン
deoxyribonucleic acid (DNA)　デオキシリボ核酸
dependent variable　従属変数
depression, low, cyclone　低気圧
derived line　派生系統
dermal system, epidermal system　表皮系
dermal toxicity　経皮毒性
derris, *Derris* spp. (*D. elliptica* (Roxb.) Benth. など)　デリス
Derris spp.【*D. elliptica* (Roxb.) Benth. など】, derris　デリス
desalinization　除塩
descendant, progeny, offspring　後代
desert plant, eremophyte　砂漠植物
desertification　砂漠化
desiccant　乾燥剤
desiccating agent　乾燥剤
desiccation　乾燥, 脱水
desiccator　デシケーター
design for fertilizer application　施肥設計
design of experiment, experimental design　実験計画
desmodium, *Desmodium* spp.　デスモディウム
Desmodium spp., desmodium　デスモディウム
desmotuble　デスモ小管
detached leaf　切断葉
detasseling　雄穂 (ゆうすい) 除去
detection of outlier　外れ値の検出

deterioration　劣化
determinate inflorescence　有限花序
determinate type　有限伸育型
deterministic model　決定論的モデル
developer　現像液
development　1) 発育, 発生　2) 現像
developmental phase　発育相
developmental physiology　発育生理
developmental stage　発育段階
devernalization　春化消去
deviation　偏差
devil's beggarticks, *Bidens frondosa* L.　アメリカセンダングサ
dew　露 (つゆ)
dew-point hygrometer　露点計
dew-point temperature　露点温度
dew retting　雨露 (うろ) 法
dextrin　デキストリン
diakinesis stage　ディアキネシス期, 移動期【減数分裂】
diallel cross　ダイアレル交配
dialysis　透析
diameter　直径
diaphragm　絞り【光学機器】
diaphragm pump　ダイアフラムポンプ
diatomaceous earth　ケイ (珪) 藻土
diazomethane (CH_2N_2)　ジアゾメタン
dibble planting　穴植え
dibbling　穴播き
2,4-dichlorophenoxy acetic acid (2,4-D)　2,4-ジクロロフェノキシ酢酸
dichogamy　雌雄異熟
dichotomous branching, dichotomy　二叉 (にさ) 分枝, 二叉 (ふたまた) 分枝
diclinism, monoecism　雌雄異花同株 [性]
dicot　双子葉植物
dicotyledon　双子葉植物
dicotyledonous　双子葉の
dictyosome　ディクチオゾーム
dictyostele　網状中心柱
dielectric constant　誘電率

differential centrifugation 分画遠心
differential display ディファレンシャルディスプレイ
differential species 識別種, 判別種
differential type 差働型【ガス分析】
differential variety 識別品種, 判別品種
differentiation 分化
diffraction 回折
diffraction grating 回折格子
diffused solar radiation 散乱太陽放射, 散乱日射
diffused light, scattered light 散乱光
diffusible auxin 拡散型オーキシン
diffusible hormone 拡散型ホルモン
diffusion 拡散
diffusion conductance 拡散伝導度
diffusion process 拡散過程
diffusion resistance 拡散抵抗
digenomic species 二基種
digestibility 消化率
digestible crude protein (DCP) 可消化粗タンパク質
digestible dry matter 可消化乾物
digestible energy (DE) 可消化エネルギー
digestible nutrient 可消化養分
digestible organic matter 可消化有機物
digestion 消化
digger 掘取り機
digging and pulling hoe 打引ぐわ (鍬)
digging hoe 打ぐわ (鍬)
digital デジタル
digital-analog (D-A) conversion D-A変換
digital elevation map 数値標高地図
digital map 数値地図
digital national land information 国土数値情報
digitalis, foxglove, *Digitalis purpurea* L. ジギタリス
Digitalis purpurea L., digitalis, foxglove ジギタリス
Digitaria adscendens (H. B. K.) Henr. (= *D. ciliaris* (Retz.) Koeler), crabgrass メヒシバ
Digitaria eriantha Steud., pangola grass パンゴラグラス
Digitaria exilis (Kippist) Stapf, fonio, fundi, hungry rice フォニオ
Digitaria violascens Link, violet crabgrass アキメヒシバ
digitizer ディジタイザー
dihybrid 両性雑種, 二遺伝子雑種
dill, *Anethum graveolens* L. イノンド, ディル
diluent 希釈剤
dilution 希釈
diluvial land 洪積地
diluvial soil 洪積土
diluvium 洪積層
dinitrogen monoxide (N_2O) 一酸化二窒素
dinkel wheat 普通系コムギ
dioecism 雌雄異株[性]
Dioscorea alata L., greater yam, water yam, winged yam ダイジョ (大薯)
Dioscorea bulbifera L., aerial yam カシュウイモ
Dioscorea cayenensis Lam., yellow guinea yam ギニアヤム【黄肉】
Dioscorea esculenta (Lour.) Burk., lesser yam, potato yam トゲドコロ
Dioscorea japonica Thunb. ex Murray, Japanese yam ヤマノイモ
Dioscorea opposita Thunb. (= *D. batatas* Decne.), Chinese yam ナガイモ (薯蕷)
Dioscorea pentaphylla L., five-leaved yam ゴヨウドコロ
Dioscorea rotundata Poir., white guinea yam ギニアヤム【白肉】
Diospyros kaki Thunb., kaki, Japanese persimon カキ (柿)
diploid, diplont 二倍体, 複相[体]
diploid plant 二倍体植物
diploidy 二倍性

diplotene stage　複糸期, ディプロテン期【減数分裂】
Dipsacus fullonum L., teasel　チーゼル
direct gene transfer, direct transformation　直接遺伝子導入
direct light　直達光
direct planting　直播【サツマイモ】
direct solar radiation　直達日射, 直達太陽放射
direct sowing (seeding)　直播, 直播き
direct sunlight　直達光
direct transformation, direct gene transfer　直接遺伝子導入
direction for safe use　安全使用基準【農薬】
disaccharide　二糖
disbudding, bud picking　摘芽
disc waterhyssop, *Bacopa rotundifolia* (Michx.) Wettst.　ウキアゼナ
discoloration　変色
discontinuous distribution　不連続分布
discontinuous variation　不連続変異
discrete distribution　離散分布
discrete variable　離散変数
discrete variate　離散変量
discriminant analysis　判別分析
discriminant function　判別関数
disease　1) 病気　2) り (罹) 病
disease and insect damage　病虫害
disease forecasting, forecasting of disease outbreak　病害発生予察
disease garden, test field for disease-tolerance [evaluation]　耐病性検定圃
disease injury　病害
disease resistance　耐病性, 病害抵抗性
disease tolerant variety　耐病性品種
disinfection　消毒
disk plow　ディスクプラウ
dispersed irrigation　分散灌漑
dispersing agent　分散剤
dissecting microscope　解剖顕微鏡
dissecting pan　解剖皿
dissecting scalpel　メス
dissecting scissors　解剖ばさみ
dissemination　1) 伝播　2) 散布【種子】
dissimilation, catabolism　異化 [作用]
dissociation　解離【化学】
dissolved oxygen (DO)　溶存酸素
distichous opposite　二列対生
distilled water　蒸留水
distribution　分布
distribution function　分布関数
distribution parameter　分布パラメータ
disturbance　かく (攪) 乱
ditcher, ridger, lister　うね (畝) 立て機
diurnal change　日変化
diurnal range　日較差
diurnal variation　日変化
divergence　開度【葉序】
division, suckering　株分け
division of root stock, root splitting　根分け
DNA (deoxyribonucleic acid)　デオキシリボ核酸
DNA ligase　DNA リガーゼ
DNA polymerase　DNA ポリメラーゼ
DO (dissolved oxygen)　溶存酸素
Dolichos lablab L. (= *Lablab purpureus* (L.) Sweet), lablab, hyacinth bean　フジマメ【鵲豆】
Dolichos uniflorus Lam. (= *Macrotyloma uniflorum* (Lam.) Verdc.), horsegram　ホースグラム
domain　ドメイン【タンパク質】
domestic animal, livestock, farm animal　家畜
domestic consumption　国内消費
domestic production　国内生産
domestic rice　国内産米
domestic tobacco　在来種【タバコ】
domestication　1) 栽培化　2) 家畜化
dominance　優性
dominant effect　優性効果
dominant mutation　優性突然変異
dominant species　優占種
dominant weed　優占雑草

doner parent 供与親
donor 供与体
dopatrium, *Dopatrium junceum* (Roxb.) Buch.-Ham. ex Benth. アブノメ
Dopatrium junceum (Roxb.) Buch.-Ham. ex Benth., dopatrium アブノメ
dormancy, rest 休眠
dormancy awakening 休眠覚醒
dormancy breaker 休眠覚醒剤
dormancy breaking 休眠打破
dormant bud, resting bud 休眠芽
dormant period 休眠期[間]
dormant season 休眠期[間]
dormant seed 休眠種子
dorsal 背側の
dorsal side 背面
dorsiventrality 背腹性
dosage, dose 用量
dose-response curve 用量-反応曲線
dot 果点
double cropping 1) 二期作 2) 二毛作
double cross 複交雑, 複交配
double crossing-over 二重乗換え
double fertilization 重複受精
double flower 八重咲き
double mutation 二重突然変異
double recessive 二重劣性
double staining 二重染色
doubled haploid 倍加半数体, 半数体倍加系統
doubling of chromosome, chromosome doubling 染色体倍加
dough パン生地
dough ripe 糊熟(こじゅく)
dough[-ripe] stage 糊熟期
downy mildew 1) 黄化萎縮病【イネ】 2) べと病
Dr. Martius' wood-sorrel, *Oxalis corymbosa* DC. ムラサキカタバミ
draft animal 役畜(えきちく)
drag, comb harrow 馬ぐわ(鍬)(まぐわ)
drainable rice nursery 乾用苗代

drainage 排水【耕地】
drainage after sprouting 芽干し
drainage of residual water, ponding water release 落水
drawing prism, camera lucida カメラルシダ
dressing furrow 施肥溝
dried corm slice 荒粉(あらこ)【コンニャク】
dried soil, oven-dry soil 乾土
drill 条播機, すじ播き機, 施肥播種機
drilling, row sowing (seeding) 条播(じょうは)
drip-free film 無滴フィルム
drip irrigation, trickle irrigation 点滴灌漑
dropping bottle 摘びん(瓶)
dropping pipet, Pasteur pipet 駒込ピペット
dropwort, *Oenanthe javanica* (Blume) DC. セリ
drought かんばつ(旱魃)
drought avoidance かんばつ(旱魃)回避性, 乾燥回避性
drought damage (injury) 干害(旱害)
drought escape かんばつ(旱魃)逃避性, 乾燥逃避性
drought resistance 耐乾性
drought tolerance かんばつ(旱魃)耐性, 乾燥耐性
drupe 核果
dry-bulb temperature 乾球温度
dry climate, arid climate 乾燥気候
dry farming 乾地農業, 乾地農法
dry fruit 乾果
dry matter 乾物
dry matter partitioning ratio 乾物分配率
dry-matter production 乾物生産, 物質生産
dry-matter yield 乾物収量
dry rot 乾腐病【コンニャク】
dry season 乾季
dry-season crop 乾季作[物]

dry-season cropping　乾季作
dry set　子球(しきゅう)【タマネギ等】
dry stem of hemp　しめそ(乾麻)
dry storage　乾燥貯蔵
dry weight　乾物重
dry wind damage (injury)　乾風害
dryer, drying machine　乾燥機
drying, desiccation　乾燥
drying machine, dryer　乾燥機
drying oil　乾性油
drying on ground　地干し
drying plant　乾燥施設
Duchesnea indica var. *japonica* Kitam. (= *D. chrysantha* (Zoll. et Mor.) Miq.), Indian strawberry　ヘビイチゴ
Duncan's multiple range test　ダンカンの多重検定
dung, feces　ふん(糞)
duplicate gene　重複遺伝子
duplicated chromosome　重複染色体
durable years　耐用年数
duration of continuous snow cover　根雪期間
duration of sunshine　日照時間
durian, *Durio zibethinus* Murr.　ドリアン
Durio zibethinus Murr., durian　ドリアン
durum wheat, macaroni wheat, *Triticum durum* Desf.　デュラムコムギ, マカロニコムギ
dust　粉剤【農薬】
dust coating, dust dressing　粉衣(ふんい)
dust diluent　増量剤【農薬】
duster　散粉機
dusting　散粉
duty of [irrigation] water, irrigation requirement　用水量
dwarf　1) わい(矮)性[の] 2) 萎縮病【イネ, ダイズ】
dwarf bamboo, *Pleioblastus chino* (Franch. et Savat.) Makino var. *viridis* (Makino) S. Suzuki (= *Arundinaria pygmaea* Mitford var. *glabra* Ohwi) ネザサ
dwarf banana, *Musa cavendishii* Lamb.　サンジャクバナナ(三尺バナナ)
dwarf bean, bush bean　ツルナシインゲン(蔓無隠元)
dwarf mutants　わい(矮)性変異体
dwarfing, stunting　わい(矮)化
dwarfing rootstock　わい(矮)性台木
dwarfism, dwarfness　わい(矮)性
dye　色素【染料】
dye [stuff] crop　染料作物
dying, death, plant death　枯死
dying-off　枯上がり【イネ】
dynamic equilibrium　動的平衡

[E]

ear　1) 穂　2)雌穂(しすい)【トウモロコシ】
ear breaking　穂切れ
ear characteristics　穂相
ear emergence, heading　出穂
ear length, panicle length　穂長
ear number, panicle number　穂数
ear-number type, many-tillering type, panicle-number type　穂数型
ear plucking　穂刈り
ear reaping　穂刈り
ear-selection method, head-selection method　穂選抜法
ear tip　穂先
ear-to-row test, head-to-row test　一穂一列検定
ear type, head type, panicle type　穂型
ear weight, panicle weight　穂重
ear-weight type, heavy-panicle type, panicle-weight type　穂重型
earliness　早晩性
early flowering　早咲き
early frost　早霜
early generation　初期世代
early harvesting　早刈り, 早取り, 早掘り

early maturation　早熟
early maturing habit　早熟性
early [maturing] variety　早生 (わせ) 品種
early planting　早植え
early-planting culture　早植栽培
early sowing (seeding)　早播き
early-season culture　早期栽培
early [season] delivery rice　早場米
earth science, geoscience　地球科学
earthing up, ridging, molding　土寄せ, 培土
easily decomposable organic matter　易分解性有機物
East Indian arrowroot, *Curcuma angustifolia* Roxb.　東インドアロールート
East Indian arrowroot, Tahiti arrowroot, *Tacca leontopetaloides* (L.) Kuntze (= *T. pinnatifida* Forst.)　タシロイモ
East Indian lemongrass, Malabar grass, *Cymbopogon flexuosus* (Nees ex Steud.) Wats.　東インドレモングラス
eastern brackenfern, bracken, *Pteridium aquilinum* (L.) Kuhn var. *latiusculum* (Desv.) Underw. ex A.Heller　ワラビ (蕨)
eastern red cedar, pencil cedar, *Sabina virginiana* (L.) Antoine (= *Juniperus virginiana* L.)　エンピツビャクシン
eating quality, palatability, taste　食味
eating quality test　食味試験
ecesis, establishment　定着, 土着
Echinochloa colonum (L.) Link, jungle rice　ワセビエ
Echinochloa crus-galli (L.) Beauv., cockspur grass, barnyard grass　ケイヌビエ
Echinochloa crus-galli (L.) Beauv. var *oryzicola* (Vasing.) Ohwi, barnyard grass　タイヌビエ
Echinochloa crus-galli (L.) Beauv. var. *crus-galli*, barnyard grass　イヌビエ

Echinochloa crus-galli (L.) P. Beauv. var. *formosensis* Ohwi, barnyard grass　ヒメタイヌビエ
Echinochloa crus-galli (L.) P. Beauv. var. *praticola* Ohwi, barnyard grass　ヒメイヌビエ
Echinochloa frumentacea (Roxb.) Link, Indian barnyard millet, billion dollar grass　インドビエ
Echinochloa utilis Ohwi et Yabuno, Japanese millet, barnyard millet　ヒエ (稗, 穆)
eclipta, false daisy, *Eclipta prostrata* L. (= *E. alba* L.)　タカサブロウ
Eclipta prostrata L. (= *E. alba* L.), eclipta, false daisy　タカサブロウ
ecological isolation　生態的隔離
ecological [pest] control　生態的防除
ecology　生態学
economic character (trait)　実用形質
economic yield　経済的収量
ecophysiology　生態生理学
ecospecies　生態種
ecosystem　生態系
ecotype　生態型
ectomycorrhiza, ectotrophic mycorrhiza　外生菌根
edaphic factor　土壌要因
edaphology　エダホロジー
eddoe, *Colocasia esculenta* (L.) Schott var. *antiquorum* Hubbard & Rehder　サトイモ (里芋)
Edgeworthia chrysantha Lindl. (= *E. papyrifera* Sieb. et Zucc.), mitsumata　ミツマタ (三椏)
edible burdock, *Arctium lappa* L.　ゴボウ (牛蒡)
edible canna, purple arrowroot, Queensland arrowroot, achira, *Canna edulis* Ker-Gawl.　食用カンナ
edible flower　食用花
effective heritability　有効遺伝率
effective population size　有効な集団の

大きさ
effective soil depth 有効土層
effective temperature 有効温度
efficiency 有効性【統計】
efficiency of light energy conversion 光エネルギー転換効率
efficiency of light energy utilization 光エネルギー利用効率
efficiency of solar energy utilization 太陽エネルギー利用効率
efflux, runoff, outflow 流去,流出
Egeria densa Planch., Brazilian elodea オオカナダモ
egg, ovum 卵(らん)
egg cell 卵細胞
egg nucleus 卵核
eggplant, *Solanum melongena* L. ナス(茄子)
Egyptian clover, Berseem clover, *Trifolium alexandrinum* L. エジプシャンクローバ
Egyptian cotton, *Gossypium barbadense* L. エジプトメン(エジプト棉)【カイトウメンの一系統】
Eh (oxidation-reduction potential), redox potential 酸化還元電位
Eichhornia crassipes (Mart.) Solms-Laub., water hyacinth ホテイアオイ(布袋葵)
eigenvalue 固有値
eigenvector 固有ベクトル
einkorn wheat 一粒系コムギ
Elaeis guineensis Jacq., oil palm, African oil palm アブラヤシ(油椰子)
Elatine triandra Schkuhr ミゾハコベ
elecampane, *Inula helenium* L. オグルマ
electric conductivity 電気伝導率
electric hotbed 電熱温床
electromagnetic wave 電磁波
electron density 電子密度
electron microscope 電子顕微鏡
electron transport chain 電子伝達系
electron transport system 電子伝達系
electrophoresis 電気泳動
electroporation エレクトロポレーション,電気穿孔法
elementary species 基本種
Eleocharis acicularis (L.) Roem. et Schult., needle spikerush, slender spikerush マツバイ
Eleocharis congesta D. Don ssp. *japonica* (Miq.) T. Koyama ハリイ
Eleocharis dulcis (Burm. f.) Trin. ex Hensch. var. *tuberosa* (Roxb.) T. Koyama, Chinese water chestnut オオクログワイ
Eleocharis kuroguwai Ohwi クログワイ
elephant foot, konjak, *Amorphophallus konjac* K. Koch コンニャク(蒟蒻)
elephant grass, napiergrass, *Pennisetum purpureum* Schumach. ネピアグラス
elephant yam, *Amorphophallus paeoniifolius* (Dennst.) Nicolson (= *A. campanulatus* Bl.) ゾウコンニャク
Elettaria cardamomum (L.) Maton, cardamon ショウズク(小豆蔲),カルダモン
Eleusine coracana (L.) Gaertn., African millet, finger millet シコクビエ(龍爪稷)
Eleusine indica (L.) Gaertn., goosegrass, wiregrass オヒシバ
elevation, altitude 高度
ELISA (enzyme-linked immunosorbent assay) 固相酵素免疫検定法
elite plant 選択株
ellipse 楕円
Elodea nuttallii (Planch.) St. John, western elodea コカナダモ
elongation 伸長
Elsholtzia ciliata (Thunb.) Hylander ナギナタコウジュ
eluviation, leaching 溶脱

Elymus canadensis L., Canada wild rye　カナダワイルドライ
Elymus junceus Fisch. (= *Psathyrostachys juncea* (Fisch.) Nevski), Russian wildrye　ロシアワイルドライ
emasculation, castration　除雄
embedding　包埋(ほうまい)
embedding agent　包埋剤
embryo　胚
embryo culture　胚培養
embryo factor　胚成長因子
embryo transplanting　胚移植
embryogenesis　胚発生, 胚形成
embryoless seed　無胚種子
embryology　発生学
embryosac　胚のう(囊)
embryosac mother cell (EMC)　胚のう(囊)母細胞
emergence　1) 出芽　2) 抽出【発育】　3) 毛状体
emergency crop　救荒作物, 備荒作物
emergent weed　抽水雑草
emersed weed　抽水雑草
Emerson effect　エマーソン効果
emission concentration regulation　濃度規制
emission spectrochemical analysis　発光分光分析
Emmer, *Triticum dicoccum* Schubl.　エンマーコムギ
emmer wheat　二粒系コムギ
empirical model　経験的モデル
empty glume, sterile glume　空穎(くうえい)
empty grain, abortive grain　しいな(粃, 秕)
emulsifiable concentrate　乳剤
emulsifier　乳化剤
emulsifying agent　乳化剤
end product　最終産物
end product inhibition　最終産物阻害
end wall　端壁(たんぺき)

end-season pollination　末期受(授)粉
endocarp　内果皮
endocrine disrupting chemicals (EDC)　外因性内分泌かく乱物質, environmental hormones　環境ホルモン
endodermis　内皮
endogenous　内生の
endogenous auxin　内生オーキシン
endogenous rhythm　内生リズム
endomycorrhiza, endotrophic mycorrhiza　内生菌根
endoplasmic reticulum (ER)　小胞体
endosperm　内[胚]乳, 胚乳【慣用】
endosperm type　もちうるち(糯粳)性
endosymbiosis　細胞内共生
energy balance, heat balance, heat budget　熱収支
energy efficiency　エネルギー効率
enhancer　エンハンサー
enokitake fungus, *Flammulina velutipes* (Fr.) Karst.　エノキタケ
enriched rice　強化米
Ensete ventricosum (Welw.) Cheesman, Abyssinian banana　アビシニアバショウ
ensilage　エンシレージ
ensiling, silage making　サイレージ調製
entomology　昆虫学
entomophilae　虫媒植物
entomophilous flower　虫媒花
entomophilous plant　虫媒植物
entomophily, insect pollination　虫媒
entropy　エントロピー
environment　環境
environmental biology　環境生物学
environmental capacity, carrying capacity　環境収容力, 環境容量
environmental conservation　環境保全
environmental correlation　環境相関
environmental education　環境教育
environmental factor　環境要因
environmental green space　環境緑地

environmental impact assessment 環境アセスメント, 環境影響評価
environmental indicator 環境指標
environmental mutagen 環境[突然]変異源
environmental pollutant 環境汚染物質
environmental pollution, public hazards 公害
environmental quality standard 環境基準
environmental science 環境科学
environmental stress 環境ストレス
environmental variance 環境分散
environmental variation 環境変異
enzyme 酵素
enzyme activity 酵素活性
enzyme immunoassay (EIA) 酵素免疫測定法
enzyme-linked immunosorbent assay (ELISA) 固相酵素免疫検定法
ephemeral 短命の
epiblast エピブラスト
epicarp 外果皮
epicotyl 上胚軸
epidermal system, dermal system 表皮系
epidermis 表皮
epigeal cotyledon 地上子葉
epinasty 上偏成長
equatorial plane 赤道面【細胞】
equatorial plate 赤道板【細胞】
equilibrium 平衡状態
Equisetum arvense L., field horsetail スギナ
ER (endoplasmic reticulum) 小胞体
eradication 根絶, 撲滅
eragrostis, *Eragrostis multicaulis* Steud. ニワホコリ
Eragrostis abyssinica (Jacq.) Link (= *E. tef* Trotter), teff [grass] テフ
Eragrostis curvula (Schrad.) Nees, weeping lovegrass ウィーピングラブグラス
Eragrostis ferruginea (Thunb.) P. Beauv. カゼクサ
Eragrostis multicaulis Steud., eragrostis ニワホコリ
Erechtites hieracifolia (L.) Raf. ex DC., American burnweed ダンドボロギク
erect ear 立ち穂
erect leaf 直立葉
erect-leaved variety 直立葉型品種
erect stem, upright stem 直立茎
erect type, upright type 直立型
erectophyll 垂直葉型の
eremophyte, desert plant 砂漠植物
Erigeron annuus (L.) Pers. (= *Stenactis annuus* (L.) Cass.), annual fleabane ヒメジョオン(姫女苑)
Erigeron bonariensis L. (= *Conyza bonariensis* (L.) Cronq.), hairy fleabane アレチノギク
Erigeron canadensis L., horseweed ヒメムカシヨモギ
Erigeron philadelphicus L., Philadelphia fleabane ハルジオン(春紫苑)
Erigeron sumatrensis Retz. (= *Conyza sumatrensis* (Retz.) Walker), tall fleabane オオアレチノギク
Eriobotrya japonica (Thunb.) Lindl., loquat, Japanese medlar ビワ(枇杷)
Eriocaulon cinereum R.Br. ホシクサ
Eriocaulon miquelianum Koern. イヌノヒゲ
Eriocaulon robustius (Maxim.) Makino ヒロハイヌノヒゲ
Erlenmeyer flask 三角フラスコ
erosion 侵食
erosion control 侵食防止
error 誤差
error of the first kind 第一種の過誤【統計】
error of the second kind 第二種の過誤【統計】
error variance 誤差分散

erucic acid　エルカ酸, エルシン酸
Erythronium japonicum Decne., Japanese dog's tooth violet　カタクリ (片栗)
escape　1) 回避　2) 野生化
escaped plant　逸出植物
escaped species　逸出種
Escherichia coli　大腸菌
essential amino acid　必須アミノ酸
essential elements　必須元素
essential oil　精油
establishment　1) 苗立ち　2) 定着, 土着
estate agriculture　農園農業
esterase　エステラーゼ
estimate　推定値
estimated value　推定値
estimation　推定
estimator　推定量
estragon, tarragon, *Artemisia dracunculus* L.　タラゴン
ethephon　エテフォン, 2-chloroethyl phosphonic acid (CEPA)　2-クロロエチルホスホン酸
ether extract　エーテル抽出物
ethyl alcohol　エチルアルコール
ethylene　エチレン
etiolation, yellowing　黄化
etioplast　エチオプラスト
eucalyptus, *Eucalyptus* spp.【*E. citriodora* Hook., *E. macarthurii* Decne. et Maiden, *E. globulus* Labill. など】ユーカリ
Eucalyptus spp.【*E. citriodora* Hook., *E. macarthurii* Decne. et Maiden, *E. globulus* Labill. など】, eucalyptus　ユーカリ
eucaryote　真核生物
Euchlaena mexicana Schrad., teosinte　テオシント
Eugenia aromatica Kuntze (= *E. caryophyllata* Thunb., *E. caryophyllus* Spreng., *Syzygium aromaticum* (L.) Merr. et Perry), clove　チョウジ (丁子, 丁字)
eulalia grass, Japanese plume-grass, *Miscanthus sinensis* Andersson　ススキ (薄)
Euodia ruticarpa (Juss.) Benth.　ゴシュユ (呉茱萸)
Euphorbia humifusa Willd. (= *E. pseudochamaesyce* Fisch., Mey. et Lallem.)　ニシキソウ
Euphorbia supina Raf., prostrate spurge　コニシキソウ
European chestnut, *Castanea sativa* Mill.　セイヨウグリ
eustele　真正中心柱
Eutrema japonica (Miq.) Koidz. (= *E. wasabi* Maxim.), wasabi　ワサビ (山葵)
eutrophication　富栄養化
evaporation　蒸発
evaporation pan, evaporimeter, atmometer　蒸発計
evaporation suppressor　蒸発抑制剤
evaporimeter, evaporation pan, atmometer　蒸発計
evapotranspiration　蒸発散
evapotranspiration rate　蒸発散速度
evasion culture　回避栽培
evening primrose, *Oenothera erythrosepala* Borbás　オオマツヨイグサ
event　事象
ever-blooming, perpetual flowering　四季咲きの
ever-flowering, perpetual flowering　四季咲きの
evergreen　常緑の
evergreen tree　常緑樹
evolution　進化
evolution theory　進化論
exalbuminous seed　無胚乳種子
exarch　外原型
excess absorption　過剰吸収

excess damage (injury)　過剰障害
excess-moisture injury, wet injury　湿害
excess symptom　過剰症[状]
excess water tolerance, wet endurance　耐湿性
excessive vine growth　つる(蔓)ぼけ
exchange acidity　交換酸度
exchange capacity　交換容量
exchange of seeds　種子交換
exchangeable base　交換性塩基
exchangeable cation　交換性陽イオン
excited state　励起状態
exclosure, nongrazing area (plot)　禁牧区
excreta　排泄物
excretion　排出
exhibition field　見本園
exhibition garden　見本園
exodermis　外皮
exogamy, outbreeding　異系交配
exogenous　外生の, 外与の
exon　エキソン
exotic plant　外来植物
exotic species　外来種
exotic weed　外来雑草
expectation　期待値
expected frequency　期待頻度
expected value　期待値
expellent, repellent　駆除剤
experimental design, design of experiment　実験計画
experimental error　実験誤差
experimental field　試験圃[場]
experimental plot, plot　試験区
expert system　エキスパートシステム
explanatory variable　説明変数【回帰分析】
explant　外植体, 外植片
exponential distribution　指数分布
exponential growth　指数[型]成長
exponential growth curve　指数[型]成長曲線
exponential growth phase　指数成長期
extender　増量剤【授粉】
extensibility　伸展性, 伸展性
extension agent　農業改良普及員
extensive crop　粗放作物
extensive cultivation　粗放栽培
extinction coefficient　吸光係数
extraction　抽出【化学】
extrapolation　外挿[法]
extreme value　極値
extremely early [maturing] variety　極早生品種
extremely late [maturing] variety　極晩生品種
exudate　分泌物, 出液【物質】
exudation, bleeding　出液【現象】
eye　目【ジャガイモなど】
eye cutting　一芽挿し
eye judgement　肉眼鑑定
eye spot　眼紋病【コムギ】
eyepiece, ocular　接眼レンズ【顕微鏡】

[F]

F (fluorine)　フッ素
F_1 (first filial generation)　雑種第一代
F_1 hybrid　F_1雑種, 一代雑種
F-distribution　F分布
F-statistic　F統計量
F-table　F表
F-test　F検定
F-value　F値
FAA (formalin acetic alcohol)　ホルマリン酢酸アルコール
FACE (Free-Air CO_2 Enrichment)　フェイス
factor　要因, 因子
factorial　階乗
factorial design　要因設計, 因子設計
factorial experiment　要因実験, 因子実験
Fagopyrum cymosum Meisn., perennial buckwheat　シャクチリソバ(赤地利

蕎麦)
Fagopyrum esculentum Moench, buckwheat　ソバ(蕎麦)
Fagopyrum tataricum (L.) Gaertn., Tartary buckwheat, Kangra buckwheat　ダッタンソバ(韃靼蕎麦)
fall (autumn) crop　秋作[物]
fall (autumn) cropping　秋作
fall (autumn) planting　秋植え
fall dressing, autumn manuring　秋肥(あきごえ)
fall flowering　秋咲き
fall panicum, *Panicum dichotomiflorum* Michx.　オオクサキビ
fall sowing (seeding), autumn sowing (seeding)　秋播き
fallow　休閑, 休閑地
fallow crop　休閑作物
fallowed field　休閑地
fallowing　休閑耕
false daisy, eclipta, *Eclipta prostrata* L. (= *E. alba* L.)　タカサブロウ
false fertilization, pseudogamy　偽受精
false fruit, pseudocarp　偽果
false hybrid　偽雑種
family　1) 科　2) 家系
family analysis　家系分析
family selection　家系選抜
famine　飢饉(ききん)
FAO (Food and Agriculture Organization of the United Nations)　国際連合食糧農業機関
far-related species　遠縁種
farm animal, livestock, domestic animal　家畜
farm crops, field crops　農作物
farm household　農家
farm implements, agricultural implements　農具
farm management　農業経営, 営農
farm practices, farm working　農作業
farm-size expansion, scale expansion　規模拡大
farming　営農
farming machines and implements　農機具
farming system　農法
farming under structure, greenhouse agriculture, protected cultivation　施設農業
farmland consolidation　圃場整備
farmyard manure　きゅう(厩)肥
fasciation　帯化(たいか)
fascicular cambium, intrafascicular cambium　維管束内形成層
fasciculate　束生(そくせい)の, そう(叢)生の
fast green　ファストグリーン【染色】
fat　脂肪
fat and oil　油脂
Fatoua villosa (Thunb.) Nakai　クワクサ
fatty acid　脂肪酸
feces, dung　ふん(糞)
feed composition　飼料成分
feed [stuff]　飼料
feedback　フィードバック
feedback inhibition　フィードバック阻害
feeder root　1) 吸収根　2) 細根【チャ】
feeding　給餌, 飼養
feeding standard　飼養標準
feeding value, forage value　飼料価[値]
Fehling's reaction　フェーリング反応
female　雌[の]
female flower, pistillate flower　雌花(しか, めばな)
female plant　雌株(めかぶ)
female sterility　雌性不稔
fence cropping　周囲作
fennel, *Foeniculum vulgare* Mill.　ウイキョウ(茴香)
fermentation　発酵
fermented rice　発酵米
fermented soybeans　納豆(なっとう)
ferric hydroxide　水酸化第二鉄

ferrous iron 二価鉄
fertile land 肥沃地
fertile soil 肥沃土
fertility 稔性
fertilization 1) 受精, 授精 2) 施肥
fertilizer, manure 肥料
fertilizer application, fertilization, manuring 施肥
fertilizer applicator, fertilizing machine 施肥機
fertilizer distributor 肥料散布機
fertilizer drill, drill 施肥播種機
fertilizer effect 肥効
fertilizer injury, burning 肥焼け (こえやけ)
fertilizer response 肥料反応
fertilizer test 肥料試験
fertilizing machine, fertilizer applicator 施肥機
Festuca arundinacea Schreb., tall fescue, tall meadow fescue トールフェスク
Festuca ovina L., sheep fescue シープフェスク
Festuca pratensis Huds., meadow fescue メドーフェスク
Festuca rubra L., red fescue レッドフェスク
Feulgen's reaction フォイルゲン反応
fiber 繊維
fibrous root 1) ひげ根 2) 細根【チャ】
fibrous root system ひげ根型根系
Ficus carica L., fig イチジク (無花果)
Ficus elastica Roxb., Indian rubber, Assam rubber インドゴム
fidelity 適合度, 群落適合度【生態】
field 畑, 圃場, 本圃
field beet, fodder beet, mangold, *Beta vulgaris crassa* L. var. *alba* DC. 飼料用ビート
field capacity 圃場容水量
field crop 畑作物, 普通作物
field crop cultivation 畑作

field crops, farm crops 農作物
field day 公開日【農場等】
field experiment, field trial 圃場試験
field for seed production 種畑 (たねばた)
field grafting 居接ぎ (いつぎ)
field heaping, stack 野積み
field horsetail, *Equisetum arvense* L. スギナ
field maturing process 熟畑化
field reclamation 開畑
field resistance 圃場抵抗性
field survey 野外調査
field test 圃場検定
field trial, field experiment 圃場試験
fig, *Ficus carica* L. イチジク (無花果)
figleaved goosefoot, *Chenopodium ficifolium* Smith (= *C. serotinum* L.) コアカザ
filament 花糸
file ファイル【コンピュータ】
filling capacity 膨こう性【タバコ】
filling fiber 充てん (填) 用繊維
filling period, grain filling period, ripening period 登熟期間
filling power 膨こう性【タバコ】
filter funnel ろうと (漏斗)
filter paper ろ (濾) 紙
filtrate ろ (濾) 液
filtration ろ (濾) 過
Fimbristylis dichotoma (L.) Vahl, forked fringerush テンツキ
Fimbristylis miliacea (L.) Vahl, globe fringe-rush ヒデリコ
final dressing, last topdressing 止め肥 (とめごえ)
final puddling 植代 (うえしろ)
fine earth 細土
fine granule 微粒剤
fine root, rootlet 細根
fine sand 細砂
fine soil 細土
finger millet, African millet, *Eleusine*

coracana (L.) Gaertn. シコクビエ (龍爪稷)
finishing ratio, yielding percentage 歩留り
finite population 有限母集団
fire burnt kernel 熱損粒
fire curing 火乾 (かかん)
fired dryer, heated-air dryer 火力乾燥機
firing, burning 火入れ
first crop, primary canopy 一番草【牧草】
first crop of tea 一番茶
first cropping 一期作
first filial generation (F_1) 雑種第一代
first flower さきがけ花【タバコ】
first frost 初霜
first heading time 出穂始期
first plowing, coarse plowing 荒起し
first puddling, coarse puddling 荒代 (あらしろ)
first setting [of fruit] 一番成り
first weeding 一番除草
first year crop 一年子【コンニャク】
fish toxicity 魚毒性
fitness 適応度
five-leaved yam, *Dioscorea pentaphylla* L. ゴヨウドコロ
fixation 1) 固定 2) 定着【写真】
fixative 1) 固定液 2)定着液
fixative solution 固定液
fixed [effects] model 固定 [効果] 模型, 定数 [効果] 模型
fixed line 固定系統
flag leaf 止葉
flake, variegation 絞り【模様】
flame photometer 炎光光度計
flame weeding 火炎除草
Flammulina velutipes (Fr.) Karst., enokitake fungus エノキタケ
flat break 平耕 (ひらうち)
flat breaking 平面耕
flat level tea field 平坦地茶園

flat seedbed, level seedbed 平床 (ひらどこ)
flattening 葉のし【タバコ】
flavor and taste 風味
flax, *Linum usitatissimum* L. アマ (亜麻)
flexuous bittercress, *Cardamine flexuosa* With. タネツケバナ
flint corn, *Zea mays* L. var. *indurata* Bailey フリントコーン
flinty, glassy ガラス質の, 硝子質 (しょうししつ)の
floating leaved weed 浮葉雑草
floating nursery 浮苗床
floating rice 浮稲 (うきいね)
floating seedling 浮苗
floating weed 浮水雑草 (ふすいざっそう)
flood damage 水害
flood tolerance 冠水耐性
flooded irrigation 湛水灌漑
flooded paddy field, flooded ricefield 湛水田
flooded soil, submerged soil, waterlogged soil 湛水土壌
flooding 湛水, 冠水
floppy disk フロッピィーディスク
flora フロラ, 植物相
floral axis, rachis 花軸
floral diagram 花式図
floral differentiation, flower bud initiation (differentiation) 花芽分化
floral organ 花器
floral sterility caused by low temperature 障害型冷害【イネ】
Florence fennel, sweet anise, *Foeniculum vulgare* Mill. var. *dulce* (Mill.) Batt. et Trab. (= *F. dulce* Mill.) イタリアウイキョウ
floret 小花
floriculture, flower gardening, ornamental horticulture 花き (卉) 園芸

Florida arrowroot, coontie, *Zamia floridana* A. DC.　フロリダアロールート

florigen　フロリゲン

floss　綿絮（めんじょ）

flour mill　製粉機

flour milling percentage　製粉歩合，製粉歩留り

flour yield　製粉歩合，製粉歩留り

flow chart　流れ図

flow diagram　流れ図

flow irrigation　掛流し灌漑

flower　花

flower abscission, flower shedding　落花

flower bean, scarlet runner bean, *Phaseolus coccineus* L.　ベニバナインゲン（紅花隠元）

flower bearing, flower setting　着花

flower bed　花壇

flower bud　花芽（かが，はなめ）

flower-bud abscission　落らい（蕾）

flower bud appearing stage　着らい（蕾）期

flower bud differentiation, floral differentiation　花芽分化

flower bud formation　花芽形成

flower bud initiation, floral differentiation　花芽分化

flower cluster　花房

flower formation, flowering　花成

flower gardening, floriculture, ornamental horticulture　花き（卉）園芸

flower induction　催花（さいか）

flower picking, defloration　摘花

flower setting, flower bearing　着花

flower shedding, flower abscission　落花

flower stalk, scape　花茎

flower stalk development, bolting　抽だい（苔）

flower thinning, defloration　摘花

flower vegetables　花菜類

flowering　1）花成　2）開花

flowering habit　開花習性

flowering hormone　花成ホルモン

flowering order　開花順位，開花順序

flowering regulation, regulation of flowering　開花調節

flowering stage, flowering time, blooming season　開花期

flowers and ornamental plants　花き（卉）

fluctuation　彷徨（ほうこう）変異

flue-cured tobacco　黄色種【タバコ】

flue-curing　黄色（おうしょく）乾燥【タバコ】

fluorescence　蛍光

fluorescence microscope　蛍光顕微鏡

fluorescence quenching　蛍光消光

fluorescent antibody technique　蛍光抗体法

fluorine (F)　フッ素

flux　フラックス，流束

flying out flour　飛粉（とびこ）【コンニャク】

flyings, primings　下葉（したは）【タバコ】

focal depth　焦点深度

focus　焦点

fodder, forage　茎葉飼料【畜産】

fodder beet, mangold, field beet, *Beta vulgaris crassa* L. var. *alba* DC.　飼料用ビート

fodder crop, forage crop　飼料作物

fodder shrub　飼料木

fodder tree　飼料木

foehn　フェーン

Foeniculum vulgare Mill., fennel　ウイキョウ（茴香）

Foeniculum vulgare Mill. var. *dulce* (Mill.) Batt. et Trab. (= *F. dulce* Mill.), Florence fennel, sweet anise　イタリアウイキョウ

fog damage　霧害

foggage　立枯れ草【飼料】

foliage　茎葉[部]，葉群

foliage leaf　本葉(ほんよう), 普通葉
foliar absorption　葉面吸収
foliar application　葉面散布
foliar application of fertlilizer　葉面施肥
foliar bud, leaf bud　葉芽
foliar intake　葉面吸収
foliar uptake　葉面吸収
foliar spray　葉面散布
fonio, fundi, hungry rice, *Digitaria exilis* (Kippist) Stapf　フォニオ
food additive　食品添加物
Food and Agriculture Organization of the United Nations (FAO)　国際連合食糧農業機関
Food and Fertilizer Technology Center (FFTC)　食糧・肥料技術センター
food chain　食物連鎖
food crop　食用作物
food self-sufficiency rate　食糧(食料)自給率
food web　食物網
foot, base　鏡脚【顕微鏡】
foot furrow of rice nursery　踏切溝
foot thresher　足踏み脱穀機
footprinting　フットプリント法
forage, fodder　茎葉飼料【畜産】
forage crop, fodder crop　飼料作物
forage crop field　飼料畑
forage establishment, sward establishment　草地造成
forage feeding, fresh forage feeding, [green] soiling　青刈り給与, 青刈り利用
forage grass　イネ科牧草
forage harvester　フォーレージハーベスタ
forage intake, herbage intake, grazing intake　採食[草]量, 食草量
forage legume　マメ科牧草
forage species　草種【牧草】
forage value, feeding value　飼料価[値]
foraging, grazing　食草
forced air dryer, power dryer　通風乾燥機

forceps　ピンセット
forcing　促成
forcing culture　促成栽培
forcing of sprouting, hastening of germination　催芽
forecasting of disease outbreak, disease forecasting　病害発生予察
forecasting of occurrence　発生予察
foreign grain　異種穀粒
foreign matter, impurity　夾雑物, 異物
forked fringerush, *Fimbristylis dichotoma* (L.) Vahl　テンツキ
formalin acetic alcohol (FAA)　ホルマリン酢酸アルコール
formula feed　配合飼料
formulated concentrate　原液
Fortunella spp., kumquats　キンカン(金柑)
foundation seed, breeder's stock　原原種
foundation seed farm, breeder's stock farm　原原種圃
foundation stock, stock seed, registered seed, original seed　原種
four-angled bean, winged bean, goa bean, asparagus pea, *Psophocarpus tetragonolobus* (L.) DC.　シカクマメ
Fourier series　フーリエ級数
Fourier transform　フーリエ変換
four-prong digging hook　四本ぐわ(鍬)
fourseeded vetch, sparrow vetch, *Vicia tetrasperma* (L.) Schreb.　カスマグサ
foxglove, digitalis, *Digitalis purpurea* L.　ジギタリス
foxtail millet, Italian millet, *Setaria italica* (L.) P. Beauv.　アワ(粟)
fractal　フラクタル
fractal dimension　フラクタル次元
fraction　画分, フラクション, 分画
fraction collector　フラクションコレクター

fractionation 分別
fraction of photosynthetically active radiation absorbed by a canopy (fAPAR) 光合成有効放射吸収率
Fragaria × *ananassa* Duch., strawberry イチゴ(苺)
fragment chromosome 断片染色体, 破片染色体
fragmentation, breakage 切断【染色体】
fragrance, aroma, scent 香り, 香気
frame, hotbed 温床(おんしょう)
frame experiment 枠試験
frame test 枠試験
Free-Air CO_2 Enrichment (FACE) フェイス
free auxin 遊離オーキシン
free nitrogen 遊離窒素
free radical, radical 遊離基
free stock 共台(ともだい)
free stone 離核【果実】
free water 自由水【土壌】
freeze-drying 凍結乾燥
freeze-etching フリーズエッチング, 凍結エッチング【電顕】
freeze-fracture フリーズフラクチャ, 凍結割断【電顕】
freezer 冷凍庫
freezing damage (injury) 凍害
freezing hardiness 耐凍性
freezing point 氷点
freezing storage 冷凍貯蔵
French bean, kidney bean, *Phaseolus vulgaris* L. インゲンマメ(隠元豆)
frequency 頻度, 度数
frequency distribution 頻度分布, 度数分布
fresh forage, green forage, soilage, green chop 生草(せいそう), 青刈り飼料
fresh forage feeding, forage feeding, [green] soiling 青刈り給与, 青刈り利用
fresh forage yield 生草収量

fresh herbage yield 生草収量
fresh leaf, green leaf 生葉(せいよう)
fresh paddy 生籾(なまもみ)
fresh weight 生体重
Fritillaria verticillata Willd. var. *thunbergii* (Miq.) Baker (= *F. thumbergii* Miq.) バイモ
frost damage (injury) 霜害, 凍霜害
frost hardiness 耐霜性
frost heaving 凍上(とうじょう)
frost-heaving damage 凍上害
frost pillars 霜柱
frost protection 防霜, 霜除け
frost resistance 耐霜性
frostless period 無霜期間
fructan フルクタン
fructification, seed-setting 結実
fructosan フルクトサン
fructose 果糖, フルクトース
fructose-1,6-bisphosphatase フルクトース-1,6-ビスホスファターゼ
fructose-1,6-bisphosphate フルクトース-1,6-ビスリン酸
fruit 実, 果実
fruit and seed crop 需実作物
fruit-bearing branch (shoot), bearing branch (shoot) 結果枝(けっかし)
fruit cluster, bunch 果房
fruit cracking 実割れ
fruit drop, fruit shedding 落果
fruit juice 果汁
fruit science, pomology 果樹学, 果樹園芸学
fruit shape 果形
fruit shedding, fruit drop 落果
fruit thinning 摘果
fruit tree 果樹
fruit vegetables 果菜類
fruitful culm (stem), productive culm (stem) 有効茎
fruitful flower 有効花
fruiting, bearing 結果【果実】
fruiting habit, bearing habit 結果習性

fruitless flower　無効花
fuchsine　フクシン【染色】
"Fuke" rice, absidia diseased rice　ふけ米
full bloom　満開
full bloom stage　満花期
full heading time　穂揃期
full maturity　完熟
full ripe　完熟
full-ripe stage　完熟期
full-time farmer　専業農家
fully fermented compost　完熟堆肥
fumigant　くん(燻)蒸剤
fumigation　くん(燻)蒸
function　1) 関数　2) 機能
fundamental meristem, ground meristem　基本分裂組織
fundamental organ　基本器官
fundamental tissue, ground tissue　基本組織
fundi, fonio, hungry rice, *Digitaria exilis* (Kippist) Stapf　フォニオ
fungi (*sing*. fungus)　菌類
fungicide　殺菌剤【真菌】
fungicide resistance　殺菌剤耐性
funicle, funiculus, ovule stalk　珠柄(しゅへい)
Furcraea gigantea (D.Dietr.) Vent., Mauritius hemp　モーリシャスアサ(モーリシャス麻)
furrow　1) うね(畝)間　2) 腹溝(ふくこう), 縦溝【ムギ類】
furrow bottom　れき(壢)底, れき(壢)溝底
furrow ditch　れき(壢)溝
furrow irrigation　けい(畦)間灌漑, うね(畝)間灌漑
furrow pan, plow sole, plow pan　耕盤
furrow planting　溝植え
furrow slice　れき(壢)条
furrow sowing (seeding)　溝播き
furrow width　溝幅
Fusarium blight, scab　赤かび病【ムギ類】
fused magnesium phosphate　溶性リン肥
fusicoccin　フシコクシン
fuzz, linters　短毛, 地毛(じもう)【ワタ】

[G]

G (guanine)　グアニン
G_0 phase　G_0 期【細胞周期】
G_1 phase　G_1 期【細胞周期】
G_2 phase　G_2 期【細胞周期】
Galinsoga quadriradiata Ruiz et Pav., hairy galinsoga　ハキダメギク
Galium spurium L. var. *echinospermon* (Wallr.) Hayek, catchweed, bedstraw　ヤエムグラ
gall　虫えい(癭)
gama grass, *Trispsacum dactyloides* L.　ガマグラス
gambir, *Uncaria gambir* Roxb.　ガンビール
gamete　配偶子
gametic lethal　配偶子致死
gametic ratio　配偶子比
gametophyte　配偶体
Garcinia mangostana L., mangosteen, mangis　マンゴスチン
garden crop, horticultural crop　園芸作物
garden pea, pea, *Pisum sativum* L.　エンドウ(豌豆)
garden poppy, opium poppy, *Papaver somniferum* L.　ケシ(芥子)
garden species, horticultural species　園芸種
garden thyme, common thyme, *Thymus vulgaris* L.　タイム
garden variety　園芸品種
Gardenia jasminoides Ellis, cape jasmine　クチナシ
garlic, *Allium sativum* L.　ニンニク(葫,

大蒜)
gas chromatography (GC) ガスクロマトグラフィー
gas chromatography-mass spectrometry (GC-MS) ガスクロマトグラフィー質量分析法
gas exchange ガス交換
gas-liquid chromatography (GLC) ガス液体クロマトグラフィー
gas phase 気相
gaseous pollutant ガス状汚染物質
gathering plowing, throw-in plowing 内返し耕
GC-MS (gas chromatography-mass spectrometry) ガスクロマトグラフィー質量分析法
Geiger-Müller counter ガイガーミュラーカウンター
geitonogamy, neighboring pollination 隣花受粉
gel ゲル
gel filtration ゲルろ過
gel shift assay ゲルシフト分析
gelatin ゼラチン
gelatinization 糊化(こか)
gelatinization property 糊化特性
gelatinization temperature 糊化温度
gene 遺伝子
gene cloning 遺伝子クローニング
gene expression 遺伝子発現
gene frequency 遺伝子頻度
gene library 遺伝子ライブラリー
gene manipulation 遺伝子操作
gene map 遺伝子地図
gene mutation 遺伝子突然変異
gene recombination, recombination of genes 遺伝子組換え
genealogical tree, phylogenetic tree 系統樹
genecology 品種生態学
general combining ability 一般組合せ能力
generation 世代
generation interval 世代間隔
genetic algorithm 遺伝的アルゴリズム
genetic analysis 遺伝分析
genetic background 遺伝の背景
genetic code 遺伝暗号
genetic correlation 遺伝相関
genetic diversity 遺伝的多様性
genetic engineering 遺伝子工学
genetic gain 遺伝獲得量
genetic marker 遺伝標識
genetic recombination 遺伝的組換え
genetic resources 遺伝資源
genetic variance 遺伝分散
genetic variation, hereditary variation 遺伝変異
genetic vulnerability 遺伝的ぜい(脆)弱性
genetics 遺伝学
genge, Chinese milkvetch, *Astragalus sinicus* L. レンゲソウ(蓮華草, 紫雲英)
genome ゲノム
genome analysis ゲノム分析
genome constitution ゲノム構成
genome size ゲノムサイズ
genotype 遺伝子型
genotype-environment interaction 遺伝子型 - 環境交互作用
genotypic selection 遺伝子型選抜
genotypic variance 遺伝子型分散
gentian violet ゲンチアナバイオレット
Gentiana lutea L., yellow gentian ゲンチアナ
genus 属
genus cross, intergeneric crossing 属間交雑
genus hybrid, intergeneric hybrid 属間雑種
geocarpa bean, *Macrotyloma geocarpum* (Harms) Maréchal et Baudet (= *Kerstingiella geocarpa* Harms) ゼオカルパマメ
geocarpy 地下結実

geographic information 地理情報
geographic information system (GIS) 地理情報システム
geographical distribution 地理的分布
geometric mean 幾何平均
geoscience, earth science 地球科学
geranium, *Pelargonium* spp.【*P. graveolens* L'Her., *P. radura* L'Her. など】ゼラニウム
germ nucleus 生殖核
German camomile (chamomile), *Matricaria chamomilla* L. カミツレ
germinability, viability of seed 発芽力
germinated grain 発芽粒
germination 発芽
germination ability, viability of seed 発芽力
germination bed 発芽床(はつがしょう)
germination inhibitor 発芽抑制物質
germination percentage, percentage of germination 発芽率, 発芽歩合
germination rate 発芽勢(はつがぜい)【≠発芽率】
germination stimulator 発芽促進物質
germination temperature 発芽温度
germination test 発芽試験
germinator [chamber] 発芽試験器
germplasm 生殖質
germplasm preservation 生殖質保存
giant duckweed, *Spirodela polyrhiza* (L.) Schleid. ウキクサ
giant foxtail, *Setaria faberi* Herrm. アキノエノコログサ
giant granadilla, square-stalked passion flower, *Passiflora quadrangularis* L. オオミノトケイソウ
giant ragweed, *Ambrosia trifida* L. オオブタクサ, クワモドキ
giant swamp taro, swamp taro, *Cyrtosperma chamissonis* (Schott) Merr. スワンプタロ
giant taro, *Alocasia macrorrhiza* (L.) Schott インドクワズイモ
gibberellin ジベレリン
gin, cotton gin 綿繰り機
gingelly, sesame, *Sesamum indicum* L. ゴマ(胡麻)
ginger, *Zingiber officinale* Rosc. ショウガ(生姜)
gingergrass, *Cymbopogon martini* Stapf var. *sofia* ジンジャーグラス
ginned cotton, lint 繰綿(そうめん)
ginning 繰綿, 綿繰り【工程】
ginning outturn (GOT), lint percentage 繰綿歩合
ginning percentage, lint percentage 繰綿歩合
ginseng, *Panax ginseng* C. A. Mey. (= *P. schinseng* Nees) ヤクヨウニンジン(薬用人参)
girdling, ringing 環状剥皮(かんじょうはくひ)
GIS (geographic information system) 地理情報システム
glabrous 無毛の
gland 腺(せん)
glandular hair, glandular trichome 腺毛
glass sash ガラス障子
glasshouse ガラス室
glassiness ガラス質性, 硝子質性(しょうししつせい)【コムギ】
glassy, flinty ガラス質の, 硝子質(しょうししつ)の
glassy kernel 硝子質粒, ガラス粒【コムギ】
GLC (gas-liquid chromatography) ガス液体クロマトグラフィー
Glechoma hederacea L. ssp. *grandis* (A. Gray) Hara カキドオシ
Glehnia littoralis F. Schmidt et Miq. (= *Philopterus littoralis* Benth. et Hook. f.) ハマボウフウ(浜防風)
gley horizon グライ層
Gley soil グライ土
global positioning system (GPS) 全球

測位システム
globe fringe-rush, *Fimbristylis miliacea* (L.) Vahl　ヒデリコ
globe sedge, *Pycreus flavidus* (Retz.) T. Koyama (= *Cyperus globosus* All.)　アゼガヤツリ
glucose　グルコース，ブドウ糖
glucoside　配糖体
glumaceous flower　穎花 (えいか)
glume　穎 (えい)，護穎，ふ (桴)
glutamic acid (Glu)　グルタミン酸
glutamine (Gln)　グルタミン
glutaraldehyde　グルタールアルデヒド
glutelin　グルテリン
gluten　グルテン
glutinous, waxy　もち (糯) 性の
glutinous barley, waxy barley　もち (糯) オオムギ
glutinous rice, waxy rice　もち (糯) 米
glutinous wheat, waxy wheat　もち (糯) コムギ
glycerin　グリセリン
glycine (Gly)　グリシン
Glycine gracilis Skov., semi-cultured soybean　半栽培ダイズ
Glycine max (L.) Merr., soybean　ダイズ (大豆)
Glycine max (L.) Merr. ssp. *soja* (Sieb. et Zucc.) H.Ohashi, wild soybean　ツルマメ (蔓豆)
glycogen　グリコーゲン
glycolate pathway (cycle)　グリコール酸経路 (回路)
glycolipid　糖脂質
glycolysis　解糖
Glycyrrhiza spp. 【*G. uralensis* Fisch. et DC., *G. glabra* L. など】, licorice　カンゾウ (甘草)
glyoxysome　グリオキシソーム
Gnaphalium affine D. Don, cudweed　ハハコグサ
Gnaphalium japonicum Thunb., Japanese cudweed　チチコグサ

Gnaphalium pensylvanicum Willd. (= *G. purpureum* L. var. *stathulatum* (Lam.) Baker), wandering cudweed　チチコグサモドキ
goa bean, asparagus pea, winged bean, four-angled bean, *Psophocarpus tetragonolobus* (L.) DC.　シカクマメ
Golgi body, Golgi apparatus　ゴルジ体
gomuti palm, *Arenga pinnata* (Kuntze) Merr.　サトウヤシ (砂糖椰子)
good harvest, bumper crop　豊作
goodness of fit　適合度【統計】
goosegrass, wiregrass, *Eleusine indica* (L.) Gaertn.　オヒシバ
Gossypium arboreum L. および *G. herbaceum* L., Asian cotton　アジアメン (アジア棉)
Gossypium barbadense L., Egyptian cotton　エジプトメン (エジプト棉)【カイトウメンの一系統】
Gossypium barbadense L., sea-island cotton　カイトウメン (海島棉)
Gossypium hirsutum L., upland cotton　リクチメン (陸地棉)
Gossypium spp., cotton　ワタ (棉)
GPS (global positioning system)　全球測位システム
grade　等級【玄米など】
grade of maturity　熟度
grader, sorter, sizer　選別機
grading　1) 選別　2) 格付
graduated cylinder　メスシリンダー
graft affinity　接ぎ木親和性
graft compatibility　接ぎ木親和性
graft hybrid　接ぎ木雑種
graft incompatibility　接ぎ木不親和性
graftage　接ぎ木
grafting　接ぎ木
grain　子実，粒，穀実，穀粒
grain amaranth, *Amaranthus hypochondriacus* L.　ハンスイヒユ (繁穂ヒユ)
grain amaranth, *Amaranthus caudatus*

L. (= *A. edulis* Spegazzini) センニンコク
grain counter 穀粒計数機
grain crop 子実作物
grain crops 穀物
grain cutter 穀粒切断機
grain density 粒着密度
grain drill グレーンドリル
grain filling, ripening 登熟
grain filling period, filling period, ripening period 登熟期間
grain length, kernel length 粒長
grain number per head 一穂粒数
grain production 子実生産
grain quality 穀粒品質
grain screen, grain sorter 万石(まんごく)
grain separator 穀粒選別機
grain shape, kernel shape 粒形
grain shedding, shattering 実こぼれ
grain size 粒大
grain sorghum, sorghum, *Sorghum bicolor* (L.) Moench モロコシ(蜀黍)
grain sorter 粒選機, 米選機, 万石(まんごく)
grain-straw ratio もみわら(籾藁)比
grain texture, kernel texture 粒質
grain thickness 粒厚
grain weight, kernel weight 粒重
grain yield 子実収量, 穀実収量
Gramineae, Poaceae イネ科
gramineous crop イネ科作物
gramineous plant, grass イネ科植物
gramineous weed, grass weed イネ科雑草
grana (*sing*. granum) グラナ
granary 穀倉
granular endoplasmic reticulum, rough-surfaced endoplasmic reticulum 粗面小胞体
granular fertilizer 粒状肥料
granule 粒剤

granule applicator 散粒機
grape, *Vitis vinifera* L. ブドウ(葡萄)
grapefruit, *Citrus paradisi* Macf. グレープフルーツ
grass イネ科植物, 草【イネ科】
grass mulch 敷草
grass pea, chickling vetch, *Lathyrus sativus* L. ガラスマメ
grass weed, gramineous weed イネ科雑草
grassland 草原, 草地
grassland agriculture 草地農業
grassland farming 草地畜産
grassland science 草地学
Gratiola japonica Miq. オオアブノメ
graupel, snow pellets あられ(霰)
gravel れき(礫)
gravel culture 礫耕[栽培]
gravitational potential 重力ポテンシャル
gravitational water 重力水
gravitropism 重力屈性
gravity separator 比重選別機
Gray Forest soil 灰色森林土
Gray Lowland soil 灰色低地土
grazable forestland, woodland pasture, wood-pasture 混牧林, 林内草地
grazing 放牧, 食草, 採食【草本の茎葉など】
grazing intake, herbage intake, forage intake 採食[草]量, 食草量
grazing land, pasture 放牧地
greater yam, water yam, winged yam, *Dioscorea alata* L. ダイジョ(大薯)
Greco-Latin square グレコラテン方格
green algae 緑藻類
green amaranth, slender amaranth, *Amaranthus viridis* L. アオビユ
green chop, fresh (green) forage, soilage 生草(せいそう), 青刈り飼料
green dead-rice kernel 青死に米
green fallow 緑色休閑
green fluorescent protein (GFP) グリー

ン蛍光タンパク質
green forage, fresh forage, soilage, green chop　生草(せいそう), 青刈り飼料
green foxtail, *Setaria viridis* (L.) P. Beauv.　エノコログサ
green gram, mung bean, *Vigna radiata* (L.) R.Wilczek (= *Phaseolus aureus* Roxb.)　リョクトウ(緑豆)
green leaf　1) 生葉(せいよう)　2) 生葉(なまは)【タバコ】
green manure　緑肥
green manure crop　緑肥作物
green mosaic, rosette　萎縮病【ムギ類】
green panic, *Panicum maximum* Jacq. var. *trichoglume* Eyles　グリーンパニック
green rice kernel　青米
green soiling, soiling, fresh forage feeding, forage feeding　青刈り給与, 青刈り利用
green sorrel, *Rumex acetosa* L.　スイバ
green soybean　枝豆
green-sprouting [under diffused light]　浴光育芽, 浴光催芽
green tea　緑茶
green vernalization　緑体春化
greenhouse　温室
greenhouse agriculture, protected cultivation, farming under structure　施設農業
greenhouse effect　温室効果
greening　緑化【育苗】
grinding　摩砕, ひき割り【工程】
gross income　粗収入, 総収入
gross photosynthesis　総光合成
gross primary productivity　総一次生産力
gross production　総生産, 粗生産
gross revenue　粗収入, 総収入
ground application　地上散布
ground barley, wheat etc., meal　ひき割り【麦類】
ground level harvesting　株刈り
ground making, land grading　整地
ground meristem, fundamental meristem　基本分裂組織
ground seedbed　地床(じどこ)【タバコ】
ground tissue, fundamental tissue　基本組織
ground truth　グランドトルース【リモートセンシング】
ground water, groundwater　地下水
groundnut, potato bean, *Apios americana* Medik.　アメリカホドイモ
groundnut, peanut, *Arachis hypogaea* L.　ラッカセイ(落花生)
groundnut oil, peanut oil　ラッカセイ油
groundwater contamination　地下水汚染
groundwater level　地下水位
groundwater table　地下水面
group-mass selection　成群集団選抜
group selection　成群選抜法
grouped row　寄せうね(畝)
grove, orchard　果樹園
growing area　栽培面積
growing point　成長点
growing point culture　成長点培養
growing process　生育過程
growth　成長, 生育
growth analysis　成長解析
growth and development　生育
growth cabinet　グロースキャビネット
growth chamber　グロースチャンバー
growth correlation　成長相関
growth curve　成長曲線
growth diagnosis　生育診断
growth efficiency　成長効率
growth factor　成長要因, 成長因子
growth habit　成長習性
growth hormone　成長ホルモン
growth inhibitor　成長阻害物質

growth model 成長モデル
growth movement 成長運動
growth promotion, growth stimulation 生育促進
growth rate 成長速度, 成長率
growth regulation 成長調節
growth regulator 成長調節物質
growth respiration 成長呼吸, 構成呼吸
growth retardant 成長抑制剤, 成長抑制物質, わい(矮)化剤
growth stimulation, growth promotion 生育促進
growth substance 成長物質
guanine (G) グアニン
guar, cluster bean, *Cyamopsis tetragonoloba* (L.) Taub. クラスタマメ
guarana, *Paullinia cupana* Humb. et Kunth ガラナ
guard cell 孔辺細胞
guava, *Psidium guajava* L. グワバ
guayule, Mexican rubber, *Parthenium argentatum* A. Gray グァユール
Guilielma gasipaes (H. B. K.) L. H. Bailey, peach palm モモミヤシ
guineagrass, *Panicum maximum* Jacq. ギニアグラス
Guizotia abyssinica (L.f.) Cass., niger seed, ramtil ニガーシード
gully erosion ガリ侵食
gum, rubber ゴム
gum arabic, gum Senegal, *Acacia senegal* Willd. アラビアゴム
gum Senegal, gum arabic, *Acacia senegal* Willd. アラビアゴム
gunny bag, jute bag 麻袋(またい)
GUS (β-glucuronidase) β-グルクロニダーゼ
guttapercha, *Palaquium gutta* (Hook. f.) Baill. グッタペルカ
guttation 排水【植物】
gymnosperm 裸子植物

gynophore, peduncle 果柄

[H]

habitat 生育地
Hachijo plume grass, *Miscanthus sinensis* Andersson var. *condensatus* (Hack.) Makino ハチジョウススキ(八丈薄)
Haematoxylon campechianum L., logwood ロッグウッド
hail damage ひょう(雹)害
hairy fleabane, *Conyza bonariensis* (L.) Cronq. (= *Erigeron bonariensis* L.) アレチノギク
hairy galinsoga, *Galinsoga quadriradiata* Ruiz et Pav. ハキダメギク
hairy root 毛状根
hairy vetch, wooly vetch, *Vicia villosa* Roth ヘアリベッチ
half-life 半減期
half-milled rice 五分づき[精]米, 半つ(搗)き米
half-polished rice 五分づき[精]米, 半つ(搗)き米
halloysite ハロイサイト
halophyte 塩生植物
hand cutting 手刈り
hand picking 手摘み【ワタなど】
hand plucking 手摘み【チャなど】
hand pollination, artificial pollination 人工授粉
hand section 徒手切片
hand sowing (seeding) 手播き
hand weeding 手取り除草
haploid, haplont 単相[体], 半数体
haploid plant 半数体植物
haploid breeding 半数体育種
haploidy 半数性
haplont, haploid 単相[体], 半数体
hapten ハプテン

hard fiber 硬質繊維
hard flour 硬質粉
hard grain 硬粒
hard red winter wheat 硬質秋播赤コムギ
hard seed 硬実(こうじつ)
hard starch cell 硬質デンプン細胞【コムギ】
hard-textured rice 硬質米
hard wheat 硬質コムギ
hardening 硬化, 順化, ハードニング【培養】
hardening property 硬化特性
hardinggrass, *Phalaris tuberosa* L. ハーディンググラス
hardness 1) 硬度【土壌・水】 2) 硬さ【米飯】
hardpan 硬盤
hardware ハードウェア
harmful weed, hazardous weed, noxious weed 害草
harmonic mean 調和平均
harrow, clod crusher, pulverizer ハロー
harrowing, pulverization 砕土
harvest index 収穫指数
harvest time 収穫期
harvested area 収穫面積
harvester 収穫機, ハーベスタ
harvest[ing] 収穫
hastening of germination, forcing of sprouting 催芽
hausa potato, *Coleus parviforus* Benth. サヤバナ
hay 乾草
hay dryer 牧草乾燥機
hay making 乾草調製
hay rack 乾草架
haylage ヘイレージ
hazardous weed, harmful weed, noxious weed 害草
He (helium) ヘリウム
head, ear, panicle, spike 穂
head feeding combine 自脱型コンバイン
head formation 結球【葉菜類】
head [polished] rice 上白米
head row 穂系統
head-selection method, ear-selection method 穂選抜法
head-to-row test, ear-to-row test 一穂一列検定
head type, ear type, panicle type 穂型
heading, ear emergence 出穂
heading-back pruning, cutting-back pruning 切返しせん(剪)定
heading stage 出穂期
heading time 出穂期
headland まくら(枕)地【桑園】
healing ゆ(癒)合
healthiness 健全性
heart rot 心腐れ
heat balance, heat budget, energy balance 熱収支
heat balance method ヒートバランス法
heat budget, heat balance, energy balance 熱収支
heat capacity, thermal capacity 熱容量
heat damage 高温障害
heat-damaged kernel 熱損粒
heat denaturation 熱変性
heat of vaporization 気化熱
heat pulse method ヒートパルス法
heat quantity 熱量
heat shock protein 熱ショックタンパク質
heat tolerance 耐暑性, 耐熱性
heat transfer 熱伝達
heated-air dryer, fired dryer 火力乾燥機
heated soil, roasted soil 焼土(やきつち)
heavy [clay] soil 重粘土
heavy manuring culture 多肥栽培
heavy metal 重金属
heavy-panicle type, ear-weight type, panicle-weight type 穂重型

heavy pruning　強せん(剪)定, 深き(剪)り
hedge, windbreak　風除け
Hedyotis diffusa Willd.　フタバムグラ
Helianthus annuus L., sunflower　ヒマワリ(向日葵)
Helianthus tuberosus L., Jerusalem artichoke　キクイモ(菊芋)
helical structure　らせん構造
helical vessel　らせん紋導管
heliograph, sunshine recorder　日照計
heliophyte, sun plant　陽生植物
heliotropism　日光屈性
Heliotropium arborescens L. (= *H. pervianum* L.), common heliotrope　ヘリオトロープ
Heliotropium indicum L., Indian heliotrope　ナンバンルリソウ
helium (He)　ヘリウム
helminthosporium leaf spot, brown spot　ごま(胡麻)葉枯病【イネ】
hematoxylin　ヘマトキシリン【染色】
hemicellulose　ヘミセルロース
Hemistepta lyrata Bunge　キツネアザミ
hemp, *Cannabis sativa* L.　タイマ(大麻)
hemp seed　お(苧)の実
hemp seed oil　大麻子油(タイマしゆ), 麻実油(アサみゆ)
hemps　アサ(麻)【麻類】
henbane, black henbane, *Hyoscyamus niger* L.　ヒヨス
henbit, *Lamium amplexicaule* L.　ホトケノザ
henequen, *Agave fourcroydes* Lem.　ヘネケン
herb　草本, 草【一般】
herbaceous　草本の
herbaceous cotton　草棉(そうめん)
herbaceous crop　草本作物
herbaceous plant, herb　草本
herbage consumption　被食[量]【草地】
herbage intake, forage intake, grazing intake　採食[草]量, 食草量
herbage mass　草量
herbalism　本草学
herbarium specimen　おし(押)葉標本
herbicidal spectrum, weed control spectrum, weeding spectrum　殺草スペクトル
herbicide　除草剤
herbicide tolerance　除草剤耐性
herbivore　草食動物
herbivory　草食
hereditary variation, genetic variation　遺伝変異
heredity, inheritance　遺伝
heritability　遺伝率
hermaphrodite flower, bisexual flower　両性花
herring cake, herring meal　ニシン粕
Hessian cloth　黄麻布(こうまふ), ヘシアンクロース
heterophylly　異形葉性
heteroploid　異数体
heteroploidy, aneuploidy　異数性
heterosis　ヘテロシス
heterostyle　異型花柱
heterostyled flower　異型花柱花
heterostylism　異型ずい(蕊)現象
heterostyly　異型花柱性
heterotypic division　異型[核]分裂
heterozygosity　異型接合性
heterozygote　異型接合体
Hevea brasiliensis Muell.-Arg., Para rubber　パラゴム
hexaploid　六倍体
hexaploidy　六倍性
hexose　六炭糖, ヘキソース
Hg (mercury)　水銀
Hibiscus cannabinus L., kenaf, ambari hemp　ケナフ
Hibiscus esculentus L. (= *Abelmoschus esculentus* (L.) Moench), okura, lady's fingers　オクラ
Hibiscus manihot L. (= *Abelmoschus*

manihot Medik.), sunset hibiscus　トロロアオイ (黄蜀葵)

Hibiscus sabdariffa L., roselle　ロゼル

hierarchical classification　階層分類

high-analysis mixed fertilizer　高度化成肥料

high bed, high ridge　高うね (畝)

high-grade rice　上米 (じょうまい)

high-level cutting　高刈り

high order tiller　高次分げつ

high performance liquid chromatography (HPLC)　高性能液体クロマトグラフィー

high permeable paddy field, water-leaking paddy field　漏水田

high-productivity land (area)　高位生産地

high protein rice　高タンパク質米

high speed liquid chromatography　高速液体クロマトグラフィー

high-yielding ability　多収性

high-yielding culture　多収穫栽培

high-yielding paddy field　高位収穫田

high-yielding variety　多収性品種, 高収性品種

highland agriculture　高地農業

hill　1) 株　2) くらつき (鞍築)

hill distance, interhill space　株間

hill plot　一株区

Hill reaction　ヒル反応

hill sowing (seeding)　点播 (てんぱ)

hill spread　株張り

hillside farm, sloping field　傾斜畑

hillside farming　傾斜地農業

hilum, navel　へそ (臍)

histidine (His)　ヒスチジン

histochemistry　組織化学

histogenesis　組織形成

histogram　ヒストグラム

histology　組織学

histone　ヒストン

Hoagland's solution　ホーグランド液

hog millet, [common] millet, proso millet, *Panicum miliaceum* L.　キビ (黍)

Holcus lanatus L., common velvetgrass, Yorkshire fog　ベルベットグラス

hollow heart　中心空洞

home-consuming crop, crop for home consumption　自給作物

home seed　本場種子

home seed-raising　自家採種

homeobox　ホメオボックス

homeodomain　ホメオドメイン

homeostasis　ホメオスタシス

homeotic mutation　ホメオチック突然変異

homogamy　雌雄同熟

homogeneity　均一性

homogeneity test, test of homogeneity　均一性の検定【分散等】

homogeneity trial, uniformity trial　一様性試験, 均一性試験【圃場等】

homogenizer　ホモジナイザー

homologous chromosome　相同染色体

homologous organ　相同器官

homology　相同 [性]

homomorphous flower　同 [花] 柱花

homotypic division　同型分裂

homozygosis　同型接合, ホモ接合

homozygosity　同型接合性, ホモ接合性

homozygote　同型接合体, ホモ接合体

honey crop　蜜源作物

hooded awn　三叉芒 (さんさぼう), 僧帽芒 (そうぼうぼう)

hoof cultivation　蹄耕法 (ていこうほう)

hooked planting　釣針植え, 釣針挿し【サツマイモ】

hop, *Humulus lupulus* L.　ホップ (忽布)

Hordeum agriocrithon Åberg, six-rowed wild barley　六条野生オオムギ

Hordeum distichon L., two-rowed barley　二条オオムギ

Hordeum spontaneum K. Koch,

two-rowed wild barley　二条野生オオムギ
Hordeum vulgare L., barley　オオムギ (大麦)
Hordeum vulgare L., six-rowed barley　六条オオムギ
Hordeum vulgare L., naked barley　ハダカムギ (裸麦)
Hordeum vulgare L., hulled barley, covered barley　カワムギ (皮麦)
horizontal distribution　水平分布
horizontal layering, continuous layering　しゅ(撞)木取り
horizontal planting　水平植え, 水平挿し【サツマイモ】
horizontal resistance　水平抵抗性
hormonal herbicide　ホルモン型除草剤
hormone　ホルモン
hormone mutants　ホルモン変異体
hornwort, *Ceratophyllum demersum* L.　マツモ (松藻)
horse nettle, *Solanum carolinense* L.　ワルナスビ
horse plowing　馬耕 (ばこう)
horsegram, *Macrotyloma uniflorum* (Lam.) Verdc. (= *Dolichos uniflorus* Lam.)　ホースグラム
horseradish, *Armoracia rusticana* Gaertn., Mey. et Scherb. (= *Cochlearia armoracia* L.)　ワサビダイコン (山葵大根)
horseweed, *Erigeron canadensis* L.　ヒメムカシヨモギ
horticultural crop, garden crop　園芸作物
horticultural science　園芸学
horticultural species, garden species　園芸種
horticulture　園芸
horticulture under structure, protected horticulture　施設園芸
host　宿主
host plant　寄主植物
hot water disinfection method (treatment)　温湯浸法 (おんとうしんぽう)
hot water emasculation [method]　温湯除雄 [法]
hot-wire anemometer　熱線風速計
hotbed, frame　温床 (おんしょう)
Hotelling's T^2 [-statistics]　ホテリングの T^2 [統計量]
housing　舎飼
Houttuynia cordata Thunb.　ドクダミ
HPLC (high performance liquid chromatography)　高性能液体クロマトグラフィー
hull　籾殻, 殻, ふ(稃), 穀皮, 包皮
hull-cracked paddy　割れ傷籾
hulled barley, covered barley, *Hordeum vulgare* L.　カワムギ (皮麦)
hulled rice, brown rice, husked rice　玄米
huller　脱ぷ(稃)機, 籾す(摺)り機
hulling, husking　脱ぷ(稃), 籾す(摺)り
humic soil　腐植質土壌
humid climate　湿潤気候
humidity　湿度
Humulus japonicus Sieb. et Zucc. (= *H. scandens* (Lour.) Merrill), Japanese hop　カナムグラ
Humulus lupulus L., hop　ホップ (忽布)
humus　腐植
hunger sign, deficiency symptom　欠乏症 [状]
hungry rice, fonio, fundi, *Digitaria exilis* (Kippist) Stapf　フォニオ
husk　籾殻, 殻, ふ(稃), 穀皮, 包皮
husked barely　玄麦 (げんばく)
husked rice, brown rice, hulled rice　玄米
husker, huller　籾す(摺)り機
husking, hulling　脱ぷ(稃), 籾す(摺)り
husking ratio　籾す(摺)り歩合

hyacinth bean, lablab, *Lablab purpureus* (L.) Sweet (= *Dolichos lablab* L.)　フジマメ (鵲豆)
hybrid　雑種
hybrid clone　雑種クローン
hybrid population　雑種集団
hybrid rice　ハイブリッドライス
hybrid variety　一代雑種品種
hybrid vigor　雑種強勢
hybrid weakness, pauperization　雑種弱勢
hybridization　交雑, 雑種形成, ハイブリダイゼーション
hybridization breeding, cross breeding　交雑育種
hydathodal cell　排水細胞
hydathodal hair　排水毛
hydathode　排水構造
Hydrangea macrophylla (Thunb. ex Murray) Ser. var. *oamacha* Makino　アマチャ (甘茶)
Hydrangea serrata (Thunb. ex Murray) Ser. var. *thunbergii* (Sieb.) H.Ohba (= *H. macrophylla* Ser. var. *thunbergii* Makino)　コアマチャ (小甘茶)
hydraulic conductance　通導効率
saturated hydraulic conductivity　飽和透水係数【土壌】
hydraulic resistance, resistance to water flow　通導抵抗
hydrilla, *Hydrilla verticillata* (L.f.) Casp.　クロモ
Hydrilla verticillata (L.f.) Casp., hydrilla　クロモ
hydrocarbon　炭化水素
Hydrocharis dubia (Blume) Backer　トチカガミ
hydrochloric acid　塩酸
Hydrocotyle maritima Honda　ノチドメ
Hydrocotyle ramiflora Maxim.　オオチドメ
Hydrocotyle sibthorpioides Lam., lawn pennywort　チドメグサ (血止草)
hydrogen ion concentration　水素イオン濃度
hydrogen peroxide　過酸化水素
hydrogen sulfide (H_2S)　硫化水素
hydrolase　加水分解酵素
hydrology　水文学
hydrolysis　加水分解
hydrophilic　親水性の
hydrophobic　疎水性の
hydrophyte, aquatic plant　水生植物
hydroponics　水耕栽培, 養液栽培
hydrotropism　水分屈性
hygrometer　湿度計
hygromycin　ハイグロマイシン
hygrophyte, hygrophytic plant　湿生植物
hygrophytic weed　湿生雑草
hygroscopic water　吸湿水
hygrothermograph　自記温湿度計
Hyoscyamus niger L., henbane, black henbane　ヒヨス
hyperbola　双曲線
hypha (*pl.* hyphae)　菌糸
hypocotyl　胚軸, 下胚軸
hypodermis　下皮
hypogeal cotyledon　地下子葉
hyponasty　下偏成長
hypostomatous leaf　片面気孔葉
hypothesis (*pl.* hypotheses)　仮説
hypoxia　低酸素状態
Hyptis suaveolens Poir., wild spikenard, wild basil　ニオイイガクサ
hysteresis　履歴現象, ヒステリシス

[I]

I (iodine)　ヨウ素
IAA (indoleacetic acid)　インドール酢酸
ICARDA (International Center for Agricultural Research in the Dry Areas)　国際乾燥地農業研究センター

ice climate 永久凍結気候
ICRISAT (International Crop Research Institute for the Semi-Arid Tropics) 国際半乾燥熱帯作物研究所
ideotype 理想型
idioblast 異型(形)細胞
idioecology 個生態学
IITA (International Institute of Tropical Agriculture) 国際熱帯農業研究所
ilang-ilang, ylang-ylang, *Canangium odoratum* Hook. f. et Thoms. (= *Cananga odorata* Baill.) イランイラン
Ilex paraguayensis A. St. Hil., mate マテチャ
ill-drained paddy field 湿田
Illicium verum Hook. f., star anise ダイウイキョウ(大茴香・八角茴香)
illipe butter イリッペ脂
illuminance 照度
illuminometer, photometer 照度計
image analysis 画像解析
image processing 画像処理
immature grain (kernel) 未熟粒
immature soil 未熟土壌
immaturity 未熟
immediate sowing (seeding) after harvest 取播き
immersion 液浸
immobilization 有機化【土壌窒素など】
immune variety 免疫性品種
immunity 免疫
immunoassay 免疫検定法, イムノアッセイ
immunoblotting 免疫ブロット法
immunofluorescence technique 免疫蛍光法
immunoglobulin 免疫グロブリン
impediment in ripening 稔実障害
Imperata cylindrica (L.) P. Beauv., cogongrass チガヤ(茅)
imperfect flower 不完全花
imperfect rice kernel 不完全米

impermeability 不透性
implement for puddling 代か(搔)き機
improved grassland (pasture), modified grassland (pasture) 改良草地
improved variety 改良品種
improvement of soil fertility 耕土培養
impurity, foreign matter 夾雑物, 異物
in situ インシチュ
in situ hybridization インシチュ・ハイブリダイゼーション
in vitro インビトロ, 生体外で
in vivo インビボ, 生体内で
inactivation 不活性化
inarching 1) 寄せ接ぎ 2) 根接ぎ【果樹】
inbred line, inbred strain 近交系
inbreeding 近親交配, 近交, 同系繁殖
inbreeding coefficient, coefficient of inbreeding 近交係数
inbreeding depression 近交弱勢
incident light 入射光
incipient wilting 初発しお(萎)れ
incompatibility 不和合性
incomplete block design 不完備[型]ブロック設計
incomplete dominance 不完全優性
incomplete leaf 不完全葉
incubation period, latent period 潜伏期[間]
incubator 定温器, 恒温器
indeterminate inflorescence 無限花序
indeterminate type 無限伸育型
Indian barnyard millet, billion dollar grass, *Echinochloa frumentacea* (Roxb.) Link インドビエ
Indian corn, corn, maize, *Zea mays* L. トウモロコシ(玉蜀黍)
Indian dwarf wheat, *Triticum sphaerococcum* Perc. インド矮性コムギ
Indian field cress, *Rorippa indica* (L.) Hochr. イヌガラシ
Indian heliotrope, *Heliotropium indicum*

L. ナンバンルリソウ
Indian jointvetch, *Aeschynomene indica* L. クサネム(草合歓)
Indian lettuce, *Lactuca indica* L. var. *dracoglossa* Kitam. リュウゼツサイ(竜舌菜)
Indian lotus, sacred lotus, *Nelumbo nucifera* Gaertn. ハス(蓮)
Indian mallow, China jute, *Abutilon avicennae* Gaertn. ボウマ(茼麻)
Indian mustard, leaf mustard, *Brassica juncea* (L.) Czern. et Coss. セイヨウカラシナ
Indian pennywort, *Centella asiatica* (L.) Urban ツボクサ
Indian rice, wildrice *Zizania aquatica* L. および *Z. palustris* L. アメリカマコモ
Indian rubber, Assam rubber, *Ficus elastica* Roxb. インドゴム
Indian strawberry, *Duchesnea chrysantha* (Zoll. et Mor.) Miq. (= *D. indica* var. *japonica* Kitam.) ヘビイチゴ
Indian toothcup, *Rotala indica* (Willd.) Koehne var. *uliginosa* (Miq.) Koehne キカシグサ
indica type インド型【イネ】
indicator 指示薬
indicator plant, plant indicator 指標植物
indigenous, native, spontaneous 自生の
indigenous variety, native variety, local variety 在来品種
indigo tree, common indigo, *Indigofera tinctoria* L. キアイ(木藍)
Indigofera suffruticosa Mill., West Indian indigo ナンバンコマツナギ
Indigofera tinctoria L., indigo tree, common indigo キアイ(木藍)
indirect nuclear division 間接核分裂
individual 個体
individual selection 個体選抜

individual test 個体検定
individual variation 個体変異
indoleacetic acid (IAA) インドール酢酸
indoor-grafting, bench-grafting 揚げ接ぎ
indoor rasing of seedling 室内育苗
induced mutation 誘発突然変異
inducer 誘発因子
inductive inference 帰納的推論
inductively coupled plasma spectrometry ICP分光分析,誘導結合高周波プラズマ分光分析
industrial crop 工芸作物,特用作物
inert ingredient 不活性成分
infection 感染
infectious disease 伝染病
infectivity 感染性
inferior ovary 下位子房
inferior spikelet 弱勢頴花
infertile land やせ(痩)地,不毛地
infinite population 無限母集団
inflorescence 花序
inflow, influx 流入
information 情報
information processing 情報処理
information retrieval 情報検索
information system 情報システム
infrared gas analyzer 赤外線ガス分析計
infrared radiation 赤外線
infructescence 果序
ingrowth, protuberance 突起
inheritance, heredity 遺伝
inhibiting gene, repressor [gene], inhibitory gene 抑制遺伝子
inhibition 阻害,抑制
inhibitor 阻害剤
inhibitory factor 抑制因子
inhibitory gene, repressor [gene], inhibiting gene 抑制遺伝子
initial cell 始原細胞
initial condition 初期条件
initial value 初期値

inland climate 内陸気候
inner glume 内苞穎(ないほうえい)
inner seed coat, internal seed coat 内種皮
innoxious weed 無害雑草
inoculation 接種
inoculum 接種原
inorganic fertilizer 無機質肥料
inorganic soil amendment 土壌改良[資]材
insect-borne virus, insect-transmitted virus 昆虫伝搬ウイルス
insect flower, insectpowder plant, Dalmatian chrysanthemum, *Pyrethrum cinerariifolium* Trevir. (= *Chrysanthemum cinerariaefolium* Visiani) ジョチュウギク(除虫菊)
insect injury 虫害
insect pest 害虫
insect pest control 害虫防除
insect pollination, entomophily 虫媒
insect pollinator 媒介昆虫【花粉】
insect resistance 耐虫性
insect-transmitted virus, insect-borne virus 昆虫伝搬ウイルス
insect transmission 昆虫伝搬
insect vector 媒介昆虫【病原体】
insecticide 殺虫剤
insectpowder plant, insect flower, Dalmatian chrysanthemum, *Pyrethrum cinerariifolium* Trevir. (= *Chrysanthemum cinerariaefolium* Visiani) ジョチュウギク(除虫菊)
insensitivity 非感受性
inspection grade 検査等級
instrument screen 百葉箱
instrument shelter 百葉箱
intact 健全な, 無傷の
intake 摂取[量]
integrated control 総合防除
integrated pest management (IPM) 総合的病害虫管理
integument 珠皮

intensive cultivation 集約栽培
interaction 1) 相互作用 2) 交互作用【統計】
interannual variation, year-to-year variation 年次変異
intercalary growth 介在成長, 部間成長
intercalary meristem 介在分裂組織, 部間分裂組織
intercellular carbon dioxide concentration (Ci) 細胞間隙二酸化炭素濃度
intercellular layer, middle lamella 中層, 中葉(ちゅうよう)
intercellular space 細胞間隙
intercepted radiation 受光量
intercrop, catch crop 間作物
intercropping, catch cropping 間作
interface インターフェース
interfascicular cambium 維管束間形成層
interference 干渉
interference microscope 干渉顕微鏡
intergeneric crossing, genus cross 属間交雑
intergeneric hybrid, genus hybrid 属間雑種
interkinesis, interphase 間期, 中間期【細胞分裂】
intermediate ecotype 中間型【ソバ】
intermediate host 中間宿主
intermediate hybrid 中間雑種
intermediate plant 中間植物
intermediate stock, interstock 中間台木
intermediate wheatgrass, *Agropyron intermedium* Beauv., *A. trichophorum* (Link) K. Richt. および *A. pulcherrimum* Grossh. インターミーディエイト・ホィートグラス
intermittent irrigation 間断灌漑
internal factor 内部要因
internal seed coat, inner seed coat 内種皮

International Board for Soil Research and Management (IBSRAM) 国際土壌研究・管理評議会
International Center for Agricultural Research in the Dry Areas (ICARDA) 国際乾燥地農業研究センター
International Center for Living Aquatic Resources Management (ICLAM) 国際水産資源管理センター
International Center for Tropical Agriculture, Centro Internacional de Agricultura Tropical (CIAT) 国際熱帯農業研究センター
International Center of Insect Physiology and Ecology (ICIPE) 国際昆虫生理生態センター
International Council for Research in Agroforestry (ICRAF) 国際アグロフォレストリー研究センター
International Crop Research Institute for the Semi-Arid Tropics (ICRISAT) 国際半乾燥熱帯作物研究所
International Fertilizer Development Center (IFDC) 国際肥料開発センター
International Food Policy Research Institute (IFPRI) 国際食糧政策研究所
International Institute of Tropical Agriculture (IITA) 国際熱帯農業研究所
International Livestock Research Institute (ILRI) 国際畜産研究所
International Maize and Wheat Improvement Center, Centro Internacional de Mejoramiento de Maiz y Trigo (CIMMYT) 国際トウモロコシ・コムギ改良センター
International Plant Genetic Resources Institute (IPGRI) 国際植物遺伝資源研究所
International Potato Center, Centro Internacional de Papa (CIP) 国際バレイショセンター
International Rice Research Institute (IRRI) 国際稲研究所
International Service for National Agricultural Research (ISNAR) 各国農業研究国際サービス
International Union for the Protection of New Varieties of Plants (UPOV) 植物新品種保護国際同盟
International Water Management Institute (IWMI) 国際水管理研究所
internet インターネット
internode 節間
internode elongation 節間伸長
internode elongation stage 節間伸長期
interphase, interkinesis 間期, 中間期【細胞分裂】
interplanting 間植
interpolation 内挿法
interrow space, row distance 条間
intersex 間性
interseeding 中播き
intersowing 中播き
interspecific competition 種間競争
interspecific crossing, species cross 種間交雑
interspecific difference 種間差
interspecific hybrid, species hybrid 種間雑種
interstock, intermediate stock 中間台木
intertillage 中耕
intertillage crop 中耕作物
interval estimation 区間推定
intervening sequence 介在配列
interview survey 聞取り調査
intolerant tree, sun tree 陽樹
intrafascicular cambium, fascicular cambium 維管束内形成層
intra-row intercropping けい(畦)内混作
intraspecific competition 種内競争
introduced variety 導入品種
introduction 導入

introgression　1) 移入交雑, 浸透交雑, 導入交雑　2) 遺伝子移入
introgressive hybridization　移入交雑, 浸透交雑, 導入交雑
intron　イントロン
intrusion, invasion　侵入【雑草】
intrusive growth　割込み成長, 侵入成長
Inula helenium L., elecampane　オグルマ
inulin　イヌリン
inundation damage　冠水害
invasion, intrusion　侵入【雑草】
inverse binomial distribution　逆二項分布
inverse matrix　逆行列
inversion　1) 転化【糖】　2) 逆位【染色体】
inversion layer　逆転層【気象】
invert sugar　転化糖
invertase　インベルターゼ
involucre　総苞 (そうほう)
iodine (I)　ヨウ素
iodine value　ヨウ素価
iodo-starch reaction　ヨウ素-デンプン反応
ion electrode　イオン電極
ion exchange resin　イオン交換樹脂
ionic pump　イオンポンプ
IPGRI (International Plant Genetic Resources Institute)　国際植物遺伝資源研究所
ipil-ipil, white popinac, *Leucaena leucocephala* (Lam.) De Wit　ギンゴウカン (銀合歓)
Ipomoea aquatica Forsk., swamp cabbage, water convolvulus, water spinach　ヨウサイ (蕹菜)
Ipomoea batatas (L.) Lam., sweet potato　サツマイモ (薩摩芋)
Ipomoea nil (L.) Roth (= *Pharbitis nil* (L.) Choisy), Japanese morning glory　アサガオ (朝顔)
Irish potato, potato, *Solanum tuberosum* L.　ジャガイモ
iron (Fe)　鉄
iron containing material　含鉄資材
iron spade　金鋤 (かなすき)
irradiation　照射
irreversible reaction　非可逆反応
IRRI (International Rice Research Institute)　国際稲研究所
irrigated grassland　灌漑草地
irrigation　灌漑 (かんがい), 灌水
irrigation association, water utilization association　水利組合
irrigation at flowering stage　花水 (はなみず)【イネ】
irrigation canal, irrigation ditch　用水路
irrigation requirement, duty of [irrigation] water　用水量
irrigation water　灌漑水
irrigation, drainage and reclamation engineering　農業土木
Isachne globosa (Thunb.) O. Kuntze　チゴザサ
Isodon japonicus Hara (= *Rabdosia japonica* (Burm. f.) H. Hara)　ヒキオコシ (引起)
isoelectric point　等電点
isogenic line　同質遺伝子系統
isolated seed production　隔離採種
isolation　単離, 隔離
isolation field (plot)　隔離圃場
isoleucine (Ile)　イソロイシン
isoline　等値線
isoline map　等値線マップ
isomer　異性体
isotherm　等温線
isotonic solution　等張液
isotope　アイソトープ, 同位元素, 同位体
isotope dilution method　同位体希釈法
isotope effect　同位体効果
isozyme　アイソザイム, イソ酵素
Italian millet, foxtail millet, *Setaria*

italica (L.) P. Beauv. アワ(粟)
Italian ryegrass, *Lolium multiflorum* Lam. イタリアンライグラス
itch grass, *Rottboellia exaltata* (L.) L. f. ツノアイアシ
ivy-leaf speedwell, *Veronica hederifolia* L. フラサバソウ
Ixeris debilis (Thunb. ex Murray) A. Gray (= *I. japonica* (Burm.) Nakai) オオジシバリ
Ixeris dentata (Thunb. ex Murray) Nakai ニガナ

[J]

Jack bean, *Canavalia ensiformis* (L.) DC. タチナタマメ(立刀豆)
jackfruit, *Artocarpus heterophyllus* Lam. パラミツ(波羅蜜)
jackknife estimation ジャックナイフ推定
Japan International Cooperative Agency (JICA) 国際協力機構(旧 国際協力事業団)
Japan International Research Center for Agricultural Sciences (JIRCAS) 国際農林水産業研究センター
Japanese Agricultural Standards (JAS) 日本農林規格
Japanese apricot, mume, *Prunus mume* Sieb. et Zucc. ウメ(梅)
Japanese bellflower, Chinese bellflower, balloonflower, *Platycodon grandiflorum* (Jacq.) A. DC. キキョウ(桔梗)
Japanese bindweed, *Calystegia hederacea* Wall. コヒルガオ
Japanese brome, *Bromus japonicus* Thunb. ex Murray スズメノチャヒキ
Japanese bulrush, *Schoenoplectus juncoides* (Roxb.) Palla ssp. *hotarui* (Ohwi) Soják (= *Scirpus juncoides* Roxb. var. *hotarui* Ohwi) ホタルイ
Japanese butterbur, *Petasites japonicus* (Sieb. et Zucc.) Maxim. フキ(蕗)
Japanese chestnut, *Castanea crenata* Sieb. et Zucc. クリ(栗)
Japanese clover, striate lespedeza, *Kummerowia striata* (Thunb. ex Murray) Schindl. ヤハズソウ
Japanese cudweed, *Gnaphalium japonicum* Thunb. チチコグサ
Japanese dock, *Rumex japonicus* Houtt. (= *R. cripus* L. ssp. *japonicus* (Houtt.) Kitam.) ギシギシ
Japanese dodder, *Cuscuta japonica* Choisy ネナシカズラ
Japanese dog's tooth violet, *Erythronium japonicum* Decne. カタクリ(片栗)
Japanese hedgeparsley, *Torilis japonica* (Houtt.) DC. ヤブジラミ
Japanese hoe 万能(まんのう)
Japanese hop, *Humulus japonicus* Sieb. et Zucc. (= *H. scandens* (Lour.) Merrill) カナムグラ
Japanese hornwort, mitsuba, *Cryptotaenia japonica* Hassk. ミツバ(三葉)
Japanese Industrial Standards (JIS) 日本工業規格
Japanese Journal of Crop Science 日本作物学会紀事【1977年から】
Japanese knotweed, *Reynoutria japonica* Houtt. (= *Polygonum cuspidatum* Sieb. et Zucc.) イタドリ
Japanese lacquer tree, *Rhus verniciflua* Stokes ウルシ(漆)【狭義】
Japanese lawngrass, *Zoysia japonica* Steud. シバ(芝)
Japanese mazus, *Mazus pumilus* (Burm. f.) V. Steenis (= *M. japonicus* (Thunb.) O. Kuntze) トキワハゼ
Japanese medlar, loquat, *Eriobotrya japonica* (Thunb.) Lindl. ビワ(枇杷)

Japanese millet, barnyard millet, *Echinochloa utilis* Ohwi et Yabuno　ヒエ(稗, 穇)

Japanese mint, *Mentha arvensis* L. var. *piperascens* Malinv. ex Holmes　ハッカ(薄荷)

Japanese morning glory, *Ipomoea nil* (L.) Roth (= *Pharbitis nil* (L.) Choisy)　アサガオ(朝顔)

Japanese mugwort, *Artemisia princeps* Pamp.　ヨモギ(蓬)

Japanese name　和名

Japanese pear, sand pear, *Pyrus pyrifolia* (Burm. f.) Nakai (= *P. serotina* Rehder)　ナシ(梨)

Japanese pearlwort, *Sagina japonica* (Sw.) Ohwi　ツメクサ

Japanese pepper, Japanese prickly ash, *Zanthoxylum piperitum* (L.) DC.　サンショウ(山椒)

Japanese persimmon, kaki, *Diospyros kaki* Thunb.　カキ(柿)

Japanese plow　和犂(わすき, わり)

Japanese plume-grass, eulalia grass, *Miscanthus sinensis* Andersson　ススキ(薄)

Japanese prickly ash, Japanese pepper, *Zanthoxylum piperitum* (L.) DC.　サンショウ(山椒)

Japanese sago palm, *Cycas revoluta* Bedd. (= *C. revoluta* Thunb.)　ソテツ(蘇鉄)【狭義】

Japanese torreya, *Torreya nucifera* Sieb. et Zucc.　カヤ(榧)

Japanese tung-oil tree, tung, *Aleurites cordata* (Thunb.) R. Br. ex Steud.　アブラギリ(油桐)

Japanese wax tree, *Rhus succedanea* L.　ハゼノキ(櫨)

Japanese yam, *Dioscorea japonica* Thunb. ex Murray　ヤマノイモ

japonica type　日本型【イネ】

JAS (Japanese Agricultural Standards) 日本農林規格

jasmine, *Jasminum* spp.　ジャスミン

Jasminum officinale L. f., poet's jasmine, common white jasmine　ソケイ(素馨), ツルマツリ(蔓莉茉)

Jasminum sambac Aiton, Arabian jasmine　マツリカ(茉莉花)

Jasminum spp., jasmine　ジャスミン

jasmonic acid　ジャスモン酸

Java apple, wax apple, *Syzygium samarangense* (Blume) Merr. et Perry レンブ

Java citronella grass, *Cymbopogon winterianus* Jowitt　ジャワシトロネラソウ

Java devilpepper, snakewood, *Rauvolfia serpentina* Benth. ex Kurtz　インドジャボク(インド蛇木)

javanica type　ジャワ型【イネ】

Jeffrey's solution　ジェフレー液

Jerusalem artichoke, *Helianthus tuberosus* L.　キクイモ(菊芋)

JICA (Japan International Cooperative Agency)　国際協力機構(旧 国際協力事業団)

jicamas, anu, *Tropaeolum tuberosum* Ruiz. et Pav.　アヌウ

Jimson weed, thorn apple, *Datura stramonium* L.　ヨウシュチョウセンアサガオ

JIRCAS (Japan International Research Center for Agricultural Sciences) 国際農林水産業研究センター

jiring, *Pithecellobium jiringa* (Jack) Prain　ジリンマメ

JIS (Japanese Industrial Standards) 日本工業規格

Job's-tears, *Coix lacryma-jobi* L. var. *ma-yuen* (Roman.) Stapf (= *C. lacryma-jobi* L. var. *frumentacea* Makino)　ハトムギ(薏苡)

Job's-tears, *Coix lacryma-jobi* L.　ジュズダマ

Johnsongrass, *Sorghum halepense* (L.) Pers.　ジョンソングラス
joint part of tuberous root　なり首, 諸梗 (しょこう)【サツマイモ】
jointhead arthraxon, *Arthraxon hispidus* (Thunb.) Makino　コブナグサ
jointing stage　茎立期【ムギ類】
jojoba, *Simmondsia chinensis* (Link) C. K. Schneid.　ホホバ
Jordan's sunshine recorder　ジョルダン日照計
Juglans regia L. var. *orientis* (Dode) Kitam., common walnut　カシグルミ
Juglans spp., walnut　クルミ(胡桃)
Juncellus serotinus (Rottb.) C. B. Clarke (= *Cyperus serotinus* Rottb.)　ミズガヤツリ
Juncus decipiens Nakai (= *J. effusus* L. var. *decipiens* Buchenau), [mat] rush　イグサ(藺草)
Juncus prismatocarpus R. Br. (= *J. leschenaultii* Gay)　コウガイゼキショウ
Juncus tenuis Willd., slender rush　クサイ
jungle rice, *Echinochloa colonum* (L.) Link　ワセビエ
Juniperus communis L., common juniper　セイヨウビャクシン(洋種杜松)
Juniperus virginiana L. (= *Sabina virginiana* (L.) Antoine), eastern red cedar, pencil cedar　エンピツビャクシン
Justicia procumbens L.　キツネノマゴ
jute, white jute, *Corchorus capsularis* L.　ジュート
jute bag, gunny bag　麻袋(またい)
juvenile form　幼形

[K]

K (potassium)　カリウム

kaa he-e, stevia, *Stevia rebaudiana* (Bertoni) Hemsl.　アマハステビア, ステビア
kaki, Japanese persimon, *Diospyros kaki* Thunb.　カキ(柿)
Kalimeris yomena (Kitam.) Kitam.　ヨメナ
kanamycin　カナマイシン
Kangra buckwheat, Tartary buckwheat, *Fagopyrum tataricum* (L.) Gaertn.　ダッタンソバ(韃靼蕎麦)
kaniwa, canihua, *Chenopodium pallidicaule* Aellen　カニウア
kaolinite　カオリナイト
kapok, *Ceiba pentandra* (L.) Gaertn.　カポック
kapok seed oil　カポック油
karyoid, nucleoid　核様体
karyokinesis, mitosis (*pl.* mitoses), caryokinesis　有糸分裂
karyology, caryology　核学
karyotype　核型
kaurene　カウレン
kelp, *Laminaria japonica* Areschoug　マコンブ
kenaf, ambari hemp, *Hibiscus cannabinus* L.　ケナフ
Kendall's rank correlation　ケンドールの順位相関
kendyr, turka, *Apocynum sibiricum* Jacq.　ケンディル, ツルカ
Kentucky bluegrass, *Poa pratensis* L.　ケンタッキーブルーグラス
kernel, grain　1) 穀実, 穀粒, 粒　2) 仁【種子】
kernel length, grain length　粒長
kernel shape, grain shape　粒形
kernel smut　墨黒穂病【イネ】
kernel texture, grain texture　粒質
kernel weight, grain weight　粒重
Kerstingiella geocarpa Harms (= *Macrotyloma geocarpum* (Harms) Maréchal et Baudet), geocarpa bean

ゼオカルパマメ
khus khus, vetiver, *Vetiveria zizanioides* Stapf (= *V. zizanioides* (L.) Nash ex Small)　ベチベル
kidney bean, French bean, *Phaseolus vulgaris* L.　インゲンマメ(隠元豆)
kikuyu grass, *Pennisetum clandestinum* Hochst. ex Chiov.　キクユグラス
kind of crop　作目
kindred, family　家系
kinetin　カイネチン
kinetochore　動原体
Kjeldahl method　ケルダール法
Klein grass, colored Guinea grass, *Panicum coloratum* L.　カラードギニアグラス
klinostat　クリノスタット
kneaded nursery bed　練り床
kneading of nursery bed　床練り
Knop's solution　クノップ液
knotgrass, *Paspalum distichum* L.　キシュウスズメノヒエ
Kodo millet, *Paspalum scrobiculatum* L.　コドラ
Kohler's illumination　ケーラー照明法
Kok effect　コック効果
kok-saghyz, Russian dandelion, *Taraxacum kok-saghyz* L. E. Rodin　ゴムタンポポ
kola, cola　1) *Cola nitida* (Vent.) Schott et Endl.　コーラ　2) *C. acuminata* (P. Beauv.) Schott et Endl.　ヒメコーラ
Kolmogorov-Smirnov test　コルモゴロフ-スミルノフ検定
konjak, elephant foot, *Amorphophallus konjac* K. Koch　コンニャク(蒟蒻)
konjak mannan　コンニャクマンナン
Korean bush-clover, Korean lespedeza, *Kummerowia stipulacea* (Maxim.) Makino　マルバヤハズソウ
Kranz anatomy　クランツ構造
Krebs cycle　クレブス回路

kudzu, kudzu-vine, *Pueraria lobata* (Willd.) Ohwi (= *P. thunbergiana* (Sieb. et Zucc.) Benth.)　クズ(葛)
Kummerowia stipulacea (Maxim.) Makino, Korean lespedeza, Korean bush-clover　マルバヤハズソウ
Kummerowia striata (Thunb. ex Murray) Schindl., striate lespedeza, Japanese clover　ヤハズソウ
kumquats, *Fortunella* spp.　キンカン(金柑)
kurram santonica, *Artemisia kurramensis* Quazilbash　クラムヨモギ
kurtosis　尖度(せんど)【統計】
Kyllinga brevifolia Rottb. (= *Cyperus brevifolius* (Rottb.) Hassk. var. *leiolepis* (Franch. et Savat.) T. Koyama)　ヒメクグ

[L]

labelled compound　標識化合物
lablab, hyacinth bean, *Lablab purpureus* (L.) Sweet (= *Dolichos lablab* L.)　フジマメ(鵲豆)
Lablab purpureus (L.) Sweet (= *Dolichos lablab* L.), lablab, hyacinth bean　フジマメ(鵲豆)
labor cost, labor expense　労働費
labor-intensive　労働集約的
labor productivity　労働生産性
labor-saving　省力化, 省力的, 労働節約的
labor-saving cultivation　省力栽培
lacquer　うるし【製品】
lacquer tapping　うるし掻き
lacquer tree, varnish tree, *Rhus* spp.　ウルシ(漆)【広義】
lactic acid　乳酸
lactic acid fermentation　乳酸発酵
Lactuca indica L. var. *dracoglossa* Kitam., Indian lettuce　リュウゼツサ

イ(竜舌菜)
Lactuca indica L. var. *laciniata* (O. Kuntze) Hara　アキノノゲシ
Lactuca sativa L., lettuce　レタス
LAD (leaf area duration)　葉積
ladino clover, *Trifolium repens* L. var. *giganteum*　ラジノクローバ
lady's fingers, okura, *Abelmoschus esculentus* (L.) Moench (= *Hibiscus esculentus* L.)　オクラ
lady's-thumb, *Persicaria vulgaris* Webb. et Moq. (= *Polygonum persicaria* L.)　ハルタデ
LAI (leaf area index)　葉面積指数
Lamarckism　ラマルク説
λ (lambda) phage　λファージ
lamella (*pl.* lamellae)　ラメラ
lamina, leaf blade　葉身
lamina joint　葉関節, 葉節
lamina joint test　葉関節試験
laminar flow　層流【気象】
Laminaria japonica Areschoug, kelp　マコンブ
Lamium album L. var. *barbatum* (Sieb. et Zucc.) Franch. et Savat. (= *L. barbatum* Sieb. et Zucc.)　オドリコソウ
Lamium amplexicaule L., henbit　ホトケノザ
Lamium purpureum L., purple deadnettle　ヒメオドリコソウ
land　土地
land classification　土地分類, 土地分級
land consolidation　1) 区画整理　2) 交換分合
land grading, ground making　整地
land improvement　土地改良
land leveler　地ならし機
land productivity　土地生産性
land reclamation, clearing　開墾
land slide, slip erosion　地滑り侵食
land use, land utilization　土地利用
landscape　景観

landscape architecture　造園学, 緑地学
lane cropping, strip cropping　帯状間作, 帯状栽培
lanolin　ラノリン
Lapsana apogonoides Maxim.　コオニタビラコ
Lapsana humilis (Thunb.) Makino　ヤブタビラコ
LAR (leaf area ratio)　葉面積比
large grain variety　大粒品種
large vascular bundle　大維管束
larva (*pl.* larvae)　幼虫
laser　レーザー
last productive-tiller emergence stage　有効分げつ終止期
last topdressing, final dressing　止め肥
last weeding　止め草
late blight　疫病【ナス科】
late cutting, late harvesting　晩刈り
late emerging head　遅れ穂
late frost　晩霜(ばんそう), 遅霜(おそじも)
late frost damage (injury)　晩霜害
late goldenrod, *Solidago virgaurea* L. ssp. *gigantea* (Nakai) Kitam. (= *S. gigantea* Ait. var. *leiophylla* Fern.)　オオアワダチソウ
late harvesting, late cutting　晩刈り
late maturation　晩熟
late maturing　晩生(ばんせい, おくて)の
late [maturing] variety　晩生(ばんせい, おくて)品種
late maturity　晩熟性
late planting　晩植
late-planting culture　晩植栽培
late raising　抑制栽培
late-season culture　晩期栽培, 晩化栽培
late sowing (seeding)　晩播き
late-summer buckwheat, autumn buckwheat　秋ソバ
late-summer ecotype, autumn ecotype　秋型【ソバ】

late-summer soybean, autumn soybean　秋ダイズ
late variety of rice　晩稲(ばんとう)
latent heat　潜熱
latent period, incubation period　潜伏期[間]
lateral branch　側枝
lateral bud　側芽
lateral flower　側生花
lateral root　側根, branch root　分枝根
lateral vein　側脈
laterite　ラテライト
latex　乳液
latex duct (tube, vessel)　乳管
Lathyrus sativus L., grass pea, chickling vetch　ガラスマメ
laticifer　乳管
laticiferous cell　乳細胞
Latin square [design]　ラテン方格[設計]
latitude　緯度
Latosol　ラトソル
lattice design　格子型設計
laurel, bay laurel, *Laurus nobilis* L.　ゲッケイジュ(月桂樹)
laurel forest　照葉樹林
laurilignosa　照葉樹林
Laurus nobilis L., laurel, bay laurel　ゲッケイジュ(月桂樹)
Lavandula angustifolia Mill. (= *L. officinalis* Chaix, *L. vera* DC.), lavender, true lavender　ラベンダー
Lavandula latifolia Medik. (= *L. spica* DC.), spike lavender, broadleaved lavender　スパイクラベンダー
lavender, true lavender, *Lavandula angustifolia* Mill. (= *L. officinalis* Chaix, *L. vera* DC.)　ラベンダー
laver, *Porphyra tenera* Kjellman　アサクサノリ
law of diminishing returns　収量漸減の法則, 報酬漸減の法則
law of large numbers　大数(たいすう)の法則
law of minimum　最少律【植物栄養】
lawn, turf　芝生
lawn pennywort, *Hydrocotyle sibthorpioides* Lam.　チドメグサ(血止草)
lax head　疎穂(そすい)
layerage　取り木
layering　取り木
layer-mixing tillage　混層耕
laying-in　伏込み【サツマイモ】
layout　割付け, 配置【試験区】
LD_{50} (lethal dose 50%), median lethal dose　半[数]致死薬量
leaching, eluviation　溶脱
leading variety, main variety　主要品種
leaf　1) 葉　2) 本葉(ほんぱ)【タバコ】
leaf abscission, leaf fall, defoliation　落葉
leaf age　葉齢【葉自身】
leaf analysis　葉分析
leaf angle　受光角度
leaf area　葉面積
leaf area duration (LAD)　葉積
leaf area index (LAI)　葉面積指数
leaf area meter　葉面積計
leaf area ratio (LAR)　葉面積比
leaf axil　葉腋
leaf blade, lamina　葉身
leaf blight　1) すす紋病【トウモロコシ】　2) 斑点病【サツマイモ】　3) 褐紋病【ソバ】
leaf bud, foliar bud　葉芽
leaf-bud cutting　葉芽挿し
leaf burn　葉焼け
leaf crop　需葉作物
leaf cushion, pulvinus　葉枕(ようちん)
leaf cutting　1) 葉挿し　2) せん(剪)葉
leaf disc method　リーフディスク法
leaf dying　葉枯れ
leaf emergence, leaf unfolding　出葉
leaf emergence rate　出葉速度

leaf fall, leaf abscission, defoliation　落葉

leaf gap　葉隙(ようげき)

leaf hopper　ヨコバイ

leaf margin　葉縁

leaf mold　腐葉土

leaf mustard, Indian mustard, *Brassica juncea* (L.) Czern. et Coss.　セイヨウカラシナ

leaf number index　葉齢指数

leaf position on stem　葉位

leaf primordium　葉原基

leaf-punch method　打抜法

leaf rust, brown rust　赤さび(銹)病【コムギ】

leaf shape　葉形

leaf sheath　葉鞘

leaf skirt　葉裾(ようきょ)【タバコ】

leaf spot　1)ごま(胡麻)葉枯病【トウモロコシ, アワ】 2)斑点病【サツマイモ】

leaf stripping　葉もぎ【タバコ】

leaf temperature　葉温

leaf tip　葉先

leaf tobacco　葉タバコ

leaf trace　葉跡(ようせき)

leaf unfolding, leaf emergence　出葉

leaf vegetables　葉菜類

leaf water potential　葉の水ポテンシャル

leaflet　小葉

leakage, seepage　漏水

least significant difference (LSD)　最小有意差

least-squares estimator　最小二乗推定量

least-squares method　最小二乗法

Leersia hexandra Sw., barect grass, southern cutgrass, tiger's-tongue grass　タイワンアシカキ

Leersia japonica Makino　アシカキ

Leersia oryzoides (L.) Sw. ssp. *japonica* (Hack.) T. Koyama (= *L. oryzoides* Sw. var. *sayanuka* Ohwi)　サヤヌカグサ

Leersia oryzoides (L.) Sw. ssp. *oryzoides*, rice cutgrass　エゾノサヤヌカグサ

legume　1)マメ科植物 2)豆果(とうか)

Leguminosae　マメ科

leguminous crop　マメ, マメ科作物

leguminous crops, pulses, pulse crops　マメ類

leguminous plant, legume　マメ科植物

lemma　外穎(がいえい)

Lemna aoukikusa Beppu et Murata (= *L. paucicostata* Hegelm.)　アオウキクサ

Lemna paucicostata Hegelm. (= *L. aoukikusa* Beppu et Murata)　アオウキクサ

lemon, *Citrus limon* (L.) Burm. f.　レモン(檸檬)

lemongrass　1) *Cymbopogon flexuosus* (Nees ex Steud.) Wats. (東インドレモングラス) 2) *C. citratus* (D. C. ex Nees) Stapf (西インドレモングラス)　レモングラス

Lens culinaris Medik (= *L. esculenta* Moench), lentil　ヒラマメ(扁豆)

lenticel　皮目(ひもく)

lentil, *Lens culinaris* Medik (= *L. esculenta* Moench)　ヒラマメ(扁豆)

Lentinus edodes (Berk.) Sing., shiitake fungus　シイタケ

Lepidium virginicum L., Virginia pepperweed, peppergrass　マメグンバイナズナ

Lepironia articulata (Retz.) Domin (= *L. mucronata* Rich.), Chinese mat rush　アンペラソウ

Leptochloa chinensis (L.) Nees, Chinese sprangletop　アゼガヤ

leptotene stage　レプトテン期, 細糸(ほそいと)期【減数分裂】

lesion　病斑

Lespedeza cuneata (Dum. Cours.) G. Don, serisea lespedeza　メドハギ

lesser yam, potato yam, *Dioscorea esculenta* (Lour.) Burk.　トゲドコロ
lethal concentration　致死濃度
lethal dose 50% (LD_{50}), median lethal dose　半[数]致死薬量
lethal factor　致死因子
lethal gene　致死遺伝子
lettuce, *Lactuca sativa* L.　レタス
Leucaena leucocephala (Lam.) De Wit, white popinac, ipil-ipil　ギンゴウカン(銀合歓)
leucine (Leu)　ロイシン
leucoplast　白色体
levee　あぜ(畦), くろ(畔), 畦畔(けいはん)
levee building　あぜ(畦)作り
levee coating　あぜ(畦)塗り, くろ(畔)塗り
levee planting　あぜ(畦)作
level　水準
level culture　平作(ひらさく)
level of significance, significance level　有意水準
level planting　平植え
level row　平うね(畝)
level seedbed, flat seedbed　平床
level sowing (seeding)　平播き
levelling　均平
ley　輪作草地
ley farming　穀草式農法, 輪換式農法
Li (lithium)　リチウム
liana, vine　つる(蔓)
Liberian coffee, *Coffea liberica* W. Bull. ex Hiern　リベリアコーヒー[ノキ]
Licania rigida Benth., oiticica　オイチシカ
licorice, *Glycyrrhiza* spp.【*G. uralensis* Fisch. et DC., *G. glabra* L. など】カンゾウ(甘草)
life cycle　生活環
life duration, longevity　寿命
life history　生活史

life science　生命科学
lifting up of vines　つる(蔓)揚げ
light break, light interruption　光中断
light compensation point　光補償点
light culture, cultivation under lightening　電照栽培
light-extinction coefficient　光消散係数
light germinater, photoblastic seed　光発芽種子
light germination　光発芽
light green　ライトグリーン【染色】
light growth reaction　光成長反応
light-harvesting complex (LHC)　集光性複合体【光合成】
light harvesting pigment　集光性色素
light intensity　光強度
light interception　受光量
light-intercepting characteristics, stand geometry　受光態勢
light interruption, light break　光中断
light microscope　光学顕微鏡
light period　明期
light pruning　浅剪り(あさぎり)
light reaction　明反応
light saturation　光飽和
light soil　軽しょう(鬆)土
light transmittance　光透過率
light trap　予察燈
light trimming of canopy　浅刈り【チャ】
light use efficiency　光利用効率
lignification　木化
lignin　リグニン
ligule　葉舌(ようぜつ), 小舌(しょうぜつ)
likelihood　尤度(ゆうど)
likelihood function　尤度関数
likelihood ratio test　尤度比検定
Lima bean, butter bean, *Phaseolus lunatus* L. (= *P. limensis* Macf.)　ライマメ
lime, sour lime, *Citrus aurantifolia* (Christm.) Swingle　ライム

lime 石灰
lime sower 石灰散布機
lime spreader 石灰散布機
lime sulfur 石灰硫黄合剤
limiting factor 制限因子, 限定要因
Limnocharis flava (L.) Buchenau, yellow velvetleaf キバナオモダカ
limnophila, *Limnophila sessiliflora* Blume キクモ
Limnophila sessiliflora Blume, limnophila キクモ
Lindernia angustifolia (Benth.) Wettst. アゼトウガラシ
Lindernia dubia (L.) Penn., low false-pimpernel アメリカアゼナ
Lindernia procumbens (Krock.) Philcox (= *L. pyxidaria* L.), common false pimpernel アゼナ
Lindernia pyxidaria L. (= *L. procumbens* (Krock.) Philcox), common false pimpernel アゼナ
line, strain, pedigree, stock 系統
LINE (long interspersed repetitive sequence) 広範囲散在反復配列
line breeding 系統育成
line cross 系統間交配
line intersection method ライン交さ(叉)[点]法
line selection, pedigree selection 系統選抜
line separation 系統分離
line test 系統検定
linear combination 線形結合
linear model 線形模型
linear programming 線形計画法
linear regression 線形回帰
linear transformation 線形変換
linearity 線形性
linkage 連鎖
linkage group 連鎖群
Linnean species リンネ種
linneon リンネ種
linoleic acid リノール酸

linolenic acid リノレン酸
linseed meal 亜麻仁粕 (アマにかす)
linseed oil 亜麻仁油 (アマにゆ)
lint 1) 綿花【狭義】, 綿毛 (めんもう) 2) 繰綿 (そうめん)【製品】
lint index 繰綿 (そうめん) 指数
lint percentage, ginning percentage, ginning outturn (GOT) 繰綿歩合
linter リンター採取機
linters, fuzz 短毛, 地毛 (じもう)【ワタ】
Linum usitatissimum L., flax アマ (亜麻)
lipase リパーゼ
lipid 脂質
Lipocarpha microcephala (R. Br.) Kunth ヒンジガヤツリ
liquid chromatography 液体クロマトグラフィー
liquid culture 液体培養
liquid fertilizer 液肥, 液体肥料
liquid formulation 液剤
liquid nitrogen 液体窒素
liquid phase 液相
liquid scintillation counter 液体シンチレーションカウンター
LISA (low-input sustainable agriculture) 低投入持続型農業
lister, ridger, ditcher うね (畝) 立て機
litchi, *Litchii chinensis* Sonn. レイシ (荔枝)
Litchii chinensis Sonn., litchi レイシ (荔枝)
lithium (Li) リチウム
Lithospermum erythrorhizon Sieb. et Zucc. ムラサキ
litter 1) リター, 落葉落枝 2) 敷料【畜産】
little bluestem, *Schizachyrium scoparius* (Michx.) Nash リトルブルーステム
little quaking grass, *Briza minor* L. ヒメコバンソウ
live green-kerneled rice 活青米 (いき

あおまい)
live weight (LW)　生体重【家畜】
live weight gain (LWG), body weight gain　増体[量]
livestock, domestic animal, farm animal　家畜
livestock farm, ranch　牧場
livestock industry, animal industry, animal husbandry　畜産
livid amaranth, *Amaranthus lividus* L.　イヌビユ
load　負荷
loading　ローディング
loam　壌土
lobelia, *Lobelia chinensis* Lour.　アゼムシロ
local adaptability test, test for regional adaptability　系統適応性検定試験
local name　地方名
local variety　在来品種, 地方品種
locality　地域性
locational conditions, condition of site　立地条件
locus (*pl.* loci)　遺伝子座
locust bean, carob, *Ceratonia siliqua* L.　イナゴマメ
lodging　倒伏
lodging index　倒伏指数
lodging resistance　耐倒伏性, 倒伏抵抗性
lodicule　鱗被(りんぴ)
loess　黄土(こうど), レス
logarithmic (log) transformation　対数変換
logistic curve　ロジスティック曲線
lognormal distribution　対数正規分布
logwood, *Haematoxylon campechianum* L.　ロッグウッド
Lolium multiflorum Lam., Italian ryegrass　イタリアンライグラス
Lolium perenne L., perennial ryegrass　ペレニアルライグラス
Lolium rigidium Gaud., wimmera ryegrass　ウィメラライグラス
Lolium temulentum L., poison ryegrass, darnel　ドクムギ(毒麦)
long day　長日
long glume rice　長穎稲(ちょうえいとう)
long interspersed repetitive sequence (LINE)　広範囲散在反復配列
long-culmed variety　長稈品種
long-day plant　長日植物
long-day treatment　長日処理
long-grained variety　長粒品種
long-lived seed, macrobiotic seed　長命種子
long-styled flower, pin flower　長[花]柱花
long-term storage　長期貯蔵
long-wave radiation　長波放射
longevity　寿命, 生存年限
longitudinal division　縦分裂
longitudinal growth　縦成長
longitudinal section　縦断切片
loquat, Japanese medlar, *Eriobotrya japonica* (Thunb.) Lindl.　ビワ(枇杷)
loss　流亡【土壌肥料】
loss from weed, weed loss　雑草害
loss function　損失関数
loss in weight　目減り
Lotus corniculatus L. var. *corniculatus*, birdsfoot trefoil　バーズフット・トレフォイル
Lotus uliginosus Schkuhr., big trefoil　ビッグトレフォイル
low, depression, cyclone　低気圧
low bed　低うね(畝), 浅うね(畝)
low cut training　低幹仕立て, 根刈り仕立て【クワ】
low false-pimpernel, *Lindernia dubia* (L.) Penn.　アメリカアゼナ
low-input sustainable agriculture (LISA)　低投入持続型農業
low-level cutting　低刈り

low-productivity land (area)　低位生産地
low protein rice　低タンパク質米
low ridge　低うね(畝),浅うね(畝)
low-temperature germinability　低温発芽性
low-temperature injury, chilling injury　低温障害
low-temperature storage　低温貯蔵
low-temperature treatment　低温処理
low-temperature warehouse　低温倉庫
low-yielding paddy field　低位収穫田
lower leaf　下位葉
lower nodal tiller　低位分げつ
lower order tiller　低次分げつ
lowland agriculture　低地農業
lowland crop　田作物(でんさくもつ)
lowland rice, paddy rice, paddy　水稲
LSD (least significant difference)　最小有意差
lucerne, alfalfa, *Medicago sativa* L.【*M. × media* Pers.を含む】　アルファルファ
luciferase　ルシフェラーゼ
luciferin　ルシフェリン
Ludwigia decurrens Walt., winged waterprimrose　ヒレタゴボウ
Ludwigia epilobioides Maxim.　チョウジタデ
Luffa acutangula (L.) Roxb., towel gourd　トカドヘチマ
Luffa cylindrica M. Roem., sponge gourd　ヘチマ(糸瓜)
lugs　中葉(ちゅうは)【タバコ】
Lupinus albus L., white lupine　シロバナルーピン(白花ルーピン)
Lupinus angustifolius L., blue lupine　アオバナルーピン(青花ルーピン)
Lupinus luteus L., yellow lupine　キバナルーピン(黄花ルーピン)
luxury absorption　ぜいたく(贅沢)吸収
Lycopersicon esculentum Mill., tomato　トマト
Lycoris radiata (L'Her.) Herb.　ヒガンバナ(彼岸花)
Lysenko hypothesis　ルイセンコ説
lysigenous aerenchyma　破生通気組織
lysigenous intercellular space　破生細胞間隙
Lysimachia japonica Thunb.　コナスビ
lysimeter　ライシメータ
lysine (Lys)　リシン,リジン
lysosome　リソソーム

[M]

M phase　M期, mitotic phase　分裂期
macaroni wheat, durum wheat, *Triticum durum* Desf.　マカロニコムギ, デュラムコムギ
mace　メース【ニクズクの仮種皮】
maceration　解離【組織学】
macha wheat, *Triticum macha* Dek. et Men.　マッハコムギ
machine winnowing　唐み(箕)選
Macleaya cordata (Willd.) R. Br.　タケニグサ
macrobiotic seed, long-lived seed　長命種子
macroelement, macronutrient, major element　多量元素, 多量要素
Macroptilium atropurpureum (DC.) Urb., siratro　サイラトロ
Macrotyloma geocarpum (Harms) Maréchal et Baudet (= *Kerstingiella geocarpa* Harms), geocarpa bean　ゼオカルパマメ
Macrotyloma uniflorum (Lam.) Verdc. (= *Dolichos uniflorus* Lam.), horsegram　ホースグラム
madake bamboo, *Phyllostachys bambusoides* Sieb. et Zucc.　マダケ(真竹)
madder, common madder, *Rubia tinctorum* L.　セイヨウアカネ(西洋

茜)
MAFF (Ministry of Agriculture, Forestry and Fisheries) 農水省(農林水産省)
magnesium (Mg) マグネシウム
magnesium ammonium phosphate 苦土リン(燐)安
magnesium fertilizer 苦土肥料
magnetic disk [磁気]ディスク
magnetic stirrer マグネチックスターラー
magnification 倍率【光学】
magnifier ルーペ
maguey, *Agave* spp. マゲー
main axis, primary axis 主軸
main crop, major crop 主作物, 主要作物
main culm 主稈
main effect 主効果
main root, taproot 主根
main season crop[ping] 表作
main stalk ear 親穂
main stem 主茎
main variety, leading variety 主要品種
main vein 中央脈, 主脈
main vine 親づる(蔓)
maintainer 維持系統
maintenance respiration 維持呼吸
maize, corn, Indian corn, *Zea mays* L. トウモロコシ(玉蜀黍)
major crop, main crop 主作物, 主要作物
major element, macroelement, macronutrient 多量元素, 多量要素
major gene 主働遺伝子
Majorana hortensis Moench (= *Origanum majorana* L.), sweet majoram マヨラナ
Malabar grass, East Indian lemongrass, *Cymbopogon flexuosus* (Nees ex Steud.) Wats. 東インドレモングラス
male 雄[の]
male flower, staminate flower 雄花(お

ばな, ゆうか)
male nucleus 雄核, sperm nucleus 精核
male plant 雄株
male sterility 雄性不稔
malformation, deformity, deformation, terata 奇形
malic acid リンゴ酸
mallet cutting しゅ(撞)木挿し
malt 麦芽(ばくが)
malting barley 1) ビール麦 2) 醸造用オオムギ
malting quality 麦芽品質
maltose 麦芽糖, マルトース
Malus pumila Mill., apple リンゴ(林檎)
Malvastrum coromandelianum (L.) Garcke エノキアオイ
management 管理
manganese (Mn) マンガン
Mangifera indica L., mango マンゴー
mangis, mangosteen, *Garcinia mangostana* L. マンゴスチン
mango, *Mangifera indica* L. マンゴー
mangold, fodder beet, field beet, *Beta vulgaris* L. var. *alba* DC. 飼料用ビート
mangosteen, mangis, *Garcinia mangostana* L. マンゴスチン
Manihot esculenta Crantz (= *M. utilissima* Pohl), cassava, manioc, tapioca plant キャッサバ
Manihot glaziovii Muell. -Arg., manihot rubber, ceara rubber マニホットゴム
manihot rubber, ceara rubber, *Manihot glaziovii* Meull. -Arg. マニホットゴム
Manila hemp, abaca, *Musa textilis* Née マニラアサ(マニラ麻)
Manilkara bidentata (A. DC.) A.Chev., balata バラタ
Manilkara zapota (L.) P. Royen (= *Achras zapota* L.), sapodilla,

naseberry　サポジラ
manioc, cassava, tapioca plant, *Manihot esculenta* Crantz (= *M. utilissima* Pohl)　キャッサバ
manipulator　マニピュレーター
mannan　マンナン
Mannit　マンニット
mannitol　マンニトール
Mann-Whitney test　マン‐ホイットニー検定
manometry　検圧法
manure　肥料, 肥(こえ)
manure-heated seedbed　醸熱温床
manure pool, night-soil reservoir　肥溜(こえだめ)
manure spreader　マニュアスプレッダ, 堆肥散布機
manuring, fertilization, fertilizer application　施肥
manuring irrigation　肥培灌漑
manuring practice　肥培管理
many-tillering type, ear-number type, panicle-number type　穂数型
maple sugar　カエデ糖
Maranta arundinacea L., arrowroot, West Indian arrowroot　アロールート
margarine　マーガリン
marginal effect　周辺効果
marginal growth　周縁成長
marginal meristem　周縁分裂組織
marginal sectorial chimera　周縁区分キメラ
marker gene　標識遺伝子
marketable tuber　上いも【ジャガイモ】
marketing standard variety, branded variety, costly registered variety　銘柄品種
Markov chain　マルコフ連鎖
Markov process　マルコフ過程
marsh, swamp　低湿地
marsh dayflower, *Murdannia keisak* (Hassk.) Hand.-Mazz. (= *Aneilema keisak* Hassk.)　イボクサ

marsh yellowcress, *Rorippa islandica* (Oeder) Borb.　スカシタゴボウ
Marsilea quadrifolia L., pepperwort, water clover　デンジソウ
mass emasculation, bulk emasculation　集団除雄
mass method of breeding, bulk method [of breeding]　集団育種法
mass pollination　集団受(授)粉, 混合受(授)粉
mass production of seed　集団採種
mass seed production　集団採種
mass selection [method of breeding]　集団選抜[法]
mass spectrometer　質量分析計
mass spectrometry (MS)　質量分析
mat bean, moth bean, *Vigna aconitifolia* (Jacq.) Maréchal (= *Phaseolus aconitifolius* Jacq.)　モスビーン
MAT gene　MAT遺伝子
mat rush, rush, *Juncus effusus* L.var. *decipiens* Buchenau (= *J. decipiens* Nakai)　イグサ(藺草)
mate, *Ilex paraguayensis* A. St. Hil.　マテチャ
material cycle, cycle of matter　物質循環
maternal effect　母性効果
maternal inheritance　母性遺伝
maternal line selection　母系選抜[法]
mathematical model　数学モデル
mating, cross, crossing　交配
matric potential　マトリックポテンシャル
Matricaria chamomilla L., German camomile (chamomile)　カミツレ
matrix　基質, 礎質【細胞】
matrix of sum of squares and products　平方和積和行列
matsutake fungus, *Trichloma matsutake* Sing.　マツタケ
maturation, maturity　成熟
mature leaf　成葉

mature paddy field　熟田
mature tea field　成園【チャ】
mature [upland] field　熟畑
maturity　1) 成熟　2) 熟性
Mauritius hemp, *Furcraea gigantea* (D. Dietr.) Vent.　モーリシャスアサ (モーリシャス麻)
maximun and minimum thermometer　最高最低温度計
maximum-likelihood estimator　最尤推定量 (さいゆうすいていりょう)
maximum-likelihood method　最尤法 (さいゆうほう)
maximum potential yield　最大限界収量
maximum temperature　最高温度
maximum tiller number stage　最高分げつ期
maximum water holding capacity　最大容水量, 飽和水分量
mayweed chamomile, *Anthemis cotula* L.　カミツレモドキ
Mazus aponicus (Thunb.) O. Kuntze (= *M. pumilus* (Burm. f.) V. Steenis), Japanese mazus　トキワハゼ
Mazus miquelii Makino　サギゴケ (鷺苔)
meadow　採草地, 草地
meadow fescue, *Festuca pratensis* Huds.　メドーフェスク
meadow foxtail, *Alopecurus pratensis* L.　メドーフォックステール
meadow saffron, colchicum, autumn crocus, *Colchicum autumnale* L.　イヌサフラン
meal　荒粉 (あらこ)【麦・豆など】
mealy　粉質の
mealy kernel, chalky kernel　粉状質粒【ムギ類】
mean, average　平均
mean air temperature　平均気温
mean square　平均平方
mean-square error　平均二乗誤差
measure, scale　尺度

measured value　測定値
measurement　測定, 測定値【統計】
measuring pipette　メスピペット
mechanical feeding thresher, automatic thresher, self-feeding thresher　自動脱穀機
mechanical isolation　機械的隔離
mechanical picking　機械摘み【ワタなど】
mechanical plucking　機械摘み【チャなど】
mechanical [weed] control　機械的 [雑草] 防除
mechanical weeding　機械除草
mechanism of action　作用機構
mechanistic model　機構的モデル
mechanization　機械化
median　中央値, メディアン
median lethal dose (LD_{50}), lethal dose 50%　半 [数] 致死薬量
median section　正中断 (せいちゅうだん)【形態】
Medicago lupulina L., black medic　ブラックメディック
Medicago polymorpha L. (= *M. hispida* Gaertn.), bur clover　バークローバ
Medicago sativa L.【*M.* × *media* Pers. を含む】, alfalfa, lucerne　アルファルファ
medicinal crop　薬用植物
medicinal rhubarb, *Rheum officinale* Baill.　ダイオウ (大黄)
medicine spoon　薬 [さ] じ
medium cut training　中幹仕立て, 中刈り仕立て【クワ】
medium-grained variety　中粒品種
medium [maturing] variety　中生 (ちゅうせい, なかて) 品種
medium pruning　中切り (ちゅうぎり)【チャ】
medium root　中根 (ちゅうこん)【チャ】
medullary cavity, pith cavity　髄腔 (ずいこう)

meiosis 減数分裂
meiosis stage, reduction division stage 減数分裂期
Melanorrhoea usitata Wall., Burmese varnish tree ビルマウルシ
Melilotus albus Medik., white sweetclover シロバナスィートクローバ(白花スィートクローバ)
Melilotus officinalis (L.) Lam., yellow sweetclover キバナスィートクローバ(黄花スィートクローバ)
mellow kernel 豊軟粒, 膨軟粒
mellow soil 膨軟土
melon, *Cucumis melo* L. メロン
memory メモリ
Mendelian inheritance メンデル性遺伝
Mendelian population メンデル集団
Mendelism メンデル[学]説
Mendel's law メンデルの法則
Mentha × piperita L., peppermint ペパーミント
Mentha arvensis L. var. *piperascens* Malinv. ex Holmes, Japanese mint ハッカ(薄荷)
Mentha pulegium L., pennyroyal mint ペニーロイヤルミント
Mentha spicata L. (= *M. viridis* L.), spearmint スペアミント
mercaptoethanol メルカプトエタノール
mercury (Hg) 水銀
mercury thermometer 水銀温度計
meristem 分裂組織
mesh climatic data メッシュ気候値
mesh data メッシュデータ
mesobiotic seed 常命種子(じょうみょうしゅし)
mesocarp 中果皮
mesocotyl 中胚軸, 中茎(ちゅうけい)
mesocotylar root 中茎根
mesophyll 葉肉
mesophyll conductance 葉肉コンダクタンス, 葉肉伝導度
mesophyll resistance 葉肉抵抗

mesophyte 中生(ちゅうせい)植物
mesophytic weed 中生雑草
messenger RNA (mRNA) 伝令 RNA, メッセンジャー RNA
mestom[e] sheath メストムシース
metabolic inhibitor 代謝阻害剤
metabolic intermediate 中間代謝物質
metabolic pathway 代謝経路
metabolism 代謝
metabolite 代謝産物
metal ion 金属イオン
metamorphosis 変態
metaphase 中期[細胞分裂]
metaphloem 後生篩部
metaplasm 後形質
metaxenia メタキセニア
metaxylem 後生木部
meteorological disaster 気象災害
methane (CH_4) メタン
methane fermentation メタン発酵
methionine (Met) メチオニン
method of fertilizer application 施肥法
methyl alcohol メチルアルコール
methylgreen メチルグリーン【染色】
Metroxylon rumphii Mart., spiny sago palm, prickly sago palm トゲサゴ[ヤシ](刺サゴ[椰子])
Metroxylon sagu Rottb., sago palm, non-spiny sago palm サゴヤシ(サゴ椰子)
Mexican rubber, guayule, *Parthenium argentatum* A. Gray グァユール
Mexican tea, *Chenopodium ambrosioides* L. ケアリタソウ
Mg (magnesium) マグネシウム
micell[e] ミセル
Michaelis constant ミカエリス定数
microanalysis 微量分析
microautoradiography ミクロオートラジオグラフィー
microbe, microorganism 微生物
microbial breakdown 微生物分解
microbial control 微生物的防除

microbial pesticide 微生物農薬
microbiotic seed, short-lived seed 短命種子
microbody ミクロボディ
microclimate 微気候
microdissection 顕微解剖
microelement, minor element, trace element 微量元素
microfibril ミクロフィブリル
micromanipulator 顕微解剖器
micrometeorology 微気象[学]
micrometer マイクロメーター
micronutrient 微量要素
micronutrient fertilizer 微量要素肥料
microorganism, microbe 微生物
microphotography 顕微鏡写真
micropyle 珠孔(しゅこう)
microscope 顕微鏡
microsome ミクロソーム
microspectrophotometry 顕微分光法
microtome ミクロトーム
microtubule 微小管
microwave マイクロ波
middle heading time 出穂盛期
middle lamella, intercellular layer 中層, 中葉(ちゅうよう)
midparent [value] 両親平均, 中間親
midrib 1) 中肋(ちゅうろく) 2) 中骨(ちゅうこつ)【タバコ】
midseason drainage 中干し(なかぼし)
milk production 産乳[量]
milk thistle, sow thistle, *Sonchus oleraceus* L. ノゲシ
milk[ripe] stage 乳熟期
milky white rice kernel 乳白米
milled rice, polished rice 精米, 白米
milled rice with embryo 胚芽米
millet, common millet, proso millet, hog millet, *Panicum miliaceum* L. キビ(黍)
millets, miscellaneous cereals 雑穀
milling 1) とう(搗)精, 精白 2) 製粉 3) 粉砕

milling loss つ(搗)き減り
milling percentage とう(搗)精歩合
milling quality 製粉性
milling test 製粉試験
mine pollution 鉱害
mine pollutant, mineral pollutant 鉱毒
mineral composition 無機組成
mineral deficiency 要素欠乏
mineral nutrition 無機栄養
mineral pollutant, mine pollutant 鉱毒
mineralization 無機化
minimax principle ミニマックス原理
minimum temperature 最低温度
minimum tillage 簡易耕起, ミニマムティレージ
minimum tillage seeding 簡易整地播き
mining pollution 鉱害
Ministry of Agriculture, Forestry and Fisheries (MAFF) 農林水産省(農水省)
minor crop, side crop 副作物
minor element, microelement, trace element 微量元素
minor gene 微働遺伝子
mioga ginger, *Zingiber mioga* (Tumb. ex Murray) Roscoe ミョウガ(茗荷)
mirror 反射鏡【光学】
Miscanthus sinensis Andersson, Japanese plume-grass, eulalia grass ススキ(薄)
Miscanthus sinensis Andersson var. *condensatus* (Hack.) Makino, Hachijo plume grass ハチジョウススキ(八丈薄)
miscellaneous cereals, millets 雑穀
misclassification 誤判別
missing plant, vacant hill 欠株
missing value 欠測値, 欠損値
mist blower ミスト機, 送風式噴霧機
mist culture, aeroponics 噴霧耕
mitochondrion (*pl.* mitochondria) ミトコンドリア
mitosis (*pl.* mitoses), karyokinesis,

caryokinesis 有糸分裂
mitotic apparatus 分裂装置
mitotic phase 分裂期, M phase M期
mitsuba, Japanese hornwort, *Cryptotaenia japonica* Hassk. ミツバ (三葉)
mitsumata, *Edgeworthia chrysantha* Lindl. (= *E. papyrifera* Sieb. et Zucc.) ミツマタ (三椏)
mix-sowing (-seeding) 混播 (こんぱ, こんぱん)
mixed cropping 混作
mixed fertilizer 配合肥料, 複合肥料
mixed line 混系
mixed model 混合模型【統計】
mixed pasture 混播草地
mixed planting, companion planting 混植
mixed pollination, mass pollination 混合受(授)粉
mixed sowing (seeding) 混播 (こんぱ, こんぱん)
mixed sward 混播草地
mixer ミキサー
mixoploidy 混数性, 混倍数性
Mn (manganese) マンガン
Mo (molybdenum) モリブデン
mobilization 易動化 (いどうか)
mode モード, 最頻値
mode of inheritance 遺伝様式
mode of reproduction, reproductive system, breeding system 繁殖様式
model モデル, 模型
model validation モデルの検証
modeling モデリング
modified grassland, improved grassland 改良草地
modified pasture, improved pasture 改良草地
modifier 変更遺伝子
modifying gene 変更遺伝子
moisture 水分, 湿気
moisture content, water content 水分含量, 含水量
moisture weight percentage 含水比
molarity モル濃度
mold かび (黴)
moldboard plow はつ (撥) 土板プラウ
molding, earthing up, ridging 土寄せ, 培土
mole drain モグラ暗きょ (渠), 弾丸暗きょ (渠)
mole ratio モル比
molecular biology 分子生物学
molecular breeding 分子育種
molecular cell biology 分子細胞生物学
molecular diffusion 分子拡散
molecular formula 分子式
molecular marker 分子マーカー
molecular sieve 分子ふるい (篩)
molecular structure 分子構造
molecular weight 分子量
Mollugo pentaphylla L. ザクロソウ
Mollugo verticillata L., common carpetweed クルマバザクロソウ
molybdenum (Mo) モリブデン
moment モーメント, 積率【統計】
Momordica charantia L., bitter gourd, balsam pear ニガウリ (苦瓜)
monitoring system モニタリングシステム
monkey-bread tree, baobab, *Adansonia digitata* L. バオバブ
monochoria, *Monochoria vaginalis* (Burm. f.) C. Presl コナギ
Monochoria korsakowii Regel et Maack ミズアオイ
Monochoria vaginalis (Burm. f.) C. Presl, monochoria コナギ
monocot 単子葉植物
monocotyledon 単子葉植物
monocotyledonous 単子葉の
monoculture 単一栽培, 単作
monoecism 雌雄同株[性], 雌雄異花同株[性]
monogenomic species 一ゲノム種, 一

基種
monogerm seed　単胚種子
monophyletic　一元性の
monopodial branching　単軸分枝
monosaccharide　単糖
monosomic plant　一染色体植物
monsoon　季節風
Monte Carlo simulation　モンテカルロ・シミュレーション
montmorillonite　モンモリロナイト
moon's age, age of the moon　月齢
moor, bog　泥炭地
mordant　媒染剤
morphogenesis　形態形成
morphological character　形態形質
morphology　形態学
mortar　乳鉢
Morus alba L. (= *M. bombycis* Koidz.), mulberry　クワ(桑)
mosaic　モザイク
Moso bamboo, *Phyllostachys heterocycla* (Carrière) Matsum. f. *pubescens* (Mazel ex Houz.) D. C. McClint.　モウソウチク(孟宗竹)
"Mosu" rice, penicillium diseased rice　もす米
moth bean, mat bean, *Vigna aconitifolia* (Jacq.) Maréchal (= *Phaseolus aconitifolius* Jacq.)　モスビーン
mother corm　親いも
mother plant　母本, 親株
mother tree　母樹
mother tuber　親いも
motor cell, bulliform cell　機動細胞
mound　盛土, くらつき(鞍築)
mountain bromegrass, *Bromus marginatus* Nees ex Steud.　マウンテンブロムグラス
mountain dairy　山地酪農
mounting agent　封入剤
moutan, tree p[-a-]eony, *Paeonia suffruticosa* Andr. (= *P. moutan* Sims)　ボタン(牡丹)

moving average　移動平均
mower　草刈り機, モーア
mowing　草刈り, 刈取り
mowing height, cutting height, clipping height　刈取り高さ
mowing interval, cutting interval, clipping interval　刈取り間隔
mRNA (messenger RNA)　伝令 RNA, メッセンジャー RNA
mucigel　ムシゲル
mucilage　粘液
Muck soil　黒泥土
Mucuna pruriens (L.) DC. var. *utilis* (Wight) Burck (= *Stizolobium hassjoo* Piper et Tracy), Yokohama [velvet] bean　ハッショウマメ(八升豆)
muddy sediment, sludge deposit　ヘドロ
mulberry, *Morus bombycis* Koidz. (= *M. alba* L.)　クワ(桑)
mulberry field　桑園
mulberry shoot　条桑(じょうそう)【クワ】
mulch　マルチ
mulch culture　被覆栽培【土を覆う】
mulcher　マルチャー
mulching　1) 畦(けい)面被覆　2) 敷草【作業】　3) マルチ【作業】
multicellular organism　多細胞生物
multi-dimensional scaling method　多次元尺度法
multigerm seed　多胚種子
multihybrid, polyhybrid　多性雑種
multi-layered cropping　多層作
multiline variety (cultivar)　多系品種
multimedia　マルチメディア
multimodal distribution　多峰分布
multinomial distribution　多項分布
multiple alleles　複対立遺伝子
multiple comparisons　多重比較
multiple correlation　重相関
multiple correlation coefficient　重相関係数
multiple cropping　多毛作

multiple crosses　多系交雑
multiple crossings　多系交雑
multiple fruit, aggregate fruit, syncarp　集合果, 多花果
multiple genes, polymeric genes　同義遺伝子
multiple regression　重回帰
multiple regression analysis　重回帰分析
multiple shoot　多芽体
multiplication　増殖, 繁殖
multiplicative　相乗的
multiplicative effect　相乗効果
multistage sampling　多段[標本]抽出
multivalent chromosome, polyvalent chromosome　多価染色体
multivariate analysis　多変量解析
multivariate normal distribution　多変量正規分布
mume, Japanese apricot, *Prunus mume* Sieb. et Zucc.　ウメ(梅)
mung bean, green gram, *Vigna radiata* (L.) R. Wilczek (= *Phaseolus aureus* Roxb.)　リョクトウ(緑豆)
Murdannia keisak (Hassk.) Hand.-Mazz. (= *Aneilema keisak* Hassk.), marsh dayflower　イボクサ
Musa cavendishii Lamb., dwarf banana　サンジャクバナナ(三尺バナナ)
Musa × paradisiaca L., banana　バナナ
Musa textilis Née, abaca, Manila hemp　マニラアサ(マニラ麻)
mustard oil　カラシ油
mutable gene　易変遺伝子(いへんいでんし)
mutable plastid　易変色素体
mutagen　突然変異源
mutagenesis　突然変異誘発
mutant　突然変異体
mutation　突然変異
mutation breeding　突然変異育種[法]
mutation pressure　突然変異圧
mutual shading　相互遮へい
mutual translocation, reciprocal translocation　相互転座
mycoplasma　マイコプラズマ
mycorrhiza　菌根
mycorrhizal fungus　菌根菌
Myriophyllum aquaticum (Vell.) Verdc. (= *M. brasiliense* Camb.), parrot's-feather　オオフサモ
Myristica fragrans Houtt., nutmeg　ニクズク(肉豆蔲)

[N]

N (nitrogen)　窒素
Na (sodium)　ナトリウム
NAA (naphthalene acetic acid)　ナフタレン酢酸
NAD (nicotinamide adenine dinucleotide)　ニコチンアミドアデニンジヌクレオチド
NADP (nicotinamide adenine dinucleotide phosphate)　ニコチンアミドアデニンジヌクレオチドリン酸
naked barley, *Hordeum vulgare* L.　ハダカムギ(裸麦)
naked grain type　はだか(裸)性【オオムギ】
naked oat, *Avena nuda* L.　ハダカエンバク(裸燕麦)
nalta jute, *Corchorus olitorius* L.　シマツナソ
nameko fungus, *Pholiota nameko* S. Ito et Imai　ナメコ
naphthalene acetic acid (NAA)　ナフタレン酢酸
napiergrass, elephant grass, *Pennisetum purpureum* Schumach.　ネピアグラス
NAR (net assimilation rate)　純同化率
narrowleaf vetch, *Vicia angustifolia* L.　カラスノエンドウ
naseberry, sapodilla, *Manilkara zapota* (L.) P. Royen (= *Achras zapota* L.)　サポジラ
Nasturtium officinale R. Br. (= *Rorippa*

nasturtium-aquaticum (L.) Hayek.), water-cress, cresson　クレソン(和蘭芥)
native, indigenous, spontaneous　自生の
native grassland, range, rangeland　野草地
native land, original habitat　原生地
native pasture, range, rangeland　野草地
native species　在来種
native variety, indigenous variety, local variety　在来品種
natural crossing　自然交雑
natural disaster　自然災害
natural dormancy　自然休眠
natural drying　自然乾燥
natural enemy　天敵
natural grass, wild grass　野草【イネ科】
natural grassland　自然草地
natural hybridization　自然交雑
natural mutation　自然突然変異, 偶発突然変異
natural pasture　自然草地
natural pollination, open pollination　放任受粉
natural population　自然集団
natural seeding, self seeding　自然下種
natural selection　自然選択
natural-shaped tea bush　自然仕立て茶園
natural supply　天然供給[量]
naturalized plant　帰化植物
naturalized weed　帰化雑草
Navashin's fluid　ナワシン液【固定液】
navel, hilum　へそ(臍)
Nawashin fluid　ナワシン液【固定液】
near-isogenic line　準(近)同質遺伝子系統
nearest neighborhood effect　最近隣効果
neck internode of panicle (spike)　穂首節間
neck node of panicle (spike)　穂首節
neck of panicle (spike), panicle base　穂首

neck rot, blast　いもち(稲熱)病【イネ】
necrosis　ネクロシス, え(壊)死
nectar　蜜, 花蜜
nectary　蜜腺
needle spikerush, slender spikerush, *Eleocharis acicularis* (L.) Roem. et Schult.　マツバイ
negative binomial distribution　負の二項分布
negative photoblastic seed, dark germinater　暗発芽種子
negative staining　ネガティブ染色
negro coffee, *Senna occidentalis* Link. (= *Cassia torosa* Cav., *C. occidentalis* L.)　ハブソウ
neighboring pollination, geitonogamy　隣花受粉
Nelumbo nucifera Gaertn., Indian lotus, sacred lotus　ハス(蓮)
nematicide, nematocide　殺線虫剤
nematode　線虫
Nephelium lappaceum L., rambutan　ランブータン
nerve, vein　葉脈
Nessler's reagent　ネスラー試薬
nest seeding (sowing)　摘播(てきは)
nested classification　枝分かれ分類
net assimilation rate (NAR)　純同化率
net culture　ネット栽培
net photosynthesis　純光合成
net primary productivity　純一次生産力
net production　純生産
net radiation　純放射
netted vein, netted venation, reticulate venation　網状脈
neural network　ニューラルネットワーク【情報処理】
neutral amino acid　中性アミノ酸
neutral detergent fiber (NDF)　中性デタージェント繊維
neutral plant, day-neutral plant　中性植物
neutralization　中和

new [crop] rice　新米
new leaf　新葉
new shoot, sprouting shoot　新芽【チャ】
New Zealand hemp, *Phormium tenax* J. R. Forst. et G. Forst.　ニュージーランドアサ(ニュージーランド麻)
Newman-Keuls method　ニューマン・ケウルス法
NFT (nutrient film technique)　培養液薄膜水耕法
Nicotiana rustica L., rustica tobacco, aztec tobacco　ルスチカタバコ
Nicotiana tabacum L., tobacco　タバコ(煙草)
nicotinamide adenine dinucleotide (NAD)　ニコチンアミドアデニンジヌクレオチド
nicotinamide adenine dinucleotide phosphate (NADP)　ニコチンアミドアデニンジヌクレオチドリン酸
nicotinic acid　ニコチン酸
nif (nitrogen fixation) gene　窒素固定遺伝子
niger seed, ramtil, *Guizotia abyssinica* (L. f.) Cass.　ニガーシード
niger seed oil　ニガー種油
night soil　下肥(しもごえ)
night-soil pail　肥桶(こえおけ)
night-soil reservoir, manure pool　肥溜(こえだめ)
night temperature　夜温
nightson pail　肥桶
ninhydrine reaction　ニンヒドリン反応
nipa palm, *Nypa fruticans* Wurmb.　ニッパヤシ
nitrate　硝酸塩
nitrate ion　硝酸イオン
nitrate nitrogen　硝酸態窒素
nitrate reductase　硝酸還元酵素
nitrate reduction　硝酸還元
nitrate toxicity　硝酸[塩]中毒
nitric acid　硝酸

nitrification　硝酸化成[作用]
nitrifier　硝酸化成菌
nitrifying bacteria　硝酸化成菌
nitrite　亜硝酸塩
nitrite ion (NO_2^-)　亜硝酸イオン
nitrogen (N)　窒素
nitrogen absorption coefficient　窒素吸収係数
nitrogen assimilation　窒素同化
nitrogen cycle, cycling of nitrogen　窒素循環
nitrogen dioxide (NO_2)　二酸化窒素
nitrogen efficiency　窒素効率
nitrogen excess　窒素過剰
nitrogen fertilizer　窒素肥料
nitrogen fixation　窒素固定
nitrogen fixation (*nif*) gene　窒素固定遺伝子
nitrogen fixer　窒素固定菌
nitrogen fixing bacterium　窒素固定菌
nitrogen-free extract (NFE)　可溶無窒素物
nitrogen metabolism　窒素代謝
nitrogen monoxide (NO)　一酸化窒素
nitrogen oxides (NO_x)　窒素酸化物
nitrogen source　窒素源
nitrogen starvation　窒素飢餓
nitrogen use efficiency　窒素利用効率
nitrogenous fertilizer　窒素肥料
NMR (nuclear magnetic resonance)　核磁気共鳴
No.1 flour　一番粉【ソバ】
noble camomile, chamomile, *Anthemis nobilis* L.　ローマカミツレ
nodal diaphragm　隔壁,隔膜【節】
nodal root　節根
nodding ear　垂れ穂
node　節(せつ)
node order　節位
nomenclature　命名法
nonadditive　非相加的
nonadditive genetic effect　非相加[的]遺伝効果

non-centrifuged sugar　含蜜糖
noncompetitive inhibition　非拮抗阻害
non-cropping　休耕
non-cultivation　休耕
nondrying oil　不乾性油
nonglutinous　粳(うるち)性の
nonglutinous rice　粳米(うるちまい)
nongrazing area, exclosure　禁牧区
nongrazing plot, exclosure　禁牧区
nonheritable variation　非遺伝的変異
non-Mendelian inheritance　非メンデル式遺伝
non[-]nodulating line　根粒非着生系統【ダイズ】
nonparametric method　ノンパラメトリック法
non-photosensitive variety　非感光性品種
non-productive tiller　無効分げつ
nonprotein nitrogen　非タンパク[態]窒素
nonrecurrent parent　一回親, 非反復親
nonreducing sugar　非還元糖
nonseasonal culture　時無し栽培
nonseasonal variety　時無し品種
nonselective herbicide　非選択性除草剤
non-spiny sago palm, sago palm, *Metroxylon sagu* Rottb.　サゴヤシ(サゴ椰子)
nonstructural carbohydrate　非構造性炭水化物
non-sulfate fertilizer　無硫酸根肥料
nontillage sowing (seeding)　不耕起播き
non-topping　無摘心
non-woven fabric　不織布
noodles　麺類
normal crop, average crop　平年作
normal density　正規密度
normal [density] curve　正規[密度]曲線
normal distribution　正規分布
normal equation　正規方程式
normal population　正規母集団

normal random number　正規乱数
normal-season culture　普通期栽培
normal type　並性【オオムギ】
normal value　平年値
normals　平年値
northern blot technique　ノーザンブロット法
northern blotting　ノーザンブロット法
northern leaf blight, leaf blight　すす紋病【トウモロコシ】
notched-belly rice kernel　胴切米(どうぎれまい)
NO_x (nitrogen oxides)　窒素酸化物
noxious weed　有害雑草, 害草
NPK elements, three major nutrients　肥料三要素
nucellus　珠心(しゅしん)
nuclear substitution　核置換[法]
nuclear transplantation　核移植[法]
nuclear division　核分裂
nuclear envelope　核膜
nuclear magnetic resonance (NMR)　核磁気共鳴
nuclear membrane　核膜
nuclear pore　核孔, 核膜孔
nucleic acid　核酸
nucleoid, karyoid　核様体
nucleolus (*pl.* nucleoli)　仁, 核小体
nucleoprotein　核タンパク質
nucleosome　ヌクレオソーム
nucleotide　ヌクレオチド
nucleus (*pl.* nuclei)　核
nucleus substitution　核置換[法]
nucleus transplantation　核移植[法]
null hypothesis　帰無仮説
nullisomic plant　零染色体植物
number of days before heading　出穂前日数
number of grains per head　一穂粒数
number of leaves on the main culm　主稈葉数
numerical integration　数値積分
numerical taxonomy　数量分類[学]

numetrical aperture　開口数【顕微鏡】
nurse crop　保護作物
nurse culture　保護培養, ナース培養
nursery [bed]　苗床, 苗代
nursery bed soil, bed soil　床土 (とこつち)
nursery box　育苗箱
nursery center　育苗センター
nursery chamber　育苗器, 出芽器
nursery field　苗圃 (びょうほ)
nursery garden　苗圃
nursery mat　成形培地
nursery stock　苗木
nursery temperature　育苗温度
nursery test　苗床検定
nursery transplanting　床替え
nut　殻果, 堅果
nut grass, *Cyperus rotundus* L.　ハマスゲ
nutmeg, *Myristica fragrans* Houtt.　ニクズク (肉豆蔲)
nutriculture, hydroponics, solution culture　養液栽培
nutrient　栄養素, 養分
nutrient absorption, nutrient uptake　養分吸収
nutrient absorption ability　吸肥力
nutrient deficiency　養分欠乏, 肥切れ
nutrient deficiency symptom　養分欠乏症
nutrient disorder　栄養障害
nutrient excess　要素過剰【土壌窒素など】
nutrient film technique (NFT)　培養液薄膜水耕法
nutrient hunger (deficiency)　肥切れ
nutrient uptake, nutrient absorption　養分吸収
nutriophysiology　栄養生理
nutrition　栄養
nutritional diagnosis　栄養診断
nutritive value　栄養価 [値]
nux vomica, strychine tree, *Strychnos nux-vomica* L.　マチン
nyctinastic movement, sleep movement　就眠運動, 睡眠運動
nyctinasty, sleep movement　就眠運動, 睡眠運動
Nypa fruticans Wurmb., nipa palm　ニッパヤシ

[O]

O (oxygen)　酸素
oatmeal　オートミール
oats【通常 *pl.*】, *Avena sativa* L.　エンバク (燕麦)【≠カラスムギ】
objective　対物レンズ
oblique cutting　斜め挿し
oblique division　斜分裂
oblique planting　斜め植え
observation　観測値【統計】
oca, *Oxalis tuberosa* Mol.　オカ
Ocimum basilicum L., basil, sweet basil　バジル
ocular, eyepiece　接眼レンズ【顕微鏡】
Oenanthe javanica (Blume) DC., dropwort　セリ
Oenothera erythrosepala Borbás, evening primrose　オオマツヨイグサ
offensive smell, off-flavor　異臭
off-farm employment　農外雇用
off-farm income　農外所得
off-flavor, offensive smell　異臭
off-season crop[ping]　裏作
off-season culture, out-of-season culture　不時栽培
off-type plant　異型植物
offspring, progeny　次代, 後代
offspring test, progeny test　次代検定
oil cell　油細胞
oil content　含油量
oil crop　油料作物
oil extraction　搾油 (さくゆ)
oil immersion method　油浸法【顕微鏡】
oil meal　油粕 (あぶらかす)

oil palm, African oil palm, *Elaeis guineensis* Jacq.　アブラヤシ(油椰子)
oil paper sash　油障子
oil percentage　含油率
oil seed　油料種子
oiticica, *Licania rigida* Benth.　オイチシカ
okura, lady's fingers, *Abelmoschus esculentus* (L.) Moench (= *Hibiscus esculentus* L.)　オクラ
old [crop] rice　古米
old flower pollination　老花受粉
old fustic, *Chlorophora tinctoria* Gaud.　オールドファスチク
Olea europaea L., olive　オリーブ
oleic acid　オレイン酸
olericulture, vegetable gardening　野菜園芸
oligosaccharide　オリゴ糖, 少糖
oligotrophication　貧栄養化
olive, *Olea europaea* L.　オリーブ
olive oil　オリーブ油
one-seeded pod　一粒莢
one-sided test　片側検定
one-tailed test　片側検定
one-thousand-grain weight, 1000-grain weight　千粒重
one-way classification　一元分類
onion, *Allium cepa* L.　タマネギ(玉葱)
Onobrychis viciifolia Scop., sainfoin　セインフォイン
ontogenesis　個体発生
ontogeny　個体発生
on-year　成り年
oolong　ウーロン(烏龍)茶
opaque　不透明の
opaque rice-kernel　死米(しにまい)
open-center type　開心型
open community　疎生群落
open culture, open-field culture, outdoor culture　露地栽培
open ditch drainage　明きょ(渠)排水

open field, outdoors　露地
open-field culture, open culture, outdoor culture　露地栽培
open-field nursery　露地苗床
open-floret panicle　提灯穂
open pollination　1) 放任受粉　2) 自然受(授)粉
open system　開放系
opening of boll, blowing of boll　開絮(かいじょ)
opening time of first leaf　開葉期【チャ】
operating system (OS)　オペレーティングシステム
operation sequence　作業体系【狭義】
operon　オペロン
Ophiopogon japonicus (L. f.) Ker-Gawl.　ジャノヒゲ
opium poppy, garden poppy, *Papaver somniferum* L.　ケシ(芥子)
opposite　対生の
oppositional gene　離反遺伝子
optical sensor　光学センサー
optimum concentration　最適濃度
optimum leaf area index　最適葉面積指数
optimum picking time　摘採適期【ワタなど】
optimum plucking time　摘採適期【チャなど】
optimum temperature　最適温度, 適温
optimum value　最適値
Opuntia ficus-indica (L.) Mill. 他多種, prickly pear　ウチワサボテン
oral toxicity　経口毒性
Orbignya speciosa (Mart.) B. Rodr. (= *O. martiana* B. Rodr.), babassu　ババスヤシ(ババス椰子)
orchard, grove　果樹園
orchardgrass, cocksfoot, *Dactylis glomerata* L.　オーチャードグラス
order　目(もく)【分類】
order of tiller　分げつ次位
order statistics　順序統計量

ordinary nursery　普通苗床
ordinate　縦軸
oregano, common marjoram, *Origanum vulgare* L.　ハナハッカ
organ　器官
organ culture　器官培養
organelle　細胞[小]器官, オルガネラ
organic acid　有機酸
organic agriculture　有機農業
organic cultivation　有機栽培
organic culture　有機栽培
organic farming　有機農業
organic fertilizer　有機質肥料
organic mercury pesticide　有機水銀剤
organic phosphorus pesticide　有機リン剤
organic soil　有機質土壌
organization　体制
organogenesis　器官形成
organography　器官学
oriental senna, *Senna obtusifolia* (L.) H. S. Irwin et Barneby (= *Cassia obtusifolia* L.)　エビスグサ
Oriental tobacco　オリエント種【タバコ】
oriental water plantain, *Alisma plantago-aquatica* L. var. *orientale* Sam.　サジオモダカ
Origanum majorana L. (= *Majorana hortensis* Moench), sweet majoram　マヨラナ
Origanum vulgare L., oregano, common marjoram　ハナハッカ
original habitat, native land　原生地
original seed, stock seed, registered seed, foundation stock　原種
original seed farm, stock seed field　原種圃
original vegetation　原植生
ornamental horticulture, floriculture, flower gardening　花き(卉)園芸
ornamental plant　観賞植物
Ornithopus sativus Brot., serradella　セラデラ
orthogonal design　直交設計
orthogonal polynomial　直交多項式
Oryza glaberrima Steud., African rice　グラベリマイネ
Oryza sativa L.【ssp. *japonica* 日本型イネ, ssp. *indica* インド型イネ, ssp. *javanica* ジャワ型イネ】, rice　イネ(稲)
OS (operating system)　オペレーティングシステム
osier, *Salix koriyanagi* Kimura (= *S. purpurea* L. var. *multinervis* Fr. et Sav.)　コリヤナギ(杞柳)
osmic acid　オスミウム酸
osmiophilic globule　好オスミウム顆粒
osmium (Os)　オスミウム
osmometer　浸透[圧]計
osmosis　浸透【水分生理】
osmotic adjustment　浸透調整, 浸透圧調節
osmotic potential　浸透ポテンシャル
osmotic pressure　浸透圧
osmotic stress　浸透圧ストレス
Ottelia alismoides (L.) Pers. (= *O. japonica* Miq.)　ミズオオバコ
outbreeding, exogamy　異系交配
outcrossing, allogamy　他殖, 他家生殖
outdoor culture, open culture, open-field culture　露地栽培
outdoors, open field　露地
outer glume　外苞穎(がいほうえい)
outer seed coat　外種皮
outflow, runoff, efflux　流出, 流去
outlier　外れ値
out-of-season culture, off-season culture　不時栽培
output, production　生産高, 生産量
ovary　子房
ovary culture　子房培養
oven-dry soil, dried soil　乾土
over production　生産過剰

overdominance　超優性

overflow irrigation, runoff irrigation　いつ(溢)流灌漑

overgrazing, overstocking　過放牧

overgrowth of rootstock　台勝ち

overgrowth of the scion　台負け

overhead irrigation　頭上灌水

overhead watering　頭上灌水

overluxuriant growth, rank growth　過繁茂

over-matured leaf for plucking　こわ葉【チャ】

overripe　過熟

overseeding　追播き

oversowing　追播き

overstocking, overgrazing　過放牧

overwintering　越冬

overwintering ability　越冬性

overwintering bud, winter bud　越冬芽

overwintering leaf　越冬葉【チャ】

ovule　胚珠(はいしゅ)

ovule culture　胚珠培養

ovule stalk, funicle, funiculus　珠柄(しゅへい)

ovum, egg　卵(らん)

owner farmer　自作農

owner-operated farming　自作

oxalic acid　シュウ酸

Oxalis corniculata L., creeping woodsorrel　カタバミ

Oxalis corymbosa DC., Dr. Martius' wood-sorrel　ムラサキカタバミ

Oxalis tuberosa Mol., oca　オカ

oxaloacetic acid　オキサロ酢酸

oxidase　オキシダーゼ, 酸化酵素

oxidation　酸化

oxidation-reduction potential (Eh), redox potential　酸化還元電位

oxidative phosphorylation　酸化的リン酸化

oxidized layer　酸化層

oxygen (O)　酸素

oxygen electrode　酸素電極

oxygen evolution　酸素放出

[P]

P (phosphorus)　リン(燐)

Pachyrhizus erosus (L.) Urban, yam bean　クズイモ

pachytene stage　パキテン期, 太糸期(ふといとき)【減数分裂】

packing, tamping　鎮圧【土木】

paclobutrazol　パクロブトラゾル

paddock　牧区(ぼくく), 追込み場

paddy　水稲, 籾

paddy field　水田, 稲田, 本田

paddy field boat　田舟

paddy field converted from upland field　転換[水]田

paddy-field crop　水田作物

paddy-field cropping　水田作

paddy field reclamation　開田

paddy field under paddy-upland rotation　輪換[水]田

paddy rice, lowland rice, paddy　水稲

paddy rice-nursery　水苗代

paddy rice-seedling　水苗

paddy sheaf rack, sheaf rack　はさ, はざ(稲架)

paddy soil　水田土壌

paddy-upland rotation　田畑輪換

Paederia scandens (Lour.) Merr.　ヘクソカズラ

Paeonia moutan Sims (= *P. suffruticosa* Andr.), moutan, tree p[-a-]eony　ボタン(牡丹)

Paeonia lactiflora Pall. (= *P. albiflora* Pall.), Chinese paeony, Chinese peony　シャクヤク(芍薬)

Paeonia suffruticosa Andr. (= *P. moutan* Sims), moutan, tree p[-a-]eony　ボタン(牡丹)

PAGE (polyacrylamide gel electrophoresis)　ポリアクリルアミド電気泳動法

pair comparison 対比較
paired row sowing (seeding) 複条播き
pairing, synapsis, syndesis 対合(たいごう)【染色体】
Palaquium gutta (Hook. f.) Baill., guttapercha グッタペルカ
palatability 嗜好性, 食味
palatable 可食性の
palea 内穎(ないえい)
palisade tissue 柵状組織
palm kernel oil パーム核油【アブラヤシ】
palm oil パーム油【アブラヤシ】
palmarosa, *Cymbopogon martini* (Roxb.) Wats. パルマローザ
palmate compound leaf 掌状複葉
palmitic acid パルミチン酸
palmyra palm, *Borassus flabellifer* L. オウギヤシ(扇椰子)
palynology 花粉学
pan 盤[層], 平鉢
Panama hat palm (plant), *Carludovica palmata* Ruiz et Pav. パナマソウ
Panama rubber, Central American rubber, *Castilla elastica* Cerv. パナマゴム
Panax ginseng C. A. Mey. (= *P. schinseng* Nees), ginseng ヤクヨウニンジン(薬用人参)
Pandanus odorus Ridl., screw pine ニオイタコノキ
Pandanus spp.【*P. tectorius* Sol. ex Parkins. 他多種】, screw pine タコノキ(林投)
panel パネル
panelist, taster パネリスト, パネル構成員
pangola grass, *Digitaria eriantha* Steud. パンゴラグラス
panicle, head, ear, spike 穂
panicle (spike) neck node differentiation stage 穂首[節]分化期
panicle base, neck of panicle (spike) 穂首
panicle differentiation stage 幼穂分化期
panicle formation stage 幼穂形成期
panicle initiation stage 幼穂分化期
panicle length, ear length 穂長
panicle number, ear number 穂数
panicle-number type, ear-number type, many-tillering type 穂数型
panicle type, ear type, head type 穂型
panicle weight, ear weight 穂重
panicle-weight type, ear-weight type, heavy-panicle type 穂重型
Panicum antidotale Retz., blue panicgrass ブルーパニックグラス
Panicum coloratum L., colored Guinea grass, Klein grass カラードギニアグラス
Panicum dichotomiflorum Michx., fall panicum オオクサキビ
Panicum maximum Jacq., guineagrass ギニアグラス
Panicum maximum Jacq. var. *trichoglume* Eyles, green panic グリーンパニック
Panicum miliaceum L., millet, common millet, proso millet, hog millet キビ(黍)
Panicum repens L., torpedo grass ハイキビ
Panicum virgatum L., switchgrass スイッチグラス
papa lisas, ulluco, *Ullucus tuberosus* Caldas ウルーコ
Papaver setigerum DC. セティゲルムケシ
Papaver somniferum L., opium poppy, garden poppy ケシ(芥子)
papaya, *Carica papaya* L. パパイア
paper chromatography ペーパークロマトグラフィー
paper-making crop 製紙料作物
paper mulberry, *Broussonetia papyrifera* (L.) L'Hér. ex Vent. カジ

ノキ (梶の木)
paper mulberry, *Broussonetia kazinoki* Sieb.　コウゾ (楮)
paper pot　ペーパーポット
papilionaceous flower　蝶形花
papilla　乳頭状突起
papyrus, *Cyperus papyrus* L.　カミガヤツリ
PAR (photosynthetically active radiation)　光合成有効放射
Para rubber, *Hevea brasiliensis* Muell.-Arg.　パラゴム
parabola　放物線
paradermal section　並皮切片
paraffin method　パラフィン法
parallel evolution　平行進化
parallel vein　平行脈
parallel venation　平行脈
parallelism　平行現象
parameter　パラメータ
parasitic weed　寄生雑草
parasitism　寄生
parathion　パラチオン
parboiled rice　パーボイルドライス
parchment coffee　パーチメントコーヒー
parenchyma　柔組織
parenchyma cell　柔細胞
parenchymatous cell　柔細胞
parent　交配母本
parent material　母材
parent-offspring correlation　親子相関
parent rock　母岩
parent root　親根
parental line　親系統
parental strain　親系統
Parkia speciosa Hassk. (= *Peltogyne speciosa* Hassk.), pete　ネジレフサマメノキ
parrot's-feather, *Myriophyllum aquaticum* (Vell.) Verdc. (= *M. brasiliense* Camb.)　オオフサモ
parsley, *Petroselinum crispum* (Mill.) Nym. ex A.W.Hill.　パセリ
Parthenium argentatum A. Gray, guayule, Mexican rubber　グアユール
parthenocarpy　単為結果
parthenogenesis　単為生殖, 単為発生, 処女生殖
partial correlation　偏相関
partial correlation coefficient　偏相関係数
partial dominance　部分優性
partial pressure　分圧
partial regression　偏回帰
partial regression coefficient　偏回帰係数
partial sterility　部分不稔[性]
particle gun　パーティクルガン, 遺伝子銃
particle size　粒径
particle size distribution　粒径分布
partition chromatography　分配クロマトグラフィー
part-time farmer　兼業農家
Pascopyrum smithii (Rydb.) A. Löve, western wheatgrass　ウェスタン・ホィートグラス
Paspalum dilatatum Poir., dallisgrass　ダリスグラス
Paspalum distichum L., knotgrass　キシュウスズメノヒエ
Paspalum notatum Flugge, bahiagrass　バヒアグラス
Paspalum scrobiculatum L., Kodo millet　コドラ
Paspalum thunbergii Kunth ex Steud.　スズメノヒエ
passage cell　通過細胞
Passiflora edulis Sims, passion fruit　パッションフルーツ
Passiflora quadrangularis L., giant granadilla, square-stalked passion flower　オオミノトケイソウ
passion fruit, *Passiflora edulis* Sims　パッションフルーツ

passive absorption　受動的吸収
passive water absorption　受動的吸水
paste crop　糊料作物
Pasteur pipet, dropping pipet　駒込ピペット
pasture　牧草地, 放牧地, 草地
pasture plant　牧草
pasturing, grazing　放牧
patchouli, *Pogostemon cablin* (Blanco) Benth. (= *P. patchouli* Pell.)　パチョリ
path analysis　経路分析, パス解析
path coefficient　経路係数
path diagram　経路図
pathogen　病原体
pathogenic bacterium　病原細菌
pathogenic fungus　病原菌
pathway　経路
pattern recognition　パターン認識
pau rosa, *Aniba rosiodora* Ducke　パウローサ
Paullinia cupana Humb. et Kunth, guarana　ガラナ
PCR (polymerase chain reaction)　ポリメラーゼ連鎖反応
pea, garden pea, *Pisum sativum* L.　エンドウ(豌豆)
peach, *Prunus persica* (L.) Batch　モモ(桃)
peach palm, *Guilielma gasipaes* (H. B. K.) L. H. Bailey　モモミヤシ
peak of occurrence　発生最盛期
peanut, groundnut, *Arachis hypogaea* L.　ラッカセイ(落花生)
peanut oil, groundnut oil　ラッカセイ油
pear, *Pyrus communis* L. var. *sativa* DC.　セイヨウナシ
pearl millet, *Pennisetum americanum* (L.) K. Schum. (= *P. typhoideum* Rich.)　パールミレット
pearled barley　精麦
pearling, milling　とう(搗)精
pearling of barley　精麦【工程】

peat　泥炭
peat moss, sphagnum　ミズゴケ(水苔)
Peat soil, Bog soil　泥炭土
pectin　ペクチン
pectinase　ペクチナーゼ
pedicel　1) 小花柄　2) 小枝梗
pedigree　系統, 系譜, 血統
pedigree [chart]　系図
pedigree breeding method　系統育種法
pedigree mass selection　系統集団選抜
pedigree selection, line selection　系統選抜
pedogenesis, soil genesis　土壌生成
pedology　ペドロジー, 土壌学
peduncle　1) 花柄, 花梗　2) 果柄
peeling　剥皮(はくひ)
PEG (polyethylene glycol)　ポリエチレングリコール
Pelargonium spp.【*P. graveolens* L'Her., *P. radura* L'Her. など】, geranium　ゼラニウム
pelleted seed, coated seed　被覆種子
pellet　1) 固形飼料, ペレット　2) 沈殿物【遠心分離】
Peltogyne speciosa Hassk. (= *Parkia speciosa* Hassk.), pete　ネジレフサマメノキ
pencil cedar, eastern red cedar, *Sabina virginiana* (L.) Antoine (= *Juniperus virginiana* L.)　エンピツビャクシン
pencil-like root, cylindrical root　梗根(こうこん), ごぼう根【サツマイモ】
penetration　浸透, 透入
penetration resistance　貫入抵抗【土壌】
penicillium diseased rice, "Mosu" rice　もす米
Pennisetum alopecuroides (L.) Spreng., Chinese pennisetum　チカラシバ
Pennisetum americanum (L.) K. Schum. (= *P. typhoideum* Rich.), pearl millet　パールミレット
Pennisetum clandestinum Hochst. ex Chiov., kikuyu grass　キクユグラス

Pennisetum purpureum Schumach., napiergrass, elephant grass　ネピアグラス
pennyroyal mint, *Mentha pulegium* L.　ペニーロイヤルミント
pentad　半旬【気象】
pentaploid　五倍体
pentose　ペントース, 五単糖
penultimate leaf　止葉の前の葉
PEP (phosphoenolpyruvic acid)　ホスホエノールピルビン酸
PEP (phosphoenolpyruvate) carboxylase　ホスホエノールピルビン酸カルボキシラーゼ
pepper, *Piper nigrum* L.　コショウ (胡椒)
peppergrass, Virginia pepperweed, *Lepidium virginicum* L.　マメグンバイナズナ
peppermint, *Mentha × piperita* L.　ペパーミント
pepperwort, water clover, *Marsilea quadrifolia* L.　デンジソウ
percentage　歩合
percentage dry matter　乾物率
percentage establishment　苗立歩合
percentage of banjhi shoots to the total (P. B. S.), ratio of banjhi shoot　出開き度【チャ】
percentage of germination, germination percentage　発芽歩合, 発芽率
percentage of glassy kernel　硝子率
percentage of ripened grains　登熟歩合
percentage [of seedling] establishment　苗立歩合
percentage of selection, selection rate　選抜 (選択) 率
percentage [of] ripening　稔実歩合
percentage of productive culms (stems)　有効茎歩合
percentage sterility　不稔歩合
perception　知覚
percolation　浸透【土壌水分】

perennial　1) 多年生 [の] 2) 多年生植物
perennial buckwheat, *Fagopyrum cymosum* Meisn.　シャクチリソバ (赤地利蕎麦)
perennial crop　多年生作物
perennial forage　多年生牧草
perennial ryegrass, *Lolium perenne* L.　ペレニアルライグラス
perennial weed　多年生雑草
perfect flower, complete flower　完全花
perfect kernel　完全粒
perfect rice grain, whole rice grain　完全米
perforation　せん (穿) 孔
perforation plate　せん (穿) 孔板
performance test　1) 生産力検定 [試験] 2) 能力検定, 性能検定
performance test for recommendable varieties　奨励品種決定試験
perianth　花被
pericarp　果皮
periclinal chimera　周縁キメラ
periclinal division　並層分裂
pericycle　内鞘
periderm　周皮
perilla, *Perilla frutescens* (L.) Britton var. *japonica* (Hassk.) H. Hara (= *P. ocymoides* L.)　エゴマ (荏胡麻)
perilla, *Perilla frutescens* (L.) Britton var. *crispa* (Thunb. ex Murray) W. Decne　シソ (紫蘇)
Perilla frutescens (L.) Britton var. *crispa* (Thunb. ex Murray) W. Decne, perilla　シソ (紫蘇)
Perilla frutescens (L.) Britton var. *japonica* (Hassk.) H. Hara (= *P. ocymoides* L.), perilla　エゴマ (荏胡麻)
perilla oil　エゴマ油
period of short supply　端境期 (はざかいき)
periodicity　周期性

peripheral meristem 周辺分裂組織
perisperm 周乳
permanent grassland 永年草地, 永久草地
permanent modification 永続変異
permanent nursery 通し苗代
permanent pasture 永年草地, 永久草地
permanent preparation 永久プレパラート
permanent tissue 永久組織
permanent wilting 永久しお(萎)れ
permeability 透過性
permutation 順列
peroxidase ペルオキシダーゼ, 過酸化酵素
peroxisome ペルオキシソーム
perpetual flowering, ever-blooming, ever-flowering 四季咲きの
Persea americana Mill., avocado, alligator pear アボカド
Persian clover, *Trifolium resupinatum* L. (= *T. suaveolens* Willd.) ペルシャンクローバ
Persian speedwell, *Veronica persica* Poir. オオイヌノフグリ
Persian wheat, *Triticum carthlicum* Nevski ペルシアコムギ
Persicaria hydropiper (L.) Spach (= *Polygonum hydropiper* L.), water pepper ヤナギタデ
Persicaria lapathifolia (L.) S. F. Gray (= *Polygonum lapathifolium* L. ssp. *nodosum* Kitam.) オオイヌタデ
Persicaria longiseta (De Bruyn) Kitag. (= *Polygonum longisetum* De Bruyn), tafted knotweed イヌタデ
Persicaria nipponensis (Makino) H. Gross ヤノネグサ
Persicaria perfoliata (L.) H. Gross (= *Polygonum perfoliatum* L.) イシミカワ
Persicaria pubescens (Blume) Hara (= *Polygonum pubescens* Blume) ボントクダテ
Persicaria scabra Mold (= *Polygonum scabrum* Moench) サナエタデ
Persicaria senticosa (Franch. et Sav.) H. Gross (= *Polygonum senticosum* (Meisn.) Franch. et Sav.) ママコノシリヌグイ
Persicaria sieboldii (Meisn.) Ohki (= *Polygonum sagittatum* L. var. *sieboldii* (Meisn.) Maxim.) アキノウナギツカミ
Persicaria thunbergii (Sieb. et Zucc.) H. Gross (= *Polygonum thunbergii* Sieb. et Zucc.) ミゾソバ
Persicaria tinctoria (Aiton.) H. Gross (= *Polygonum tinctorium* Lour.), Chinese indigo アイ(藍)
Persicaria vulgaris Webb. et Moq. (= *Polygonum persicaria* L.), lady's-thumb ハルタデ
persistency 1) 生存年限 2) 永続性【牧草】
personal breeding 民間育種
pest control 病害虫防除
pesticide, agricultural chemicals 農薬
pestle 乳棒
petal 花弁
Petasites japonicus (Sieb. et Zucc.) Maxim., Japanese butterbur フキ(蕗)
pete, *Parkia speciosa* Hassk. (= *Peltogyne speciosa* Hassk.) ネジレフサマメノキ
petiolate 有柄の
petiole 葉柄
petiolule 小葉柄
petri dish ペトリ皿, シャーレ
Petroselinum crispum (Mill.) Nym. ex A. W. Hill., parsley パセリ
pF value pF値
PGA (phosphoglyceric acid) ホスホグリセリン酸
pH indicator pH指示薬

pH-meter pHメーター
pH-stat pHスタット
phage ファージ
phagemid ファージミド
Phalaris arundinacea L., reed canarygrass リードカナリーグラス
Phalaris tuberosa L., hardinggrass ハーディンググラス
phanerogams 顕花植物
Pharbitis nil (L.) Choisy (= *Ipomoea nil* (L.) Roth), Japanese morning glory アサガオ(朝顔)
phase 相
phase-contrast microscope 位相差顕微鏡
phaseic acid ファゼイン酸
Phaseolus aconitifolius Jacq. (= *Vigna aconitifolia* (Jacq.) Maréchal), moth bean, mat bean モスビーン
Phaseolus acutifolius A. Gray var. *latifolius* Freem., tepary bean テパリビーン
Phaseolus angularis L. (= *Vigna angularis* (Willd.) Ohwi et Ohashi), adzuki bean, azuki bean, small red bean アズキ(小豆)
Phaseolus aureus Roxb., *Vigna radiata* (L.) R. Wilczek, green gram, mung bean リョクトウ(緑豆)
Phaseolus calcaratus Roxb. (= *Vigna umbellata* (Thunb.) Ohwi et Ohashi), rice bean タケアズキ
Phaseolus coccineus L., scarlet runner bean, flower bean ベニバナインゲン(紅花隠元)
Phaseolus lunatus L. (= *P. limensis* Macf.), Lima bean, butter bean ライマメ
Phaseolus mungo L. (= *Vigina mungo* (L.) Hepper), black gram, urd, black matpe ケツルアズキ(毛蔓小豆), ブラックマッペ
Phaseolus pendulus Makino (= *Vigna umbellata* (Thunb.) Ohwi et Ohashi) ツルアズキ(蔓小豆)
Phaseolus vulgaris L., kidney bean, French bean インゲンマメ(隠元豆)
phasmid ファスミド
phellem, cork tissue コルク組織
Phellodendron amurense Rupr., Amur cork-tree キハダ
phellogen, cork cambium コルク形成層
phenocopy 表現型模写
phenol フェノール
phenology 生物季節学
phenolphthalein フェノールフタレイン
phenotype 表現型
phenotypic correlation 表現型相関
phenotypic variance 表現型分散
phenylalanine (Phe) フェニルアラニン
Philadelphia fleabane, *Erigeron philadelphicus* L. ハルジオン(春紫苑)
Philopterus littoralis Benth. (= *Glehnia littoralis* F. Schmidt et Miq.) ハマボウフウ(浜防風)
Phleum pratense L., timothy チモシー
phloem 篩部(しぶ)
Phoenix dactylifera L., date palm ナツメヤシ(棗椰子)
Phoenix sylvestris Roxb., sugar date palm サトウナツメヤシ
Pholiota nameko S. Ito et Imai, nameko fungus ナメコ
Phormium tenax J.R. Forst. et G. Forst., New Zealand hemp ニュージーランドアサ(ニュージーランド麻)
phosphatase ホスファターゼ
phosphate リン酸塩
phosphate absorption coefficient リン酸吸収係数
phosphate buffer リン酸緩衝液
phosphate retention capacity リン酸保持容量
phosphatic fertilizer リン酸肥料

phosphoenolpyruvic acid (PEP)　ホスホエノールピルビン酸
phosphoenolpyruvate (PEP) carboxylase　ホスホエノールピルビン酸カルボキシラーゼ
phosphoglyceric acid (PGA)　ホスホグリセリン酸
phospholipid　リン脂質
phosphoric acid　リン(燐)酸
phosphorus (P)　リン(燐)
phosphorylase　ホスホリラーゼ, リン酸化酵素
photoblastic seed, light germinater　光発芽種子
photochemical oxidant　光化学オキシダント
photochemical reaction　光化学反応
photochemical smog　光化学スモッグ
photochemical system, photosystem　光化学系
photodecomposition　光分解
photoinhibition　光阻害, 強光阻害, 光障害
photolysis　光分解
photometer, illuminometer　照度計
photomorphogenesis　光形態形成
photon, quantum　光量子
photon flux density (PFD)　光量子束密度
photoperiod　光周期
photoperiod sensitive phase, photophase　感光相
photoperiodic control　日長調節
photoperiodic response　日長反応
photoperiodic sensitivity, photosensitivity　感光性
photoperiodic treatment　日長処理
photoperiodism　光周性
photophase, photoperiod sensitive phase　感光相
photophosphorylation　光リン酸化
photorespiration　光呼吸
photosensitive variety　感光性品種
photosensitivity, photoperiodic sensitivity　感光性
photosynthate, photosynthetic product　光合成産物
photosynthesis　光合成
photosynthetic ability　光合成能力
photosynthetic activity　光合成活性
photosynthetic capacity　光合成能力
photosynthetic organ　光合成器官
photosynthetic phosphorylation　光合成的リン酸化
photosynthetic photon flux density (PPFD)　光合成有効光量子束密度
photosynthetic pigment　光合成色素
photosynthetic product, photosynthate　光合成産物
photosynthetic rate　光合成速度
photosynthetic tissue　光合成組織
photosynthetically active radiation (PAR)　光合成有効放射
photosystem, photochemical system　光化学系
phototropism　光屈性
Phragmites australis (Cav.) Trin. ex Steud. (= *P. communis* Trin.), common reed　ヨシ(葭)
Phyllanthus urinaria L.　コミカンソウ
phyllochron　出葉間隔
Phyllostachys bambusoides Sieb. et Zucc., madake bamboo　マダケ(真竹)
Phyllostachys heterocycla (Carrire) Matsum. f. *pubescens* (Mazel ex Houz.) D. C. McClint., Moso bamboo　モウソウチク(孟宗竹)
Phyllostachys nigra (Lodd. ex Loudon) Munro f. *henonis* (Mitord) Stapf ex Rendle, black bamboo　ハチク(淡竹)
phyllotaxis　葉序
phyllotaxy　葉序
phylogenetic tree, genealogical tree　系統樹

phylogeny 系統学，系統発生
phylum 門【分類】
physical containment 物理的封じ込め
physical genetic map 物理的遺伝子地図
physiological active substance 生理活性物質
physiological disease 生理病
physiological disorder 生理障害
physiological ecology 生理生態学
physiological factor 生理的因子
physiological leaf spot, weather fleck 生理的斑点病【タバコ】
physiological race 生理品種，生理変種
physiological salt solution 生理[的]食塩水
physiology 生理学
phytochrome フィトクロム
phytogeography, plant geography 植物地理学
phytohormone, plant hormone 植物ホルモン
Phytolacca americana L., pokeweed, scoke ヨウシュヤマゴボウ
phytomer ファイトマー
phytometer 植物計
phytomorphology, plant morphology 植物形態学
phytopathology, plant pathology 植物病理学
phytoplasma ファイトプラズマ
phytotoxicity 植物毒性
phytotoxin 植物毒[素]
phytotron ファイトトロン
picked cotton 摘採棉(綿)
picker, plucker 摘採機
picking 1) 摘刈り(つみがり) 2) 摘採【ワタなど】 3) 葉か(掻)き【タバコ】
pigeon pea, cajan pea, *Cajanus cajan* (L.) Millsp. キマメ
pigment 色素
pilot farm 模範農場

Pimenta dioica (L.) Merr. (= *P. officinalis* Lindl.), allspice, pimento オールスパイス
pimento, allspice, *Pimenta dioica* (L.) Merr. (= *P. officinalis* Lindl.) オールスパイス
Pimpinella anisum L., anise アニス
pin flower, long-styled flower 長[花]柱花
pinchcock ピンチコック
pinching, topping, top pruning 摘心
pineapple, *Ananas comosus* (L.) Merr. パイナップル
Pinellia ternata (Thunb.) Breit. カラスビシャク
pinnate compound leaf 羽状複葉
pipe drainage, underdrainage, tile drainage 暗きょ(渠)排水
Piper nigrum L., pepper コショウ(胡椒)
pistachio, *Pistacia vera* L. ピスタチオ
Pistacia vera L., pistachio ピスタチオ
pistil 雌しべ(蕊)，雌ずい(蕊)
pistillate flower, female flower 雌花(しか，めばな)
Pisum sativum L., pea, garden pea エンドウ(豌豆)
pit 1) 壁孔 2) 室(むろ)
pith 髄(ずい)
pith cavity, medullary cavity 髄腔(ずいこう)
Pithecellobium jiringa (Jack) Prain, jiring ジリンマメ
pithiness す(鬆)入り【ダイコン等】
pithy tissue す(鬆)【ダイコン等】
pixel 画素【画像解析】
place of origin, provenance 原産地
placenta 胎座(たいざ)
planophyll 水平葉型の
plant age in leaf number 葉齢【植物体】
plant anatomy 植物解剖学
plant and wood ashes 草木灰
plant body 植物体

plant breeding 植物育種
plant bugs カメムシ
plant community 植物群落
plant death, death, dying 枯死
plant ecology 植物生態学
plant geography, phytogeography 植物地理学
plant growth regulator 植物成長調節物質
plant height 草高【≠草丈 plant length】
plant histology 植物組織学
plant hormone, phytohormone 植物ホルモン
plant husbandry, cultivation 栽培
plant indicator, indicator plant 指標植物
plant kingdom 植物界
plant length 草丈【≠草高 plant height】
plant morphology, phytomorphology 植物形態学
plant nutrition 植物栄養[学]
plant opal プラントオパール, 植物タンパク石
plant organography 植物器官学
plant pathology, phytopathology 植物病理学
plant physiology 植物生理学
plant pigment 植物色素
plant poison, phytotoxin 植物毒
Plant Production Science 【日本作物学会英文誌】
plant protection 植物保護
plant quarantine 植物検疫
plant remain 植物遺体, 植物残さ(渣)
plant residue 植物遺体, 植物残さ(渣)
plant sociology 植物社会学
plant spacing, planting distance 栽植距離
plant taxonomy 植物分類学
plant thremmatology 植物育種学
plant-to-row test 一個体一列検定, 一母本一列検定
plant type 草型
Plantago asiatica L. (= *P. major* L. var. *asiatica* (L.) Dec.), Asiatic plantain オオバコ
Plantago lanceolata L., buck-horn plantain ヘラオオバコ
plantation プランテーション
plantation white sugar 耕地白糖
planted area 作付面積
planter 植付け機, 点播機
planthopper ウンカ
planting 栽植, 植付け, 定植
planting cord 植え縄
planting density, planting rate 栽植密度
planting distance, plant spacing 栽植距離
planting furrow 植え溝
planting hole, planting pit 植え穴
planting pattern, spacial arrangement [of plants] 栽植様式
planting rate, planting density 栽植密度
planting time 栽植時期, 定植期
planting trench 植え溝
planting with wide spacing, sparse planting 疎植
plaque プラーク
plasmagene, cytoplasmic gene, cytogene 細胞質遺伝子
plasmid プラスミド
plasmodesmata (*sing.* plasmodesma) 原形質連絡
plasmolysis 原形質分離
plastic film プラスチックフィルム
plastic-tube watering チューブ灌水
plastic-tunnel culture トンネル栽培
plasticity 可塑性, 塑性
plastid プラスチド, 色素体
plastid mutation 色素体突然変異
plastochron[e] 葉間期, プラストクロン
plastoquinone プラストキノン

plate culture 平板培養
Platycodon grandiflorum (Jacq.) A. DC., Japanese bellflower, Chinese bellflower, balloonflower キキョウ(桔梗)
Pleioblastus chino (Franch. et Savat.) Makino var. *viridis* (Makino) S. Suzuki (= *Arundinaria pygmaea* Mitford var. *glabra* Ohwi), dwarf bamboo ネザサ
pleiotropism 多面発現
pleiotropy 多面作用
plerome 原中心柱
plot, experimental plot 試験区
plot-to-plot irrigation 田越灌漑(たごしかんがい)
plow 洋犂(ようり), プラウ
plow layer, top soil 作土
plow pan, furrow pan 耕盤
plow sole, furrow pan 耕盤
plowing 耕起, り(犂)耕, プラウ耕, すき(犂)起こし
plowing and harrowing 耕起砕土, 耕は(耙)
plowing by cow 牛耕
plowing-in すき(犂)込み
plucked new shoot 生葉(なまは)【チャ】
plucker, picker 摘採機
plucking 摘採【チャなど】
plucking machine, tea plucker 摘採機【チャ】
plucking surface 摘採面【チャ】
plump kernel 肥大粒
plumule 幼芽
pluviometer, rain gauge 雨量計
PMC (pollen mother cell) 花粉母細胞
Poa annua L., annual bluegrass スズメノカタビラ
Poa compressa L., Canada bluegrass カナダブルーグラス
Poa pratensis L., Kentucky bluegrass ケンタッキーブルーグラス
Poa trivialis L., rough bluegrass, rough-stalked meadow grass ラフブルーグラス
Poaceae, Gramineae イネ科
pod 莢(さや)【インゲン, エンドウ, ナタネなど】
pod corn, *Zea mays* L. var. *tunicata* St. Hil. ポドコーン
pod shedding 落きょう(莢)
podding 結莢(けっきょう)
Podzol ポドゾル
poet's jasmine, common white jasmine, *Jasminum officinale* L. f. ソケイ(素馨), ツルマツリ(蔓莉茉)
Pogostemon cablin (Blanco) Benth. (= *P. patchouli* Pell.), patchouli パチョリ
poison 毒物
poison ryegrass, darnel, *Lolium temulentum* L. ドクムギ(毒麦)
poisonous herb 毒草
poisonous weed 有毒雑草
Poisson distribution ポアソン分布
pokeweed, scoke, *Phytolacca americana* L. ヨウシュヤマゴボウ
polar nucleus 極核
polar transport 極性移動
polarity 極性
polarization 偏波
polarized light 偏光
polarizing microscope 偏光顕微鏡
polder 干拓地
polderland 干拓地
pole 極
pole climbing, viny つる(蔓)性の
pole number 極数
Polish wheat, *Triticum polonicum* L. ポーランドコムギ
polished rice 精米, 白米, 研磨米
polishing, milling 精白
pollen 花粉
pollen analysis 花粉分析
pollen culture 花粉培養

pollen grain 花粉粒
pollen gun 花粉散布器, 花粉銃, 花粉放射器
pollen mother cell (PMC) 花粉母細胞
pollen parent 花粉親
pollen sac 花粉嚢 (かふんのう)
pollen tetrad 花粉四分子
pollen tube 花粉管
pollen tube nucleus 花粉管核
pollination 受粉, 授粉
pollinator 花粉媒介者
pollinizer 授粉樹
polluted field, contaminated field 汚染圃場
polluted rice 汚染米
pollution, contamination 汚染
pollution source, source of contamination 汚染源
polyacrylamide gel electrophoresis (PAGE) ポリアクリルアミド電気泳動法
polyamine ポリアミン
polyarch 多原型
polycross[ing] 多交配
polyculture 複作
polydeoxyribonucleotide synthase ポリデオキシリボヌクレオチドシンターゼ
polyembryony 多胚現象
polyethylene film ポリエチレンフィルム
polyethylene glycol (PEG) ポリエチレングリコール
Polygala senega L., senega, seneca, snake root セネガ
Polygala senega L. var. *latifolia* Torr. et A. Gray ヒロハセネガ
polygene ポリジーン
Polygonum aviculare L., prostrate knotweed ミチヤナギ
Polygonum cuspidatum Sieb. et Zucc. (= *Reynoutria japonica* Houtt.), Japanese knotweed イタドリ
Polygonum hydropiper L. (= *Persicaria hydropiper* (L.) Spach), water pepper ヤナギタデ
Polygonum lapathifolium L. ssp. *nodosum* Kitam. (= *Persicaria lapathifolia* (L.) S.F.Gray) オオイヌタデ
Polygonum longisetum De Bruyn (= *Persicaria longiseta* (De Bruyn) Kitag.), tafted knotweed イヌタデ
Polygonum perfoliatum L. (= *Persicaria perfoliata* (L.) H. Gross) イシミカワ
Polygonum persicaria L. (= *Persicaria vulgaris* Webb. et Moq.), lady's-thumb ハルタデ
Polygonum pubescens Blume (= *Persicaria pubescens* (Blume) Hara) ボントクダテ
Polygonum sagittatum L. var. *sieboldii* (Meisn.) Maxim. (= *Persicaria sieboldii* (Meisn.) Ohki) アキノウナギツカミ
Polygonum scabrum Moench (= *Persicaria scabra* Mold) サナエタデ
Polygonum senticosum (Meisn.) Franch. et Sav. (= *Persicaria senticosa* (Franch. et Sav.) H.Gross) ママコノシリヌグイ
Polygonum thunbergii Sieb. et Zucc. (= *Persicaria thunbergii* (Sieb. et Zucc.) H. Gross) ミゾソバ
Polygonum tinctorium Lour. (= *Persicaria tinctoria* (Aiton.) H. Gross), Chinese indigo アイ (藍)
polyhybrid, multihybrid 多性雑種
polymerase chain reaction (PCR) ポリメラーゼ連鎖反応
polymeric genes, multiple genes 同義遺伝子
Polymnia sonchifolia Poepp. et Endl., yacon ヤーコン
polynomial regression 多項式回帰
polypeptide ポリペプチド

polyphyletic 多元性の
polyploid 倍数体
polyploidy 倍数性
Polypogon mouspeliensis (L.) Desf., rabbit's-foot polypogon, annual beard grass ハマヒエガエリ
polyribosome ポリリボソーム
polysaccharide 多糖類
polysome ポリソーム
polyvalent chromosome, multivalent chromosome 多価染色体
pomaceous fruit ナシ状果, 仁果
pome 仁果, ナシ状果
pomegranate, *Punica granatum* L. ザクロ(石榴)
pomology, fruit science 果樹学, 果樹園芸学
ponding water release, drainage of residual water 落水
pooled estimator こみにした推定量
poor crop, bad crop 不作, 凶作
poor harvest, bad crop 不作, 凶作
poor ripening 登熟不良
poor soil 不良土壌
poor sunshine か(寡)照
pop corn, *Zea mays* L. var. *everta* Bailey ポップコーン
poppy seed oil ケシ油
population 1) 集団, 個体群 2) 母集団 【統計】
population density 集団密度, 個体群密度
population ecology 個体群生態学
population engaged in agriculture 農業就業人口
population genetics 集団遺伝学
population improvement 集団改良
population mean 母集団平均[値], 母平均[値]
population parameter 母集団パラメータ
pore space, void 孔げき(隙)
porometer ポロメーター

porosity 孔げき(隙)率
Porphyra tenera Kjellman, laver アサクサノリ
Portulaca oleracea L., common purslane スベリヒユ
positional information 位置情報
post-fixation 後固定(こうてい)
postharvest ポストハーベスト
pot ポット, 植木鉢, 鉢
pot culture 鉢栽培
pot plant, potted plant 鉢物
Potamogeton distinctus A.W.Benn. ヒルムシロ
potash fertilizer カリ肥料
potassium (K) カリウム
potassium chloride 塩化カリ[ウム](塩加)
potassium hydroxide 水酸化カリウム
potassium sulfate 硫酸カリウム
potato, Irish potato, *Solanum tuberosum* L. ジャガイモ
potato bean, groundnut, *Apios americana* Medik. アメリカホドイモ
potato harvester ポテトハーベスタ
potato planter ポテトプランタ
potato starch ジャガイモデンプン
potato yam, lesser yam, *Dioscorea esculenta* (Lour.) Burk. トゲドコロ
potential arable land, uncultivated arable land 未耕地
potential evapotranspiration 蒸発散位, 可能蒸発散量
potential productivity 潜在生産力
potential soil fertility 潜在地力
potential soil productivity 潜在地力
potential yield ポテンシャル収量, 潜在収量
potometer ポトメーター, 吸水計
potted plant, pot plant 鉢物
potting 鉢植え, 鉢上げ
poultry manure 鶏糞きゅう(厩)肥
powder paper 薬包紙
powdered tea 抹茶

powdery mildew　うどんこ病
power　検出力【統計的検定】
power dryer, forced air dryer　通風乾燥機
power duster　動力散粉機
power huller　動力籾す(摺)り機
power spectrum　パワー・スペクトル
power sprayer　動力噴霧機
power tea plucker　可搬型摘採機【チャ】
power thresher　動力脱穀機
power tiller　耕うん(耘)機, 動力耕耘機
PPFD (photosynthetic photon flux density)　光合成有効光量子束密度
PPi (pyrophosphate)　ピロリン酸
prairie　プレーリー
prairie grass, rescuegrass, *Bromus catharticus* Vahl (= *B. unioloides* (Willd.) H.B.K.)　レスキュグラス
preceding cropping　前作
precipitation, amount of precipitation　降水量
precise culture experiment　精密栽培試験
precision agriculture　精密農業, 精密圃場管理農業
precision farming　精密圃場管理
precocious ear　走り穂
precooling　予冷
precursor　前駆物質
predicted value　予測値
prediction　予測
predictor　予測量
pregerminated seed　催芽種子
preharvest sprouting　穂発芽
preliminary performance test　生産力検定予備試験
preliminary survey　予備調査
preliminary yield trial　生産力検定予備試験
premature bolting　早期抽だい(苔)
premature heading, unseasonable heading　不時出穂
preparation　1) 調製　2) プレパラート

preprophase band (PPB)　前期前微小管束
preservation, storage　保蔵, 保存
preservation of line　系統保存
preservation of stock, clonal preservation　株保存
preservative　保存液
pressed barley, rolled barley　押麦(おしむぎ)
pressed juice　搾汁(さくじゅう)
pressure chamber method　圧ボンベ法, プレッシャーチャンバー法
pressure flow theory　圧流説
pressure potential　圧ポテンシャル
pressure probe　プレッシャープローブ【膨圧測定】
pretreatment　1) 前処理　2) 予措(よそ)【種子】
price of rice, rice price　米価(べいか)
prickly pear, *Opuntia ficus-indica* (L.) Mill. 他多種　ウチワサボテン
prickly sago palm, spiny sago palm, *Metroxylon rumphii* Mart.　トゲサゴ[ヤシ](刺サゴ[椰子])
prickly sida, *Sida spinosa* L.　アメリカキンゴジカ
primary axis, main axis　主軸
primary bed　親床(おやどこ)【タバコ】
primary branch of panicle, primary rachis-branch of panicle　一次枝梗
primary canopy, first crop　一番草【牧草】
primary [cell] wall　一次[細胞]壁
primary growth　一次成長
primary leaf　初生葉
primary meristem　一次分裂組織
primary phloem　一次篩部
primary pit-field　一次壁孔域, 初生壁孔域
primary rachis-branch of panicle, primary branch of panicle　一次枝梗
primary root　一次根
primary succession　一次遷移

primary thickening [growth]　一次肥大 [成長]
primary tiller　一次分げつ
primary tissue　一次組織
primary vascular bundle　一次維管束
primary xylem　一次木部
prime and curing　連干し乾燥【タバコ】
primer　プライマー
priming, picking　葉か(掻)き【タバコ】
primings, flyings　下葉(したは)【タバコ】
primordium (*pl.* primordia)　原基, 始原体
principal component　主成分【統計】
principal component analysis (PCA)　主成分分析
principal component score　主成分スコア
probability　確率
probability density　確率密度
probability density function　確率密度関数
probability distribution　確率分布
problem soil　問題土壌
procambium　前形成層
Proceedings of the Crop Science Society of Japan　日本作物学会紀事【1976年まで】
process-based model　過程模型, プロセスモデル
processing　加工, 調製
processing suitability　加工適性
processor　プロセッサ
procumbent　伏地性の
procumbent stem　平伏茎
producer　生産者
producer rice price　生産者米価
producing area　生産地
product inhibition　生産物阻害
production, output　生産, 生産高, 生産量
production adjustment　生産調整
production control　生産調整

production cost　生産費
production trait　生産形質
productive stem (culm), fruitful stem (culm)　有効茎
productive structure　生産構造
productive tiller, bearing tiller　有効分げつ
productivity, yield potential　生産力
proembryo　前胚
progeny, offspring　次代, 後代
progeny test　次代検定, 後代検定
program　プログラム
prolamin　プロラミン
proliferation, multiplication, propagation　増殖
proline (Pro)　プロリン
promeristem　前分裂組織
promoter　プロモーター
prop root, brace root　支根, 支持根, 支柱根
propagation　増殖, 繁殖
propagation bed, propagation bench　挿木床
propagation farm, seed [production] farm　採種圃
prophase　前期【細胞分裂】
prophyll　前葉(ぜんよう), 前出葉
proplastid　プロプラスチド, 原色素体
proportionate sampling　比例抽出[法], 比例標本抽出[法]【統計】
proso millet, [common] millet, hog millet, *Panicum miliaceum* L.　キビ(黍)
prostrate knotweed, *Polygonum aviculare* L.　ミチヤナギ
prostrate spurge, *Euphorbia supina* Raf.　コニシキソウ
prostrate type, creeping type　ほふく(匍匐)型
protandrous flower　雄ずい(蕊)先熟花
protandry, proterandry　雄ずい(蕊)先熟
protease　プロテアーゼ, タンパク[質]

分解酵素
protected cultivation, greenhouse agriculture, farming under structure　施設農業
protected horticulture, horticulture under structure　施設園芸
protected rice-nursery　保温苗代, 保護苗代
protected upland rice-nursery　保温畑苗代
protein　タンパク質
protein grain　タンパク粒
protein metabolism　タンパク質代謝
protein nitrogen　タンパク態窒素
proterandry, protandry　雄ずい(蕊)先熟
proterogyny　雌ずい(蕊)先熟
protoderm　前(原)表皮
protogyny　雌ずい(蕊)先熟
proton pump　プロトンポンプ
protophloem　原生篩部
protoplasm　原形質
protoplasmic streaming　原形質流動
protoplast　プロトプラスト, 原形質体
protostele　原生中心柱
prototype　原型
protoxylem　原生木部
protuberance, ingrowth　突起
provenance, place of origin　原産地
provisional planting, temporary planting　仮植(かしょく)
pruning　1) せん(剪)定　2) せん(剪)枝【チャ】
pruning at the base　台刈り
Prunus amygdalus Batsch, almond　アーモンド(扁桃)
Prunus armeniaca L., apricot　アンズ(杏)
Prunus avium L.他数種, cherry　オウトウ(桜桃)
Prunus mume Sieb. et Zucc., mume, Japanese apricot　ウメ(梅)
Prunus persica (L.) Batch, peach　モモ(桃)

Psathyrostachys juncea (Fisch.) Nevski (= *Elymus junceus* Fisch.), Russian wildrye　ロシアワイルドライ
pseudo color　擬似カラー
pseudocarp, false fruit　偽果
pseudocereals　偽(擬)禾穀類
pseudocompatibility　偽和合性
pseudodominance　偽優性
pseudofertility　偽稔性
pseudogamy, false fertilization　偽受精
pseudogene　偽遺伝子
Psidium guajava L., guava　グワバ
Psophocarpus tetragonolobus (L.) DC., winged bean, four angled bean, goa bean, asparagus pea　シカクマメ
psychrometer　乾湿計, サイクロメータ
Pteridium aquilinum (L.) Kuhn var. *latiusculum* (Desv.) Underw. ex A. Heller, eastern brackenfern, bracken　ワラビ(蕨)
Pterocarpus santalinus L. f., red sandalwood, red sanders　コウキシタン(紅木紫檀)
ptorescine, putrescine　プトレッシン
public hazards, environmental pollution　公害
pUC plasmid　pUC系プラスミド
puddling and levelling　代か(搔)き
Pueraria lobata (Willd.) Ohwi (= *P. thunbergiana* (Sieb. et Zucc.) Benth.), kudzu, kudzu-vine　クズ(葛)
pueraria starch　クズデンプン
pulling hoe　引きぐわ(鍬)
pulling of seedling　採苗, 苗取り
pulling-out hills from field, uprooting hills from field　株揚げ
pulp, sarcocarp　果肉
pulse, leguminous crop　マメ
pulses, leguminous crops　マメ類
pulse-chase experiment　パルス-チェイス実験
pulse crop, leguminous crop　マメ科作

物
pulse crops, leguminous crops　マメ類
pulse threshing　マメ打ち
pulses and cereals　しゅく(菽)穀類
pulverization, harrowing　砕土
pulverizer, harrow, clod crusher　ハロー
pulvinus, leaf cushion　葉枕(ようちん)
pumpkin, winter squash, *Cucurbita maxima* Duch. ex Lam.　セイヨウカボチャ
pumpkin, winter squash, *Cucurbita moschata* (Duch. ex Lam.) Duch. ex Poir.　ニホンカボチャ
pumpkin, summer squash, *Cucurbita pepo* L.　ペポカボチャ
Punica granatum L., pomegranate　ザクロ(石榴)
pupa (*pl.* pupae)　さなぎ(蛹)
purchased feed　購入飼料
pure line　純系
pure line selection　純系選抜
pure line separation　純系分離
pure line theory　純系説
pure sward　単播草地
purification　精製
purple ammannia, *Ammannia coccinea* Rottb.　ホソバヒメミソハギ
purple arrowroot, edible canna, Queensland arrowroot, achira, *Canna edulis* Ker-Gawl.　食用カンナ
purple deadnettle, *Lamium purpureum* L.　ヒメオドリコソウ
purple rice　紫稲(むらさきイネ)
purple stain of seed　紫斑病【ダイズ】
putrefaction　腐敗
putrescine, ptorescine　プトレッシン
Pycreus flavidus (Retz.) T. Koyama (= *Cyperus globosus* All.), globe sedge　アゼガヤツリ
Pycreus polystachyos (Rottb.) P. Beauv. (= *Cyperus polystachyos* Rottb.)　イガガヤツリ
Pycreus sanguinolentus (Vahl) Nees (= *Cyperus sanguinolentus* Vahl)　カワラスガナ
pyranometer, solarimeter　日射計
Pyrethrum cinerariifolium Trevir. (= *Chrysanthemum cinerariaefolium* Visiani), insectpowder plant, insect flower, Dalmatian chrysanthemum　ジョチュウギク(除虫菊)
pyrophosphate (PPi)　ピロリン酸
Pyrus communis L. var. *sativa* DC., pear　セイヨウナシ
Pyrus pyrifolia (Burm.f.) Nakai (= *P. serotina* Rehder), Japanese pear, sand pear　ナシ(梨)
pyruvic acid　ピルビン酸

[Q]

quackgrass, *Agropyron repens* (L.) P. Beauv.　シバムギ
quadrat　コドラート, 方形区, わく(枠)
quadrate plucking　枠摘み【チャ】
quadratic curve　二次曲線
qualitative analysis　定性分析
qualitative character (trait)　質的形質
quality　品質
quality inspection　品質検査
Quality of the Environment in Japan　環境白書
quality test　品質検定
quantification　数量化
quantitative analysis　定量分析
quantitative character (trait)　量的形質, 計量形質
quantity of element　成分量
quantization　量子化
quantum, photon　光量子
quantum efficiency　量子収率
quantum meter　光量子計
quantum yield　量子収量
quartet, tetrad　四分子
quartile　四分位点
quartile deviation　四分位偏差

Queensland arrowroot, edible canna, purple arrowroot, achira, *Canna edulis* Ker-Gawl.　食用カンナ
questionnaire　質問票, 聞取り調査表
quick acting fertilizer, readily available fertilizer　速効性肥料
quicklime, calcium oxide, calx　生石灰
quiescent center　静止中心
quiescent state, resting stage　静止期【細胞分裂】
quinine, cinchona, *Cinchona* spp.【*C. pubescens* Vahl など】キナ
quinoa, *Chenopodium quinoa* Willd.　キノア

[R]

rabbit's-foot polypogon, annual beard grass, *Polypogon mouspeliensis* (L.) Desf.　ハマヒエガエリ
Rabdosia japonica (Burm. f.) H. Hara (= *Isodon japonicus* Hara)　ヒキオコシ(引起)
race　レース, 病原型
raceme　総状花序
rachis　1) 中軸, 花軸, 葉軸【複葉】 2) 穂軸(すいじく)【イネ・ムギ類】
rachis branch　枝梗(しこう)
rack drying　1) はざ(稲架)干し　2) 架乾(かかん)
radial division　放射分裂
radial wall　放射壁【細胞】
radiation　放射
radiation biology　放射線生物学
radiation breeding　放射線育種
radiation cooling　放射冷却
radiation genetics　放射線遺伝学
radiation mutagenesis　放射線突然変異生成
radiation use efficiency　日射利用効率
radiative cooling　放射冷却
radical, free radical　遊離基
radical leaf　根出葉, 根生葉
radicle　幼根
radioactive isotope (RI)　放射性同位元素, 放射性同位体
radioactivity　放射能
radioimmunoassay (RIA)　放射性免疫検定法, ラジオイムノアッセイ
radioisotope (RI)　放射性同位元素, 放射性同位体
radish, *Raphanus sativus* L.　ダイコン(大根)
radius　半径
rail-tracking tea plucker　レール走行式摘採機【チャ】
rain damage　雨害
rain-fed cultivation　天水栽培
rain-fed culture　天水栽培
rain-fed paddy field　天水田
rain gauge, pluviometer　雨量計
raindrop erosion　雨滴侵食
rainy season　雨季
rainy season crop[ping], wet-season crop[ping]　雨季作, 雨期作
raised bed　揚げ床
raised seedbed　揚げ床
raising　育成
raising of seedling under structure　施設育苗
raising seedling, rearing of seedling　育苗
rake　レーキ
rakkyo, Baker's garlic, *Allium chinense* G. Don　ラッキョウ(辣韮)
rambutan, *Nephelium lappaceum* L.　ランブータン
random number　乱数
ramie, China grass, *Boehmeria nivea* (L.) Gaud.　チョマ(苧麻)
Ramsch method of breeding　ラムシュ育種法
ramtil, niger seed, *Guizotia abyssinica* (L. f.) Cass.　ニガーシード
ranch, livestock farm　牧場
random arrangement　無作為配置

rec (287)

random drift　機会的浮動
random [effects] model　変量[効果]模型
random error　偶然誤差
random mating　無作為交配
random planting　乱雑植え
random sample　無作為標本
random sampling　無作為[標本]抽出
random variable　確率変数
randomization　無作為化
randomized block design (method)　無作為ブロック設計, 乱塊法
range　1) 野草地, 牧野　2) 範囲【統計量】
range of variation　変異[の]幅
rangeland, native grassland (pasture)　野草地, 牧野
rank　順位
rank correlation　順位相関
rank growth, overluxuriant growth　過繁茂
Ranunculus quelpaertensis (Lév.) Nakai (= *R. silerifolius* Lév.), buttercup　キツネノボタン
Ranunculus muricatus L., roughseed buttercup　トゲミノキツネノボタン
Ranunculus sceleratus L., crowfoot, buttercup　タガラシ
rape, *Brassica napus* L. および *B. campestris* L.　ナタネ(菜種)
rape, *Brassica campestris* L. (= *B. rapa* L. var. *campestris* (L.) Clapham　在来種ナタネ
rape, *Brassica napus* L.　洋種ナタネ
rapeseed meal　ナタネ粕
rapeseed oil　ナタネ油
Raphanus sativus L., radish　ダイコン(大根)
raphe　せい(臍)条
raphia palm, *Raphia pedunculata* Beauv. (= *R. ruffia* Mart.)　ラフィアヤシ
Raphia pedunculata Beauv. (= *R. ruffia*

Mart.), raphia palm　ラフィアヤシ
raspberry【*Rubus idaeus* L. およびその園芸品種を含む】ラズベリー
raster data　ラスターデータ【ディスプレイ】
ratio of banjhi shoot, percentage of banjhi shoots to the total (P. B. S.)　出開き度【チャ】
ratio of contribution, contribution ratio　寄与率
ratio of good nursery plant　成苗率【チャ】
ratoon　1) ひこばえ(蘖)【イネ, サトウキビ, パイナップルなど】　2) 株出し苗【サトウキビ】　3) 刈り株苗
ratoon cropping　株出し[栽培]
ratooning　株出し[栽培]
rattan [palm], rotan, cane palm, *Calamus caesius* Blume 他多種　トウ(籐)
Rauvolfia serpentina Benth. ex Kurtz, snakewood, Java devilpepper　インドジャボク(インド蛇木)
raw cotton　生綿(きわた), 原綿(げんめん)
raw sugar　粗糖, 生砂糖(きざとう)
ray　放射組織
reaction, response　反応, 応答
reaction center　反応中心
reaction rate　反応速度
readily available fertilizer, quick acting fertilizer　速効性肥料
reagent　試薬
reagent bottle　試薬びん
reaper　刈取り機, リーパ
reaper and binder, binder　バインダ
reaping　刈取り【禾穀類】
rearing of seedling, raising seedling　育苗
receptor　受容体, 受容器, レセプター
recessive　劣性の
recessive mutation　劣性突然変異
reciprocal cross[ing]　逆交雑

reciprocal crosses (crossing)　正逆交雑, 相反交雑
reciprocal recurrent selection　相反循環選抜
reciprocal translocation, mutual translocation　相互転座
reclaimed land　開拓地
reclaimed paddy field　干拓田, 新田
reclaimed settlement, settlement [site]　入植地
reclamation in water area　干拓
recombinant DNA　組換え DNA
recombinant inbred line　組換え型自殖系統
recombination　組換え
recombination of genes, gene recombination　遺伝子組換え
recombination value　組換え価
recommended variety　奨励品種
recorder　記録計
recording chart　記録紙
recovered line　再生系統
recreation crop　嗜好料作物
rectangular nursery　短冊苗代
rectangular planting　長方形植え
recurrent parent　反復親
recurrent selection　循環選抜法
red clover, *Trifolium pratense* L.　アカクローバ
red coloring of leaves　紅葉
red fescue, *Festuca rubra* L.　レッドフェスク
red [kerneled] rice　赤米 (あかまい)
red pepper, chili, capsicum, *Capsicum annuum* L. (= *C. frutescens* L.)　トウガラシ (唐芥子)
red sandalwood, red sanders, *Pterocarpus santalinus* L. f.　コウキシタン (紅木紫檀)
Red soil　赤色土
red sorrel, *Rumex acetosella* L.　ヒメスイバ
redifferentiation　再分化

redox potential, oxidation-reduction potential (Eh)　酸化還元電位
redtop, *Agrostis gigantea* Roth　レッドトップ
reduced layer, reduction zone　還元層
reduced organ　退化器官
reducing sugar　還元糖
reduction　還元
reduction division stage, meiosis stage　減数分裂期
reduction zone, reduced layer　還元層
reductive pentose phosphate cycle　還元的ペントースリン酸回路
reed canarygrass, *Phalaris arundinacea* L.　リードカナリーグラス
refined flour　精粉 (せいこ)【コンニャク】
refined oil　白絞油 (しらしめゆ)
refined sugar　精製糖
reflectance spectrum　反射スペクトル
reflection　反射
reflection coefficient　反射係数
reflectivity　反射率
reflorescence　返り咲き
refraction index　屈折率
refractometer index　屈折計示度
refrigerated centrifuge　冷却遠心機
regenerated plant　再生個体
regeneration　1) 再生　2) 更新【チャ, クワ】
regional trial　現地試験
registered seed, stock seed, original seed, foundation stock　原種
registered seed　登録種子
registered variety　登録品種
registration　登録
registry　登録
Regosol　未熟土
regreening　回青 (かいせい)【カンキツ類】
regression　回帰
regression analysis　回帰分析
regression coefficient　回帰係数

regression curve 回帰曲線
regression equation 回帰式
regression line 回帰直線
regression model 回帰模型
regression parameter 回帰パラメータ
regrowing stage 起生期【ムギ類】
regrowth 1) 再成長 2) 再生【草地】
regrowth vigor 再生力
regular planting 正条植え
regular planting with ruler 定規植え
regulation of flowering, flowering regulation 開花調節
regulator gene, regulatory gene 調節遺伝子
Rehmannia glutinosa (Gaertn.) Libosh. ex Fisch.et C. A. Mey var. *purpurea* (Makino) アカヤジオウ (赤矢地黄)
rejection 棄却【仮説検定】
rejuvenation 若返り
rejuvenescence 若返り
related species, allied species 近縁種
relationship 類縁
relative frequency 相対頻度
relative growth 相対成長
relative growth rate (RGR) 相対成長率
relative humidity 相対湿度
relative water content 相対含水量
relay cropping つなぎ作
relay intercropping つなぎ作
relic [species] 残存種
relief map, topographic map 地形図
remaining in rosette state 座止 (ざし)
remnant method 残穂法 (ざんすいほう)
remote sensing リモートセンシング, 遠隔測定, 隔測
Rendzina レンジナ
renewal of seeds 種子更新
renewal, regeneration 更新【チャ, クワ】
renovation 更新【草地など】
repeated selection 反復選抜
repellent 忌避剤, 駆除剤
repetitive sequence 反復配列

replanting 改植, 植直し
replicated plot 反復試験区
replication 1) 反復 2) 複製【核酸】
reporter gene レポーター遺伝子
repotting 鉢替え
repressor リプレッサー, 抑制遺伝子
repressor gene 抑制遺伝子
reprocessed sugar 加工糖
reproduction 1) 生殖, 繁殖 2) 再生産
reproductive cell 生殖細胞
reproductive growth 生殖成長
reproductive organ 生殖器官
reproductive phase 生殖相
reproductive stage 生殖成長期
reproductive system, mode of reproduction, breeding system 繁殖様式
rescuegrass, prairie grass, *Bromus catharticus* Vahl (= *B. unioloides* (Willd.) H.B.K.) レスキュグラス
reseeding, resowing 再播
reserve carbohydrate 貯蔵炭水化物
reserve protein, storage protein 貯蔵タンパク質
reserve starch, storage starch 貯蔵デンプン
reserve substance 貯蔵物質
reserved seed 備蓄種子
reservoir 貯水地
residual 残差
residual activity 残効
residual effect 残効
residual herbage 残草
residual mean-square 残差平均平方
residual stalks 残幹【タバコ】
residual sum of squares 残差平方和
residual toxicity 残留毒性
residual variance 残差分散
residue 残留物
resin 樹脂
resin crop 樹脂料作物
resistance 抵抗性

resistance thermometer　抵抗温度計
resistance to breaking, breaking resistance　挫折抵抗
resistance to submergence　冠水抵抗性
resistance to water flow　通水抵抗, 通導抵抗
resistant variety　抵抗性品種
resolution　分解能
resolving power　1) 分解能　2) 解像力
resowing, reseeding　再播 (さいは)
respiration　呼吸
respiration rate　呼吸速度
respiratory activity　呼吸活性
respiratory quotient (RQ)　呼吸商, 呼吸率
respiratory rate　呼吸速度
respiratory root　呼吸根
response, reaction　反応, 応答
response curve　反応曲線, 応答曲線
rest, dormancy　休眠
resting bud, dormant bud　休眠芽
resting stage, quiescent state　静止期【細胞分裂】
restitution nucleus　復旧核
restorer　稔性回復系統
restriction enzyme　制限酵素
restriction fragment length polymorphism (RFLP)　制限断片長多型
reticulate venation, netted vein, netted venation　網状脈
reticulate vessel　網紋導管
retrogressive succession　退行遷移
retrotransposon　レトロトランスポゾン
retting　精練 (せいれん)【アサ】
return period　再現期間
return plowing　往復耕
reverse mutation, back mutation　逆突然変異, 復帰突然変異
reverse selection, adverse selection　逆選抜
reverse transcriptase　逆転写酵素
reversible reaction　可逆反応

reversion, atavism　先祖返り
Reynoutria japonica Houtt. (= *Polygonum cuspidatum* Sieb. et Zucc.), Japanese knotweed　イタドリ
RFLP (restriction fragment length polymorphism)　制限断片長多型
RGR (relative growth rate)　相対成長率
Rheum officinale Baill., medicinal rhubarb　ダイオウ (大黄)
Rhizobia, root-nodule bacteria　根粒菌
rhizome　根茎
rhizoplane, root-soil interface　根面
rhizosphere　根圏
Rhodes grass, *Chloris gayana* Kunth　ローズグラス
Rhus spp., lacquer tree, varnish tree　ウルシ (漆)【広義】
Rhus succedanea L., Japanese wax tree　ハゼノキ (櫨)
Rhus succedanea L. var. *dumoutieri* Pier.　トンキンウルシ
Rhus verniciflua Stokes, Japanese lacquer tree　ウルシ (漆)【狭義】
RI (radioisotope), radioactive isotope　放射性同位元素, 放射性同位体
Ri plasmid　Ri プラスミド
ribonucleic acid (RNA)　リボ核酸
ribosomal RNA (rRNA)　リボソーム RNA
ribosome　リボソーム
ribulose-1,5-bisphosphate (RuBP)　リブロース-1,5-ビスリン酸
ribulose-1,5-bisphosphate carboxylase/oxygenase (Rubisco)　リブロース-1,5-ビスリン酸カルボキシラーゼ/オキシゲナーゼ
Ricciocarpos natans (L.) Corda　イチョウウキゴケ
rice　米
rice, *Oryza sativa* L.【ssp. *japonica* 日本型イネ, ssp. *indica* インド型イネ, ssp. *javanica* ジャワ型イネ】イネ (稲)

rice-barley double cropping　イネ・ムギ二毛作
rice bean, *Vigna umbellata* (Thunb.) Ohwi et Ohashi (= *Phaseolus calcaratus* Roxb.)　タケアズキ
rice bran　糠(ぬか), 米糠(こめぬか)
rice bran oil　米糠油
rice bundle　稲束(いなたば)
rice cake　餅
rice center, rice processing plant　ライスセンター
rice cracker, rice flake　せんべい(煎餅)
rice cropping　稲作
rice cultivation　稲作
rice cutgrass, *Leersia oryzoides* (L.) Sw. ssp. *oryzoides*　エゾノサヤヌカグサ
rice field　本田
rice flake, rice cracker　せんべい(煎餅)
rice flatsedge, *Cyperus iria* L.　コゴメガヤツリ
rice flour　米粉(べいふん)
rice for sake brewery, brewers' rice　酒米(さかまい)
rice grain　米粒(べいりゅう)
rice harvest, rice reaping　稲刈り
rice inspection　米穀検査
rice kernel　米粒
rice mill　精米機
rice milling　精米
rice milling machine　精米機
rice oil, bran oil　糠油(ぬかゆ)
rice planting rope　田植綱
rice price, price of rice　米価(べいか)
Rice Price Council　米価審議会(米審)
rice processing plant, rice center　ライスセンター
rice quality　米質(べいしつ)
rice reaping, rice harvest　稲刈り
rice screenings　屑米(くずまい)
rice seed, seed rice　種籾
rice sorter, grain sorter　米選機
rice starch　米デンプン
rice straw　稲わら(藁)
rice stubble　稲株
rice terrace, terrace paddy field　棚田
rice threshing　稲こ(扱)き
rice transplanter　田植機
rice transplanting　田植え
rice transplanting frame, frame rule for transplanting of rice seedling　田植枠
rice-wheat double cropping　イネ・ムギ二毛作
rice year　米穀年度
ricefield, riceland, paddy field　稲田
Ricinus communis L., castor, castor bean　ヒマ(蓖麻)
ridge　うね(畝)
ridge bed, ridge-up bed　うね(畝)床
ridge bed width　うね(畝)床幅
ridge culture　うね(畝)立て栽培
ridge distance, row width, row space　うね(畝)幅
ridge height　うね(畝)高
ridge planting　うね(畝)作, うね(畝)仕立て
ridge plowing, ridging　うね(畝)立て
ridge shoulder　うね(畝)肩
ridge side　うね(畝)肩
ridge sowing (seeding)　うね(畝)播き
ridger　うね(畝)立て機, 土寄機
ridge-up bed, ridge bed　うね(畝)床
ridging　1) 土寄せ, 培土, うね(畝)立て 2) うね(畝)立て耕
ridging on tilled land　無心うね(畝)立て
ridging up　うね(畝)揚げ
ridging with untilled core　有心うね(畝)立て
riding tractor　乗用トラクタ
riding-type tea plucker　乗用型摘採機【チャ】
right crop for right land　適地適作
ring rot　輪腐病【ジャガイモ】
ringing, girdling　環状剥皮(かんじょうはくひ)
rinsing　水洗

ripeness to flower　花熟
ripening　1) 登熟, 稔実　2) 成熟【果実】
ripening period, [grain] filling period　登熟期間
risk analysis　リスク分析
river bulrush, *Bolboschoenus fluviatilis* (Torr.) T. Koyama ssp. *yagara* (ohwi) T. Koyama (= *Scirpus yagara* Ohwi)　ウキヤガラ
rivet wheat, *Triticum turgidum* L.　リベットコムギ
RNA (ribonucleic acid)　リボ核酸
RNA polymerase　RNAポリメラーゼ
roasted soil, heated soil　焼土 (やきつち)
Robinson anemometer　ロビンソン風速計
Robitzsch bimetallic actinograph　ロビッチ自記日射計
robot system　ロボットシステム
robusta coffee, Congo coffee, *Coffea robusta* Linden (= *C. canephora* Pierr. ex Froeh.)　ロブスタコーヒー
robustness　頑健性【統計】
rock wool　ロックウール
rodenticide　殺鼠剤 (さっそざい)
rod-row plot　一畦区 (いっけいく)
roguing　株抜き【混種, 罹病株などの】
rolled barley　圧扁大麦, 押麦 (おしむぎ)
rolled wheat　圧扁小麦
roller gin　ローラージン【ワタ】
rolling　圧扁 (あっぺん)
root　根
root and tuber crops　いも類
root apex, root tip　根端
root apical meristem　根端分裂組織
root axis　根軸
root box　根箱
root cap　根冠
root crop　需根作物, 根菜
root cutting　根挿し
root exudate　根の分泌物
root grafting　根接ぎ

root hair　根毛
root length density, rooting density　根長密度
root [length] scanner　ルートスキャナー
root mass　根量
root nodule　根粒
root-nodule bacteria, Rhizobia　根粒菌
root pressure　根圧
root pruning　せん (剪) 根, 断根
root rot　根腐れ [病]
root-soil interface, rhizoplane　根面
root splitting, division of root stock　根分け
root spread　根張り
root system　根系
root tip, root apex　根端
root type　結しょ (藷) 型【サツマイモ】
root vegetable, root crop　根菜
root vegetables　根菜類
root weed　根生雑草
rooted cutting　挿木苗
rooting　活着, 発根
rooting ability　活着力, 発根力
rooting density, root length density　根長密度
rooting zone　根域
rootlet　細根
rootstock, stock　台木
rope root　索根 (さくこん)【サトウキビ】
Roripa nasturtium-aquaticum (L.) Hayek. (= *Nasturtium officinale* R. Br.), water-cress, cresson　クレソン (和蘭芥)
Rorippa indica (L.) Hochr., Indian field cress　イヌガラシ
Rorippa islandica (Oeder) Borb., marsh yellowcress　スカシタゴボウ
Rorippa sylvestris (L.) Bess., yellow fieldcress　キレハイヌガラシ
Rosa damascena Mill., damask rose　ダマスクバラ
Rosa rugosa Thunb., rugosa rose　ハマ

roselle, *Hibiscus sabdariffa* L. ロゼル
rosemary, *Rosmarinus officinalis* L. ローズマリー
rosette 1) ロゼット 2) 萎縮病【ムギ類】
rosewood, *Dalbergia cochinchinensis* Pierre ex Laness. など数種 シタン(紫檀)
Rosmarinus officinalis L., rosemary ローズマリー
rosulate ロゼット状の
Rotala pusilla Tulasne (= *R. mexicana* Cham. et Schl.) ミズマツバ
Rotala indica (Willd.) Koehne var. *uliginosa* (Miq.) Koehne, Indian toothcup キカシグサ
rotan, rattan [palm], cane palm, *Calamus caesius* Blume 他多種 トウ(籐)
rotary thresher 回転脱穀機
rotary tiller ロータリ耕うん(耘)機
rotary tilling ロータリ耕
rotary weeder 回転除草機
rotational crossing 輪番交雑
rotational grassland, ley 輪作草地
rotational irrigation 輪番灌漑
Rottboellia exaltata (L.) L. f., itch grass ツノアイアシ
rough bluegrass, rough-stalked meadow grass, *Poa trivialis* L. ラフブルーグラス
rough rice, unhulled rice, paddy 籾
rough rice screenings 屑籾(くずもみ)
rough-stalked meadow grass, rough bluegrass, *Poa trivialis* L. ラフブルーグラス
rough-surfaced endoplasmic reticulum, granular endoplasmic reticulum 粗面小胞体
rough-weaving crop 組編料作物(そへんりょうさくもつ)
roughage 粗飼料
roughseed buttercup, *Ranunculus muricatus* L. トゲミノキツネノボタン
round-bottom flask 丸底フラスコ
round-grain rice 円粒種【イネ】
round ridge 丸うね(畝)
roundabout plowing 回り耕
row 条
row competition けい(畦)間競争
row distance, interrow space 条間
row planting 並木植え
row plot 列条試験区
row sowing (seeding), drilling 条播(じょうは)
row space, ridge distance, row width うね(畝)幅
row width, ridge distance, row space うね(畝)幅
RQ (respiratory quotient) 呼吸商, 呼吸率
rRNA (ribosomal RNA) リボソームRNA
rubber, gum ゴム
rubber crop ゴム料作物
Rubia tinctorum L., common madder, madder セイヨウアカネ(西洋茜)
Rubisco (ribulose-1,5-bisphosphate carboxylase/oxygenase) リブロース-1,5-ビスリン酸カルボキシラーゼ/オキシゲナーゼ
RuBP (ribulose-1,5-bisphosphate) リブロース-1,5-ビスリン酸
Rubus idaeus L., raspberry ラズベリー
ruderal plant 人里植物
rudimentary glume 副護穎
rue, common rue, *Ruta graveolens* L. ヘンルーダ
rugosa rose, *Rosa rugosa* Thunb. ハマナス
Rumex acetosa L., green sorrel スイバ
Rumex acetosella L., red sorrel ヒメスイバ
Rumex japonicus Houtt. (= *R. cripus* L. ssp. *japonicus* (Houtt.) Kitam.),

Japanese dock　ギシギシ
Rumex obtusifolius L., broadleaf dock　エゾノギシギシ
ruminant　反すう(芻)動物
run　連【順位統計量】
runner, stolon　ふく(匐)枝, ほふく(匍匐)枝, ストロン
runoff, outflow, efflux　流去, 流出
runoff irrigation, overflow irrigation　いつ(溢)流灌漑
rural　農村の
rush, mat rush, *Juncus effusus* L. var. *decipiens* Buchenau (= *J. decipiens* Nakai)　イグサ(藺草)
Russian dandelion, kok-saghyz, *Taraxacum kok-saghyz* L. E. Rodin　ゴムタンポポ
Russian wildrye, *Psathyrostachys juncea* (Fisch.) Nevski (= *Elymus junceus* Fisch.)　ロシアワイルドライ
rust　さび(錆)病
rustica tobacco, aztec tobacco, *Nicotiana rustica* L.　ルスチカタバコ
rusty rice [kernel]　茶米
Ruta graveolens L., rue, common rue　ヘンルーダ
rutabaga, Swedish turnip, swede, *Brassica napus* L. var. *napobrassica* (Mill.) Reichb.　ルタバガ
rye, *Secale cereale* L.　ライムギ

[S]

S (sulfur)　イオウ(硫黄)
S-adenosylmethionine (SAM)　S-アデノシルメチオニン
S phase　S期, synthetic phase　合成期
Sabina virginiana (L.) Antoine (= *Juniperus virginiana* L.), eastern red cedar, pencil cedar　エンピツビャクシン
saccharification　糖化
saccharophyll, sugar leaf　糖葉

Saccharum officinarum L., sugar cane　サトウキビ(砂糖黍)
Sacciolepis indica (L.) Chase ssp. *oryzetorum* (Makino) T. Koyama　ヌメリグサ
sacred datura, *Datura meteloides* Dunal. (= *D. inoxa* Mill.)　アメリカチョウセンアサガオ
sacred lotus, Indian lotus, *Nelumbo nucifera* Gaertn.　ハス(蓮)
saddle grafting　鞍接ぎ(くらつぎ)
safflower, *Carthamus tinctorius* L.　ベニバナ(紅花)
safflower oil　ベニバナ(紅花)油
saffron, *Crocus sativus* L.　サフラン(泪夫藍)
safranin　サフラニン
sage, *Salvia officinalis* L.　セージ
Sagina japonica (Sw.) Ohwi, Japanese pearlwort　ツメクサ
Sagittaria aginashi Makino　アギナシ
Sagittaria pygmaea Miq.　ウリカワ
Sagittaria trifolia L., arrowhead　オモダカ
Sagittaria trifolia L. var. *edulis* (Sieb.) Ohwi, arrowhead　クワイ
sago palm, non-spiny sago palm, *Metroxylon sagu* Rottb.　サゴヤシ(サゴ椰子)
Saigon cinnamon, *Cinnamomum sieboldii* Meisn. (= *C. loureirii* Nees)　ニッケイ(肉桂)
sainfoin, *Onobrychis viciifolia* Scop.　セインフォイン
Salacca edulis Reinw., salak　サラカヤシ
salad oil　サラダ油
salak, *Salacca edulis* Reinw.　サラカヤシ
saline soil　塩類土壌
salinity　塩分濃度【土壌】
salinity tolerance　耐塩性
Salix koriyanagi Kimura (= *S. purpurea*

L. var. *multinervis* Fr. et Sav.), osier　コリヤナギ(杞柳)
salt tolerance　耐塩性
salt accumulation　塩類集積
salt damage (injury)　塩害
salt stress　塩分ストレス, 塩類ストレス
salt-tolerant rice　塩水稲
salting　塩蔵
salting out　塩析
saltmarsh aster, *Aster subulatus* Michx.　ホウキギク
salty wind damage (injury)　潮風害
Salvia officinalis L., sage　セージ
Salvinia natans (L.) All.　サンショウモ
sample　標本
sample mean　標本平均
sample size　標本の大きさ
sample unit　標本単位
sample variance　標本分散
sampling　標本抽出, 抜取り【統計】
sampling error　標本抽出誤差
sampling inspection　抜取り検査
sampling ratio　標本抽出率
sampling survey　標本調査[法]
sampling unit　抽出単位【標本抽出】
sanago, dark flour　さなご, 末粉(すえこ)【ソバ】
sand, sandy soil　砂土
sand culture, sandponics　砂耕
sand dune　砂丘
sand pear, Japanese pear, *Pyrus pyrifolia* (Burm. f.) Nakai (= *P. serotina* Rehder)　ナシ(梨)
sandy soil, sand　砂土
sandalwood, white sandalwood, *Santalum album* L.　ビャクダン(白檀)
sandponics, sand culture　砂耕
sandy loam　砂壌土
sandy soil　砂質土壌
sansevieria, bowstring hemp, *Sansevieria nilotica* Baker 他数種　サンセベリア

Sansevieria nilotica Baker 他数種, sansevieria, bowstring hemp　サンセベリア
Santalum album L., sandalwood, white sandalwood　ビャクダン(白檀)
sap　汁液
sap fruit, berry, succulent fruit　液果, 多肉果
sapling, young tree　若木
sapodilla, naseberry, *Manilkara zapota* (L.) P. Royen (= *Achras zapota* L.)　サポジラ
sappanwood, *Caesalpinia sappan* L.　スオウ(蘇芳)
sarashina flour,　さらしな粉【ソバ】
sarcocarp, pulp　果肉
satsuma mandarin, *Citrus unshiu* Marcovitch　ウンシュウミカン(温州蜜柑)
saturated fatty acid　飽和脂肪酸
saturated solution　飽和溶液
saturation deficit　飽差
saturation vapor pressure　飽和水蒸気圧
Satureja hortensis L., summer savory　サマーサボリー
savanna　サバンナ
saw gin　ソージン【ワタ】
scab　1) 赤かび病【ムギ類】 2) そうか(瘡痂)病【ジャガイモ】
scale, measure　尺度
scale expansion, farm-size expansion　規模拡大
scaly leaf, scale leaf　鱗片葉
scanning electron microscope (SEM)　走査型電子顕微鏡
scape, flower stalk　花茎
scarification　種皮処理
scarlet runner bean, flower bean, *Phaseolus coccineus* L.　ベニバナインゲン(紅花隠元)
scatter diagram　散布図
scattered light, diffused light　散乱光
scent, aroma, fragrance　香り, 香気

scented rice, aromatic rice　香り米
Scheffe's test　シェッフェの検定
Schizachyrium scoparius (Michx.) Nash, little bluestem　リトルブルーステム
schizogenous intercellular space　離生細胞間隙
Schoenoplectus juncoides (Roxb.) Palla ssp. *hotarui* (Ohwi) Soják (= *Scirpus juncoides* Roxb. var. *hotarui* Ohwi), Japanese bulrush　ホタルイ
Schoenoplectus juncoides (Roxb.) Palla ssp. *juncoides* (= *Scirpus juncoides* Roxb. ssp. *juncoides* Roxb.)　イヌホタルイ
Schoenoplectus lacustris (L.) Palla ssp. *validus* (Vahl) T. Koyama (= *Scirpus lacustris* L.), black rush　フトイ (太藺, 莞)
Schoenoplectus triqueter (L.) Palla (= *Scirpus triqueter* L.), bulrush　タイコウイ (太甲藺)
Schultze solution　シュルツェ液
scientific name　学名
scion　穂木, 接ぎ穂, 挿し穂
scion grafting　枝接ぎ
Scirpus juncoides Roxb. ssp. *juncoides* Roxb. (= *Schoenoplectus juncoides* (Roxb.) Palla ssp. *Juncoides*)　イヌホタルイ
Scirpus juncoides Roxb. var. *hotarui* Ohwi (= *Schoenoplectus juncoides* (Roxb.) Palla ssp. *hotarui* (Ohwi) Soják), Japanese bulrush　ホタルイ
Scirpus lacustris L. (= *Schoenoplectus lacustris* (L.) Palla ssp. *validus* (Vahl) T. Koyama), black rush　フトイ (太藺, 莞)
Scirpus lineolatus Franch. et Savat.　ヒメホタルイ
Scirpus planiculmis Fr. Schm.　コウキヤガラ
Scirpus triqueter L. (= *Schoenoplectus triqueter* (L.) Palla), bulrush　タイコウイ (太甲藺)
Scirpus yagara Ohwi (= *Bolboschoenus fluviatilis* (Torr.) T. Koyama ssp. *yagara* (ohwi) T. Koyama), river bulrush　ウキヤガラ
sclerenchyma　厚壁組織
sclerenchyma cell　厚壁細胞
sclerenchymatous cell　厚壁細胞
sclerophyllous tree　硬葉樹
sclerotium (*pl.* sclerotia)　菌核
scoke, pokeweed, *Phytolacca americana* L.　ヨウシュヤマゴボウ
score　スコア, 評点
scrape sowing (seeding)　削り播き
scraping hemp fiber　麻挽き (アサびき)
screening, sieving　ふるい (篩) 分け
screenings　屑粒 (くずりゅう)
screw pine, *Pandanus* spp.【*P. tectorius* Sol. ex Parkins. 他多種】タコノキ (林投)
screw pine, *Pandanus odorus* Ridl.　ニオイタコノキ
Scutellaria baicalensis Georgi, Baical skullcap　コガネバナ
scutellum (*pl.* scutella)　胚盤
scythe　大がま (鎌)
SDS (sodium dodecyl sulphate)　ドデシル硫酸ナトリウム
sea-island cotton, *Gossypium barbadense* L.　カイトウメン (海島棉)
sea wormwood, *Artemisia maritima* L.　ミブヨモギ
seasonal prevalence　発生消長
seasonal productivity　季節 [的] 生産性
Secale cereale L., rye　ライムギ
Sechium edule (Jacq.) Swartz, chayote　ハヤトウリ
second crop　1) 二期作【作物】 2) 二番草
second cut　二番刈り【牧草】
second cutting　二番刈り

second flush 二番芽, 二番立ち
second puddling 中代(なかしろ)
second weeding 二番除草
secondary association, secondary pairing 二次対合(たいごう)
secondary bed 子床, 植替え床【タバコ】
secondary branch of panicle, secondary rachis-branch of panicle 二次枝梗
secondary [cell] wall 二次[細胞]壁
secondary crop 二次作物
secondary dormancy 二次休眠
secondary growth 二次成長
secondary meristem 二次分裂組織
secondary metabolite 二次代謝産物
secondary pairing, secondary association 二次対合(たいごう)
secondary phloem 二次篩部
secondary polyploidy 二次倍数性
secondary rachis-branch of panicle, secondary branch of panicle 二次枝梗
secondary root 二次根
secondary selection 二次選抜
secondary succession 二次遷移
secondary thickening [growth] 二次肥大[成長]
secondary tiller 二次分げつ
secondary tissue 二次組織
secondary tuber 孫芋
secondary vascular bundle 二次維管束
secondary xylem 二次木部
secretion 分泌, 分泌物
secretory tissue 分泌組織
section 節(せつ)【分類】
sectorial chimera 区分キメラ
sedge weed, cyperaceous weed カヤツリグサ科雑草
sedimentation 堆積, 沈降
Sedum bulbiferum Makino コモチマンネングサ
seed 種子
seed age 種子齢

seed and seedling 種苗
seed bank 種子バンク
seed bed 播き床
seed blower 風選機
seed bulb, seed corm 種球
seed certification 種子検定
seed cleaning 種子精選
seed coat, testa 種皮
seed coating 種子粉衣
seed contamination 混種
seed corm, seed bulb 種球
seed cotton 実棉(じつめん)
seed disinfection 種子消毒
seed disinfection in hot bath 風呂湯浸法
seed dormancy 種子休眠
seed dressing 種子粉衣
seed furrow 作条
seed furrow making 作条【作業】
seed grading, seed selection 選種
seed growing, seed production culture 採種栽培
seed home 種場(たねば)
seed identification 種子鑑別
seed impregnation 種肥(たねごえ)
seed infection, seed transmission 種子伝染
seed inspection, seed testing 種子検査
seed longevity 種子寿命
seed maturity 種子成熟
seed plant, spermatophyte 種子植物
seed preservation 種子保存
seed production 種子生産, 採種
seed production by open pollination 放任採種
seed production culture, seed growing 採種栽培
seed [production] farm, propagation farm 採種圃
seed productivity 種子生産力
seed propagation 種子繁殖
seed protein 種子タンパク質
seed quality 種子品質

seed retention [habit] 難脱粒性
seed rice, rice seed 種籾
seed selection, seed grading 選種
seed selection with salt solution 塩水選
seed-set percentage 結実率
seed-setting, fructification 結実
seed soaking 浸種
seed storage 種子貯蔵
seed tape シードテープ
seed testing, seed inspection 種子検査
seed transmission, seed infection 種子伝染
seed tuber 種イモ
seed vernalization 種子春化
seed vigor 種子活力
seed yield 種子収量
seedbed, nursery [bed] 苗床
seeder, seeding machine 播種機
seeding, sowing 播種
seeding bed 播種床
seeding density, seeding rate 播種密度
seeding depth 播種深度
seeding furrow 播き溝
seeding in hill 株播き
seeding in nursery bed 床播き
seeding method 播種法
seeding rate 播種密度, 播種量
seeding time 播種期
seeding width 播き幅
seeding machine, seeder 播種機
seedless 無核
seedling 幼植物, 実生(みしょう), 苗, 芽ばえ
seedling age 苗齢(びょうれい)
seedling infection 苗感染
seedling pot 育苗ポット
seedling raising by illuminated nursery 補光育苗
seedling-raising in box 箱育苗
seedling selection 実生選抜
seedling stage 幼苗期
seedling storage, storage of seedling 苗貯蔵

seedling test 幼植物検定
seeds 種物(たねもの)
seepage, leakage 漏水
segmental interchange 部分交換【染色体】
segregating generation 分離世代
segregation 分離【遺伝】
selection 選抜, 選択
selection by truncation, truncation selection 切断型選抜
selection experiment 選抜(選択)実験
selection index 選抜(選択)指数
selection intensity 選抜(選択)強度
selection marker 選抜(選択)マーカー
selection objective 選抜(選択)目標
selection pressure 選抜(選択)圧
selection rate, percentage of selection 選抜(選択)率
selective absorption 選択吸収
selective fertilization 選択受精
selective herbicide 選択性除草剤
selective permeability 選択的透過性
self-compatibility 自家和合性
self-feeding thresher, automatic thresher, mechanical feeding thresher 自動脱穀機
self-fertilization 自家受精, 自殖
self-fertilizing plant, autogamous plant 自殖性植物, 自家受精植物
self-incompatibility 自家不和合性
self-pollination 自家受粉
self-propagation 自殖
self-reproduction 自殖
self seeding, natural seeding 自然下種
self-sterility 自家不稔性
self-sufficiency 自給自足
self-sufficiency rate 自給率
self-supplied feed 自給飼料
self-supplied manure 自給肥料
self-supply 自給
self-topping 心止り
self-unfruitfulness 自家不結果性
selfed line 自殖系統

selfing 自家受精, 自殖
SEM (scanning electron microscope) 走査型電子顕微鏡
semi-cultured soybean, *Glycine gracilis* Skov. 半栽培ダイズ
semi-irrigated rice nursery 折衷苗代
semi-low cut training 高根刈り仕立て【クワ】
semi-regular planting 片正条植え
semiarid land 半乾燥地
semidrying oil 半乾性油
semidwarf variety 半わい(矮)性品種
seminal root 種子根
semipermeability 半透性
semipermeable membrane 半透膜
Sencha 煎茶
seneca, senega, snake root, *Polygala senega* L. セネガ
Senecio vulgaris L., common groundsel ノボロギク
senescence 老化
senna, *Senna angustifolia* Batka および *S. alexandrina* Mill. (= *Cassia senna* L.) センナ
Senna angustifolia Batka および *S. alexandrina* Mill. (= *Cassia senna* L.), senna センナ
Senna obtusifolia (L.) H. S. Irwin et Barneby (= *Cassia obtusifolia* L.), oriental senna エビスグサ
Senna occidentalis Link. (= *Cassia torosa* Cav., *C. occidentalis* L.), negro coffee ハブソウ
sensitivity 1) 感受性 2) 感度【計測】
sensitivity analysis 感度分析
sensitivity to nursery days 苗代日数感応性
sensitivity to temperature, thermosensitivity 感温性
sensor センサー
sensory test 官能試験
sepal がく(萼)片
separating funnel, separatory funnel 分液ろうと(漏斗)
Sephadex セファデックス
septoria brown spot 褐紋病【ダイズ】
sequence-tagged sites (STS) 配列標識部位
sequential cropping, continuous cropping 連作
sequential test 逐次検定
sere 遷移系列
serial correlation 系列相関
sericulture 養蚕
series of techniques, systematized techniques, system of techniques 技術体系
serine (Ser) セリン
serisea lespedeza, *Lespedeza cuneata* (Dum. Cours.) G. Don メドハギ
serological reaction 血清反応
serradella, *Ornithopus sativus* Brot. セラデラ
serrated sickle のこぎり(鋸)鎌
serration 鋸歯(きょし)
sesame, gingelly, *Sesamum indicum* L. ゴマ(胡麻)
sesame meal ゴマ粕
sesame oil ゴマ油
Sesamum indicum L., sesame, gingelly ゴマ(胡麻)
sesquioxide 三二酸化物
sessile 無柄の
sessile joy-weed, *Alternanthera sessilis* (L.) R. Br. ex Roem. et Schult. ツルノゲイトウ
Setaria faberi Herrm., giant foxtail アキノエノコログサ
Setaria glauca (L.) Beauv., yellow foxtail キンエノコロ
Setaria italica (L.) Beauv. var. *germanicum* Trin., small foxtail millet コアワ
Setaria italica (L.) P. Beauv., Italian millet, foxtail millet アワ(粟)
Setaria viridis (L.) P. Beauv., green

foxtail　エノコログサ
setting, planting　定植
settlement　1) 入植　2) 入植地
settlement site, reclaimed settlement　入植地
sewing, stringing　葉あみ【タバコ】
sex chromosome　性染色体
sex-limited inheritance　限性遺伝
sex-linked inheritance　伴性遺伝
sex ratio　性比
sexual affinity　性的親和性
sexual cell　性細胞
sexual isolation　性的隔離
sexual propagation　有性繁殖
sexual reproduction　有性生殖
shade, shading　日除け
shade culture　遮光栽培, 被覆栽培
shade leaf　陰葉
shade plant　陰生植物
shade tolerance　耐陰性
shade tolerant tree　陰樹
shade tree　1) ひ(庇)陰樹　2) 陰樹
shaded tea　おおい(覆)下茶
shading　遮光, 日除け
shadowing　シャドウイング【電顕】
shaking table sorter　揺動選別機
shallow flooding　浅水
shallow planting　浅植え
shallow rooted　浅根性の
shallow-rooted crop　浅根性作物
shallow sowing (seeding)　浅播き
shallow tillage　浅耕
shatter　花振るい【ブドウ】
shattering, grain shedding　実こぼれ
shattering habit, shedding habit, threshability　脱粒性
shea butter　シア脂
shea [butter] tree, *Vitellaria paradoxa* (A. DC.) C. F. Gaertn. (= *Butyrospermum parkii* Don Kotschy)　シアバターノキ(シアバターの木)
sheaf rack, paddy sheaf rack　はさ, はざ(稲架)
shear plucking　はさみ摘み【チャ】
sheath blight　紋枯病【イネ】
shed burn　吊り腐れ【タバコ】
shedding habit, shattering habit, threshability　脱粒性
sheep fescue, *Festuca ovina* L.　シープフェスク
sheet erosion　面状侵食
shell　1) 殻　2) 莢【ラッカセイ】
shelling, hulling, husking　籾す(摺)り
shelter belt　防風林帯
shelter hedge　防風垣
shelter tree　遮へい(蔽)樹
shepherd's-purse, *Capsella bursa-pastoris* (L.) Medik.　ナズナ
Sheppard's correction　シェパードの補正
shifting cultivation, slash-and-burn agriculture　移動耕作, 焼畑農耕
shifting field　切替え畑
shiitake fungus, *Lentinus edodes* (Berk.) Sing.　シイタケ
shock　立束(たてたば)
shocking　立干し(たてほし)
shoot, foliage　茎葉[部]
shoot apex　茎頂
shoot apex culture　茎頂培養
short-culmed variety　短桿品種
short day　短日
short-day plant　短日植物
short-day treatment　短日処理
short-day vernalization　短日春化性
short-grained variety　短粒品種
short grass　短草
short interspersed repetitive sequence (SINE)　短い散在反復配列
short-lived seed, microbiotic seed　短命種子
short-season crop　短期作物
short-styled flower, thrum flower　短[花]柱花
short-term rotation　短期輪作

short-wave radiation　短波放射
shortening of breeding cycle　世代短縮
shrub, bush　低木
shuttle vector　シャトルベクター
Si (silicon)　ケイ(珪)素
sib cross　きょうだい交配
sib test　きょうだい検定
sick soil　忌地(いやち)【土壌】
sickle　かま(鎌)
Sida spinosa L., prickly sida　アメリカキンゴジカ
side crop, minor crop　副作物
side dressing　側条施肥
side-grafting　腹接ぎ【接木】
side-looking radar, synthetic aperture radar (SAR)　合成開口レーダ
side-oats grama, *Bouteloua curtipendula* (Michx.) Torr.　サイドオートグラマ
Siegesbeckia pubescens Makino　メナモミ
sieve area　篩域(しいき)
sieve plate　篩板(しばん)
sieve pore　篩孔(しこう)
sieve tube　篩管(しかん)
sieved soil　ふるい(篩)土
sieving, screening　ふるい(篩)分け
sigmoid curve　S字形(状)曲線
sign test　符号検定
signal　信号
signal transduction　シグナル伝達
signal-to-noise ratio (SN ratio)　SN比
significance　有意性
significance level, level of significance　有意水準
significance test, test of significance　有意性検定
significant　有意な
significant difference　有意差
silage　サイレージ
silage making, ensiling　サイレージ調製
silencer　サイレンサー
silica body　珪酸体

silicate fertilizer　ケイ(珪)酸質肥料
silicic acid　ケイ(珪)酸
silicified cell　珪化細胞
silicon (Si)　ケイ(珪)素
silk　絹糸(けんし)【トウモロコシ】
silking stage　絹糸抽出期
silkworm, *Bombyx mori* L.　カイコ(蚕), 家蚕(かさん)
silo　サイロ
silt　シルト
silver staining　銀染色法
similarity　相似【図形】
Simmondsia chinensis (Link) C. K. Schneid., jojoba　ホホバ
simple correlation　単相関
simple leaf　単葉
simple sequence length polymorphism (SSLP)　単純配列長多型
simulation　シミュレーション
Sinapis alba L. (= *Brassica alba* (L.) Boiss.), white mustard　シロガラシ(白芥子)
SINE (short interspersed repetitive sequence)　短い散在反復配列
single-cropped paddy field　一毛作田
single cropping　1) 一毛作　2) 単作
single cross　単交雑
single grained structure　単粒構造
single leaf　個葉
single pedigree　単独系
single planting　一本植え
single seed descent method (SSD method)　単粒系統法
single selection　一回選抜
singling　一本立て
sink　シンク
siratro, *Macroptilium atropurpureum* (DC.) Urb.　サイラトロ
sisal, *Agave sisalana* Perr. ex Engelm.　サイザル
Sisyrinchium atlanticum Bicknell, blue-eyed grass　ニワゼキショウ
six-rowed barley, *Hordeum vulgare* L.

六条オオムギ
six-rowed wild barley, *Hordeum agriocrithon* Åberg　六条野生オオムギ
sizer, grader, sorter　選別機
skiffing　整枝【チャ】
skin abrasion of rough rice　籾ずれ
skin-abrased rice　肌ずれ米
skipped-row culture　隔畦栽培
SLA (specific leaf area)　比葉面積
slag　鉱さい(滓), スラグ
slaked lime　消石灰
slash-and-burn agriculture, shifting cultivation　焼畑農耕, 移動耕作
sleep movement, nyctinastic movement, nyctinasty　就眠運動, 睡眠運動
slender amaranth, green amaranth, *Amaranthus viridis* L.　アオビユ
slender grain　狭粒
slender rush, *Juncus tenuis* Willd.　クサイ
slender spikerush, needle spikerush, *Eleocharis acicularis* (L.) Roem. et Schult.　マツバイ
slender starwort, *Stellaria alsine* Grimm var. *undulata* (Thunb.) Ohwi　ノミノフスマ
slender wheatgrass, *Agropyron trachycautum* Link (= *A. pauciflorum* Hitchc.)　スレンダーホィートグラス
slide [glass]　スライドグラス
slip　か(搔)き苗【サツマイモ】
slip erosion, land slide　地滑り侵食
sloping field, hillside farm　傾斜畑
slow-release fertilizer, controlled release fertilizer　緩効性肥料
sludge　汚泥
sludge deposit, muddy sediment　ヘドロ
slurry　スラリー
SLW (specific leaf weight)　比葉重
small foxtail millet, *Setaria italica* (L.) Beauv. var. *germanicum* Trin.　コアワ
small fruits　小果類

small grain variety　小粒品種
small red bean, azuki bean, adzuki bean, *Vigna angularis* (Willd.) Ohwi et Ohashi (= *Phaseolus angularis* L.)　アズキ(小豆)
small vascular bundle　小維管束
small vein, veinlet　細脈
smallflower umbrella sedge, *Cyperus difformis* L.　タマガヤツリ
smear method　なすりつけ法【試料作成】
smoke crop　喫煙料作物
smooth bromegrass, *Bromus inermis* Leyss.　スムーズブロムグラス
smooth-surfaced endoplasmic reticulum (SER)　滑面小胞体
smoothing　平滑化
smut　黒穂病【コムギ, トウモロコシ】
SN ratio, signal-to-noise ratio　SN比
snake gourd, *Trichosanthes cucumerina* Buch.-Ham. ex Wall. (= *T. anguina* L.)　ヘビウリ
snake root, senega, seneca, *Polygala senega* L.　セネガ
snakewood, Java devilpepper, *Rauvolfia serpentina* Benth. ex Kurtz　インドジャボク(インド蛇木)
snow blight　雪腐病【ムギ類】
snow cover　積雪
snow damage (injury)　雪害
snow endurance　耐雪性
snow mold　雪腐病
snow pellets, graupel　あられ(霰)
SO_2 (sulfur dioxide)　二酸化イオウ
sod, sward, turf　芝地
sod culture　草生栽培
sod culture system　草生法
sod mulch　ソドマルチ
sod-mulch system　草生マルチ法
sod seeding　不耕起播き【牧草】
sodium (Na)　ナトリウム
sodium chloride　塩化ナトリウム
sodium dodecyl sulphate (SDS)　ドデシ

ル硫酸ナトリウム
soft corn, *Zea mays* L. var. *amylacea* Sturt.　ソフトコーン
soft flour　薄力粉 (はくりきこ)【コムギ】
soft grain　軟質粒
soft rot　軟腐 [病]
soft-textured rice　軟質米
soft wheat　軟質コムギ
softening　軟化
software　ソフトウェア
softwood cutting　緑枝挿し
soil　土壌
soil acidity　土壌酸性
soil aeration　土壌通気
soil amendment, soil improvement　土壌改良
soil analysis　土壌分析
soil-borne disease　土壌病害, 土壌伝染病
soil buffer action　土壌緩衝能
soil burning　焼土 (やきつち)【作業】
soil chemistry　土壌化学
soil colloid　土壌コロイド
soil compaction　土壌の締固め
soil condition　土壌条件
soil conditioner　土壌改良剤
soil conservation　土壌保全
soil contaminant, soil pollutant　土壌汚染物質
soil contamination, soil pollution　土壌汚染
soil culture　土耕
soil diagnosis　土壌診断
soil disinfectant, soil fungicide　土壌殺菌剤
soil disinfection　土壌消毒
soil dressing　客土
soil erosion　土壌侵食
soil extract　土壌浸出液
soil fertility　土壌肥沃度
soil fumigator　土壌消毒機
soil fungicide, soil disinfectant　土壌殺菌剤
soil genesis, pedogenesis　土壌生成
soil hardness　土壌硬度
soil horizon　土壌層位
soil improvement, soil amendment　土壌改良
soil injector　土壌注入機
soil insulation　間土【播種】
soil management　土壌管理
soil map　土壌図
soil microorganism　土壌微生物
soil microstructure　土壌微細構造
soil mineral　土壌鉱物
soil mixes, commercial compost　配合土
soil moisture (water)　土壌水 [分]
soil moisture (water) content　土壌水分含量
soil moisture stress　土壌水分ストレス
soil monolith　土壌モノリス, モノリス
soil mulch　土壌被覆
soil nitrogen　土壌窒素
soil organic matter　土壌有機物
soil organism　土壌生物
soil particle　土壌粒子
soil persistent pesticide　土壌残留性農薬
soil physics　土壌物理 [学]
soil-plant-atmosphere continuum (SPAC)　土壌 - 植物 - 大気連続体
soil pollutant, soil contaminant　土壌汚染物質
soil pollution, soil contamination　土壌汚染
soil pore space　土壌孔げき (隙)
soil porosity　土壌孔げき (隙) 率
soil productivity　土壌生産力
soil profile　土壌断面
soil reaction　土壌反応
soil respiration　土壌呼吸
soil science　土壌学
soil seed bank　土壌種子バンク
soil series　土壌統
soil sickness　忌地 (いやち)【現象】

soil solution 土壌溶液
soil sterilization 土壌殺菌
soil strength 土壌強度
soil structure 土壌構造
soil survey 土壌調査
soil temperature 地温
soil testing 土壌検定
soil texture 土性
soil transmission 土壌伝染
soil type 土壌型
soil‐vegetation‐atmosphere transfer (SVAT) 土壌−植物−大気伝達
soil water (moisture) 土壌水[分]
soil water (moisture) content 土壌水分含量
soil water holding agent 土壌保水剤
soil water potential 土壌[の]水ポテンシャル
soil water tension 土壌水分張力
soilage, fresh forage, green forage, green chop 生草 (せいそう), 青刈り飼料
soiling 1) 青刈り 2) 青刈り給与, 青刈り利用
soiling corn 青刈りトウモロコシ
soiling crop 青刈り作物
soiling maize 青刈りトウモロコシ
soilless culture 無土壌栽培
sol ゾル
solanine ソラニン
Solanum carolinense L., horse nettle ワルナスビ
Solanum melongena L., eggplant ナス (茄子)
Solanum nigrum L., black nightshade イヌホオズキ
Solanum tuberosum L., potato, Irish potato ジャガイモ
solar altitude 太陽高度
solar constant 太陽常数
solar radiation 太陽放射
solarimeter, pyranometer 日射計
solid bulb, corm 球茎

solid fertilizer 固形肥料
solid medium (*pl.* media) 固形培地
solid medium culture, substrate culture, aggregate culture 固形培地耕
solid phase 固相
Solidago altissima L., tall goldenrod セイタカアワダチソウ
Solidago virgaurea L. ssp. *gigantea* (Nakai) Kitam. (= *S. gigantea* Ait. var. *leiophylla* Fern.), late goldenrod オオアワダチソウ
Solonchak ソロンチャク
Solonetz ソロネッツ
solubility 溶解度
soluble carbohydrate 可溶性炭水化物
solute 溶質
solution 溶液
solution culture, hydroponics, nutriculture 養液栽培
solvent 溶剤, 溶媒
somaclonal variation 体細胞[性]変異
somaclone 体細胞由来繁殖系, ソマクローン
somatic cell 体細胞
somatic embryogenesis 体細胞不定胚形成
somatic mutation 体細胞突然変異
Sonchus asper (L.) Hill, spiny sowthistle オニノゲシ
Sonchus brachyotus DC. ハチジョウナ
Sonchus oleraceus L., milk thistle, sow thistle ノゲシ
sorbitol ソルビトール
sorghum, grain sorghum, *Sorghum bicolor* (L.) Moench モロコシ (蜀黍)
Sorghum bicolor (L.) Moench, sorghum, grain sorghum モロコシ (蜀黍)
Sorghum bicolor (L.) Moench var. *hoki* Ohwi, broom corn ホウキモロコシ (箒蜀黍)
Sorghum bicolor (L.) Moench var. *saccharatum* (L.) Mohlenbr., sweet

sorghum, sugar sorghum, sorgo　サトウモロコシ(砂糖蜀黍)
Sorghum halepense (L.) Pers., Johnsongrass　ジョンソングラス
Sorghum sudanense (Piper) Stapf, Sudan grass　スーダングラス
sorgo, sweet sorghum, sugar sorghum, *Sorghum bicolor* (L.) Moench var. *saccharatum* (L.) Mohlenbr.　サトウモロコシ(砂糖蜀黍)
sorter, grader, sizer　選別機
sorting　選別
sour apple, sour-sop, *Annona muricat* L.　トゲバンレイシ
sour lime, lime, *Citrus aurantifolia* (Christm.) Swingle　ライム
sour-sop, sour apple, *Annona muricata* L.　トゲバンレイシ
source　ソース
source of contamination, pollution source　汚染源
southern blight, stem rot　白絹病【トウモロコシ, ダイズ, インゲンマメ】
Southern blot technique　サザンブロット法
Southern blotting　サザンブロット法
southern cutgrass, bareet grass, tiger's-tongue grass, Leersia hexandra Sw.　タイワンアシカキ
southern pea, cowpea, *Vigna unguiculata* (L.) Walp. (= *V. sinensis* Endl.)　ササゲ(豇豆, 大角豆)
sow thistle, milk thistle, *Sonchus oleraceus* L.　ノゲシ
sowing, seeding　播種
sowing bed　播種床
sowing by seeder, automatic seeding　機械播き
sowing density, sowing rate　播種密度
sowing depth　播種深度
sowing furrow　播き溝
sowing in hill　株播き
sowing in nursery bed　床播き

sowing method　播種法
sowing on roughly prepared field, minimum tillage seeding　簡易整地播き
sowing rate　播種密度, 播種量
sowing time　播種期
sowing width　播き幅
sown grassland, artificial grassland　人工草地
sown pasture, artificial pasture　人工草地
SO_x (sulfur oxides)　イオウ(硫黄)酸化物
soy milk　豆乳(とうにゅう)
soy sauce　しょうゆ(醤油)
soybean, *Glycine max* (L.) Merr.　ダイズ(大豆)
soybean flour　黄粉(きなこ), 大豆粉
soybean meal　大豆粕
soybean oil　大豆油
soybean paste　味噌
soybean pod borer　マメシンクイガ
soybean pod gall midge　ダイズサヤタマバエ
soybean sauce　しょうゆ(醤油)
SPAC (soil-plant-atmosphere continuum)　土壌-植物-大気連続体
spacial arrangement [of plants], planting pattern　栽植様式
spade　すき(鋤)
sparrow vetch, fourseeded vetch, *Vicia tetrasperma* (L.) Schreb.　カスマグサ
sparse planting, planting with wide spacing　疎植
sparse sowing (seeding), thin sowing (seeding)　疎播(そは), 薄播き
spatula　へら, スパチュラ
Spearman's rank correlation　スピアマンの順位相関
spearmint, *Mentha spicata* L. (= *M. viridis* L.)　スペアミント
speciation　種分化
species　種

species biology　種生物学
species cross, interspecific crossing　種間交雑
species ecology, autecology　種生態学
species hybrid, interspecific hybrid　種間雑種
species specificity　種特異性
specific activity　比活性
specific combining ability　特定組合せ能力
specific gravity　比重
specific heat　比熱
specific humidity　比湿
specific leaf area (SLA)　比葉面積
specific leaf weight (SLW)　比葉重
specific root length　根長/根重比, 比根長
specimen　標本
specimen bottle　標本びん(瓶)
spectral analysis　スペクトル解析
spectral reflectance　分光反射率
spectrochemical analysis　分光分析
spectrophotometer　分光光度計
spectrophotometric analysis　吸光分光分析
spectro-radiometer　分光放射計
spectrum (*pl.* spectra)　スペクトル
speed sprayer　スピードスプレーヤ
spelt [wheat], *Triticum spelta* L.　スペルトコムギ
sperm nucleus　精核, male nucleus　雄核
spermatophyte, seed plant　種子植物
spermidine　スペルミジン
spermine　スペルミン
sphagnum, peat moss　ミズゴケ(水苔)
spice, condiment　香辛料
spice crop　香辛料作物
spike　穂, 穂状花序
spike (panicle) neck node differentiation stage　穂首[節]分化期
spike lavender, broadleaved lavender, *Lavandula latifolia* Medik. (= *L. spica* DC.)　スパイクラベンダー
spikelet　小穂(しょうすい)
spinach, *Spinacia oleracea* L.　ホウレンソウ
Spinacia oleracea L., spinach　ホウレンソウ
spindle body　紡錘体
spindle fiber　紡錘糸
spindly growth　徒長(とちょう)
spiny amaranth, *Amaranthus spinosus* L.　ハリビユ
spiny sago palm, prickly sago palm, *Metroxylon rumphii* Mart.　トゲサゴ[ヤシ](刺サゴ[椰子])
spiny sowthistle, *Sonchus asper* (L.) Hill　オニノゲシ
spiral phyllotaxis　らせん葉序
Spirodela polyrhiza (L.) Schleid., giant duckweed　ウキクサ
Spirogyra arcla Kutz.　アオミドロ
spiroscalate phyllotaxis　らせん階段型対生葉序
spleen amaranth, *Amaranthus patulus* Bert.　ホソアオゲイトウ
splice grafting　合わせ接ぎ
split application　分施【肥料】
split dressing　分施【肥料】
split-hull paddy　割れ籾
split-plot design　分割試験区設計
spodogram　灰像(かいぞう)
spodography　灰像法
sponge gourd, *Luffa cylindrica* M. Roem.　ヘチマ(糸瓜)
spongy tissue　海綿状組織
spontaneous, indigenous, native　自生の
spontaneous mutation　偶発突然変異
spore　胞子
sporophyte　造胞体, 胞子体
spot blotch　斑点病【コムギ】
spray irrigation　散水灌漑
spray, application　散布
sprayer　噴霧機

spraying　噴霧
spread type　開張型【チャ】
spreading　開張性の
spreading panicle　散穂 (さんすい)
spring cereals　春穀物
spring crop　春作 [物]
spring cropping　春作
spring dressing　春肥 (はるごえ)
spring flush　スプリングフラッシュ【牧草】
spring habit　春播き性
spring planting　春植え
spring plowing　春耕
spring pruning　春切【クワ】
spring shoot　春枝
spring sowing (seeding)　春播き
spring type　春播き型
spring variety　春播き品種
spring wheat　春コムギ
sprinkler　スプリンクラ
sprout sowing (seeding)　芽出し播き
sprout[ing]　萌芽
sprouted vine planting　挿苗 (そうびょう), 苗挿し【サツマイモ】
sprouting shoot, new shoot　新芽【チャ】
sprouting time, time of bud opening　萌芽期【チャ】
square planting　正方形植え
square-stalked passion flower, giant granadilla, *Passiflora quadrangularis* L.　オオミノトケイソウ
squash method　押しつぶし法
stability　安定性
stable, barn　畜舎
stable isotope　安定同位元素, 安定同位体
stable manure　きゅう (厩) 肥, 堆厩肥
Stachys sieboldii Miq. (= *S. affinis* Fresen), chorogi　チョロギ
stack　1) 野積み　2) かたい (禾堆), にお　3) しま (島)【イネ, ソバなど】
stack drying　島立て乾燥
stage　1) 段階【標本抽出】　2) ステージ【顕微鏡】
stained rice　汚染米【外観】
staining　染色
stalk-cut cotton　木採棉 (きどりわた)
stalk-cut curing　幹干し乾燥【タバコ】
stalk cutting　幹刈り【タバコ】
stalk position　着葉位置【タバコ】
stamen　雄しべ (蕊), 雄ずい (蕊)
staminate flower, male flower　雄花 (おばな, ゆうか)
staminate inflorescence　雄穂 (ゆうすい)
stamping　踏込み
stamping spade　踏すき (鋤)
stand　株立ち, 立毛 (りつもう), 群落, 植分
stand geometry, light-intercepting characteristics　受光態勢
stand observation　検見 (けみ), 立毛調査 (りつもうちょうさ)
standard application rate of fertilizer　施肥基準
standard deviation　標準偏差
standard dosage of fertilizer　標準施肥量
standard error　標準誤差
standard normal distribution　標準正規分布
standard specimen of rice kernel　検査等級標準米
standard variety　1) 標準品種　2) 基準品種【食味試験】
standing crop　現存量
staple food　主食
star anise, *Illicium verum* Hook. f.　ダイウイキョウ (大茴香・八角茴香)
star fruit, carambola, *Averrhoa carambola* L.　ゴレンシ (五斂子)
starch　デンプン
starch crop　デンプン [料] 作物
starch grain (granule)　デンプン粒
starch leaf, amylophyll　デンプン葉
starch seed　デンプン種子

starch synthase　デンプン合成酵素
starch value　デンプン価
starch yielding percentage　デンプン歩留り
starchy endosperm　デンプン質胚乳
starter　根付け肥
starvation　飢餓
state variable　状態変数
statistic　統計量
statistical genetics　統計遺伝学
statistical inference　統計的推論
statistical method　統計的方法
statistics　統計学
statocyst, statocyte　平衡細胞
statolith　平衡石
steady state　定常状態
stearic acid　ステアリン酸
stele, central cylinder　中心柱
Stellaria alsine Grimm var. *undulata* (Thunb.) Ohwi, slender starwort　ノミノフスマ
Stellaria aquatica (L.) Scop., water starwort　ウシハコベ
Stellaria media (L.) Villars, common chickweed　ハコベ
stem　1) 茎 2) 中骨(ちゅうこつ)【タバコ】
stem cutting　茎挿し, 枝挿し
stem drying stage　中骨乾燥期【タバコ】
stem length　茎長
stem rot　1) つる割病【サツマイモ】 2) 白絹病【トウモロコシ, ダイズ, インゲンマメ】
stem vegetables　茎菜類
Stenactis annuus (L.) Cass. (= *Erigeron annuus* (L.) Pers.), annual fleabane　ヒメジョオン(姫女苑)
Stephania cepharantha Hayata　タマザキツヅラフジ
steppe　ステップ
stereoscopic microscope　実体[解剖]顕微鏡
sterile　不稔の

sterile culture, aseptic culture　無菌培養
sterile flower　不稔花
sterile glume, empty glume　空穎
sterile seed　不稔種子
sterility　不稔[性]
sterilization　殺菌, 滅菌
stevia, kaa he-e, *Stevia rebaudiana* (Bertoni) Hemsl.　アマハステビア, ステビア
Stevia rebaudiana (Bertoni) Hemsl., stevia, kaa he-e　アマハステビア, ステビア
stickiness　粘り
sticky chickweed, *Cerastium glomeratum* Thuill.　オランダミミナグサ
stiff-strawed variety, strong-strawed variety　強稈品種
stifle disease　赤枯れ[病]【イネの生理病】
stigma　柱頭
stimulus (*pl*. stimuli)　刺激
stipule　托葉
Stizolobium hassjoo Piper et Tracy (= *Mucuna pruriens* (L.) DC. var. *utilis* (Wight) Burck), Yokohama [velvet] bean　ハッショウマメ(八升豆)
stochastic independence　確率的独立性
stochastic process　確率過程
stock　1) 株 2) 台木 3) 系統
stock plant　母株
stock seed, registered seed, original seed, foundation stock　原種
stock seed field, original seed farm　原種圃
stock solution　原液【培養液等】
stolon, runner　ストロン, ほふく(匍匐)枝, ふく(匐)枝
stoma (pl. stomata)　気孔
stomatal aperture　気孔開度
stomatal conductance　気孔コンダクタンス, 気孔伝導度

stomatal resistance　気孔抵抗
stomatal transpiration　気孔蒸散
stone cell　石細胞
stone fruit, drupe　核果
storage　保存, 保蔵, 貯蔵
storage ability　貯蔵性
storage feeding　貯蔵給与, 貯蔵利用【飼料】
storage loss　置減り
storage of seedling, seedling storage　苗貯蔵
storage organ　貯蔵器官
storage protein, reserve protein　貯蔵タンパク質
storage root　貯蔵根
storage starch, reserve starch　貯蔵デンプン
storage tissue　貯蔵組織
stored rice　貯蔵米
stover, straw　わら（藁）
straight cross　正交配
straight fertilizer　単肥
straight head, straighthead　青立ち【イネ】
straighthead due to drought　ひでり（旱）青立
strain, pedigree, stock, line　系統
stratification　階層構造
stratified clip method　層別刈取法
stratified sampling　層化［標本］抽出
stratum (*pl.* strata)　層【標本抽出】
straw, stover　わら（藁）
straw ash　わら（藁）灰
straw bag　俵, かます（叺）
straw mulch　敷わら（藁）
straw of barley　麦わら【オオムギ】
straw of rye　麦わら【ライムギ】
straw of wheat　麦わら【コムギ】
straw rice bag　米俵【表装のみ】
straw rice bale　米俵【内容を含む】
straw stiffness, culm stiffness　強稈性
straw weight　わら（藁）重
straw windbreak　わら（藁）立て

strawberry, *Fragaria* × *ananassa* Duch.　イチゴ（苺）
strawberry clover, *Trifolium fragiferum* L.　ストロベリクローバ
stress　ストレス
stress tolerator　ストレス耐性種
striate lespedeza, Japanese clover, *Kummerowia striata* (Thunb. ex Murray) Schindl.　ヤハズソウ
stringing, sewing　葉あみ【タバコ】
strip cropping, lane cropping　帯状間作, 帯状栽培
stripe　1) 縞葉枯病【イネ】 2) 斑葉病【オオムギ】
stroma (*pl.* stromata)　ストロマ
strong flour　強力粉【コムギ】
strong-strawed variety, stiff-strawed variety　強稈品種
strophiole, caruncle　種枕（しゅちん）
structural carbohydrate　構造性炭水化物
structural formula　構造式
structural gene　構造遺伝子
structural hybrid　構造的雑種
structural protein　構造タンパク質
struggle for existence　生存競争
strychine tree, nux vomica, *Strychnos nux-vomica* L.　マチン
Strychnos nux-vomica L., strychine tree, nux vomica　マチン
stubble　刈株【イネ科】
stubble breaking　株切り
stubble field　刈田
stubble grazing　刈跡放牧
Studentized range　スチューデント化された範囲
stump　切株
stunt, dwarf　萎縮病【イネ, ダイズ】
stunting, dwarfing　わい（矮）化
style　花柱
stylo, *Stylosanthes guianensis* (Aubl.) Sw. 他数種　スタイロ
Stylosanthes guianensis (Aubl.) Sw. 他数種, stylo　スタイロ

subarctic zone, subpolar zone 亜寒帯
subclover, subterranean clover, *Trifolium subterraneum* L. サブクローバ, サブタレニアンクローバ
subculture 継代培養
suberin スベリン
suberization スベリン化, コルク化
subfamily 亜科
subirrigation 地下灌漑
submerged soil, flooded soil, waterlogged soil 湛水土壌
submerged weed 沈水雑草
submergence, flooding 冠水
submersed weed 沈水雑草
subpolar zone, subarctic zone 亜寒帯
subroutine サブルーチン
subsample 副次標本
subsidiary cell 副細胞【気孔】
subsoil 心土, 下層土
subsoil breaker 心土破砕機
subsoil compaction 床じめ
subsoil improvement 土層改良
subsoil plowing 心土耕
subsoil puddling 盤練り
subsoiler サブソイラ, 心土プラウ
subsoiling 心土耕, 心土破砕
subspecies 亜種
substitute cropping 代作
substomatal cavity 気孔内腔
substrate 基質
substrate culture, solid medium culture, aggregate culture 固形培地耕
subsurface drainage 地下排水
subterranean clover, subclover, *Trifolium subterraneum* L. サブタレニアンクローバ, サブクローバ
subterranean part, underground part 地下部
subterranean stem 地下茎
subtropical zone 亜熱帯
subtropics 亜熱帯
subunit サブユニット
suburban gardening 近郊園芸

succeeding cropping 後作(あとさく)
succession 遷移
succinic acid コハク酸
succulent 多肉の
succulent fruit, berry, sap fruit 液果, 多肉果
succulent plant 多肉植物
succulent root 多肉根
succulent stem 多肉茎
sucker 1) 吸枝, ひこばえ, 台芽 2) わき芽【タバコ】 3) サッカー【サゴヤシ苗】
sucker control わき芽抑制【タバコ】
suckercide わき芽抑制剤【タバコ】
suckering, division 株分け
sucking root, absorbing root, feeder root 吸収根
sucrose スクロース, ショ(蔗)糖
sucrose-phosphate synthase スクロースリン酸シンターゼ
suction force 吸水力
suction pressure 吸水圧
Sudan grass, *Sorghum sudanense* (Piper) Stapf スーダングラス
sugar 糖
sugar-acid ratio 甘味比(かんみひ)
sugar apple, sweet sop, custard apple, *Annona squamosa* L. バンレイシ
sugar beet, *Beta vulgaris* L. var. *rapa* Dumort. テンサイ(甜菜)
sugar cane, *Saccharum officinarum* L. サトウキビ(砂糖黍)
sugar date palm, *Phoenix sylvestris* Roxb. サトウナツメヤシ
sugar leaf, saccharophyll 糖葉
sugar maple, *Acer saccharum* Marsh. サトウカエデ(砂糖楓)
sugar palm【サトウヤシ(gomuti palm), オウギヤシなど, 砂糖を採るヤシ類】サトウヤシ(砂糖椰子)
sugar phosphate 糖リン酸
sugar sorghum, sweet sorghum, sorgo, *Sorghum bicolor* (L.) Moench var.

saccharatum (L.) Mohlenbr. サトウモロコシ(砂糖蜀黍)
sugars 糖類
sulfur (S) イオウ(硫黄)
sulfur dioxide (SO_2) 二酸化硫黄
sulfur oxides (SO_x) イオウ(硫黄)酸化物
sum of products 積和
sum of squares 平方和
summer buckwheat 夏ソバ
summer bud 夏芽
summer cereals 夏穀物
summer depression 夏枯れ【牧草など】
summer ecotype 夏型【ソバ】
summer fallow 夏期休閑
summer planting 夏植え
summer pruning 夏切【クワ】
summer savory, *Satureja hortensis* L. サマーサボリー
summer sowing (seeding) 夏播き
summer soybean 夏ダイズ
summer squash, pumpkin, *Cucurbita pepo* L. ペポカボチャ
summer weed 夏雑草
SUMP method スンプ法
sun-curing 日干乾燥【タバコ】
sun (sunn) hemp, *Crotalaria juncea* L. サンヘンプ
sun leaf 陽葉
sun plant, heliophyte 陽生植物
sun tree, intolerant tree 陽樹
sunburn 日焼け
sunflower, *Helianthus annuus* L. ヒマワリ(向日葵)
sunflower oil ヒマワリ油
sunscald 日焼け
sunset hibiscus, *Abelmoschus manihot* Medik. (= *Hibiscus manihot* L.) トロロアオイ(黄蜀葵)
sunshine 日照
sunshine recorder, heliograph 日照計
superficial root うわ根
superior ovary 上位子房
superior spikelet 強勢穎花
supernatant [liquid] 上ずみ(澄)[液]
superphosphate 過リン(燐)酸石灰
supplement 補助飼料
supplement application, topdressing 追肥
supplemental feeding 補助飼料給与
supplementary feed 補助飼料
supplementary feeding 補助飼料給与
supplementary illumination 補光
supplementary lighting 補光
supplementary planting, complementary planting 補植
suppressor サプレッサー, 抑圧遺伝子
suppressor gene 抑圧遺伝子
surface-active agent 界面活性剤
surface application of fertilizer 表層施肥
surface area 表面積
surface drainage 地表排水
surface irrigation 地表灌漑
surface runoff 表面流出
surface soil, top soil 表[層]土
surface tillage 表土耕
surfactant 界面活性剤
survival rate, viability 生存率
susceptibility 感受性, り(罹)病性
susceptible to lodging 倒伏性の
susceptible variety り(罹)病性品種
suspension culture 懸濁培養, 浮遊培養
suspensor 胚柄
sustainable agriculture 持続型農業
suture 縫合線
swamp, marsh 低湿地
swamp cabbage, water convolvulus, water spinach, *Ipomoea aquatica* Forsk. ヨウサイ(蕹菜)
swamp taro, giant swamp taro, *Cyrtosperma chamissonis* (Schott) Merr. スワンプタロ
sward 草地, 芝地
sward canopy 草冠【草地】
sward composition 草種構成

sward establishment, forage establishment　草地造成
sward height　草高【草地】
sward management　草地管理
sward renovation　草地更新
swath, cutting width　刈り幅
swede, Swedish turnip, rutabaga, *Brassica napus* L. var. *napobrassica* (Mill.) Reichb.　ルタバガ
Swedish turnip, swede, rutabaga, *Brassica napus* L. var. *napobrassica* (Mill.) Reichb.　ルタバガ
sweet acacia, cassie, *Acacia farnesiana* (L.) Willd.　キンゴウカン（金合歓）
sweet anise, Florence fennel, *Foeniculum vulgare* Mill. var. *dulce* (Mill.) Batt. et Trab. (= *F. dulce* Mill.)　イタリアウイキョウ
sweet basil, basil, *Ocimum basilicum* L.　バジル
sweet corn, *Zea mays* L. var. *saccharata* Bailey　スイートコーン
sweet majoram, *Origanum majorana* L. (= *Majorana hortensis* Moench)　マヨラナ
sweet orange, *Citrus sinensis* Osbeck　スイートオレンジ
sweet potato, *Ipomoea batatas* (L.) Lam.　サツマイモ（薩摩芋）
sweet potato starch　サツマイモデンプン
sweet sop, sugar apple, custard apple, *Annona squamosa* L.　バンレイシ
sweet sorghum, sugar sorghum, sorgo, *Sorghum bicolor* (L.) Moench var. *saccharatum* (L.) Mohlenbr.　サトウモロコシ（砂糖蜀黍）
sweet taste　甘味【タバコ】
sweet vernalgrass, *Anthoxanthum odoratum* L.　スイートバーナルグラス
sweetness　甘味（かんみ）
swelling　膨潤
swine cress, *Coronopus didymus* (L.) J. E. Smith　カラクサナズナ
switchgrass, *Panicum virgatum* L.　スイッチグラス
sword bean, *Canavalia gladiata* (Jacq.) DC.　ナタマメ（刀豆）
symbiosis　共生
symmetry　対称，相称
sympatric　同所性の
sympatric species　同所種
sympatry　同所性
symplast　シンプラスト
symplastic growth　同調成長
sympodial branching　仮軸分枝
symptom　病徴
synapsis, syndesis　対合，接合【染色体】
syncarp, multiple fruit, aggregate fruit　集合果，多花果
synchronized culture, synchronous culture　同調培養
synchronous division　同調分裂
synchronously emerging leaf　同伸葉
synchronously emerging tiller　同伸分げつ
syndesis, synapsis　対合，接合【染色体】
synecology, community ecology　群集生態学，群落生態学
synergid　助胎細胞，助細胞
synergism　相乗作用
synergistic　相助的
synergistic effect　相助効果
synthetic aperture radar (SAR), side-looking radar　合成開口レーダ
synthetic auxin　合成オーキシン
synthetic phase　合成期，S phase　S期
synthetic species, synthetic breed, synthetic strain　合成種
synthetic variety　合成品種
system　システム，系
system of techniques, systematized techniques, series of techniques　技術体系
systematic error　系統誤差
systematic sampling　系統的[標本]抽

出

systematics, taxonomy 分類学

systematized techniques, series of techniques, system of techniques 技術体系

systemic herbicide 浸透性除草剤

Syzygium aqueum (Burm. f.) Alston, water apple ミズレンブ

Syzygium aromaticum (L.) Merr. Et Perry (= *Eugenia caryophyllata* Thunb., *E. caryophyllus* Spreng., *E. aromatica* Kuntze), clove チョウジ（丁字，丁子）

Syzygium samarangense (Blume) Merr. et Perry, wax apple, Java apple レンブ

[T]

T (thymine) チミン
t-distribution t分布
T-DNA (transferred DNA) ティーディーエヌエー
T-R ratio (top-root ratio) TR率
T-shaped tubing connector T字管
t-table t表
t-test t検定
table of random numbers 乱数表
Tacca leontopetaloides (L.) Kuntze (= *T. pinnatifida* Forst.), East Indian arrowroot, Tahiti arrowroot タシロイモ
Tacca pinnstifida Forst., Indian arrow root インディアンアロールート
tafted knotweed, *Persicaria longiseta* (De Bruyn) Kitag. (= *Polygonum longisetum* De Bruyn) イヌタデ
Tahiti arrowroot, East Indian arrowroot, *Tacca leontopetaloides* (L.) Kuntze (= *T. pinnatifida* Forst.) タシロイモ
take-all 立枯病【ムギ類】
tall fescue, tall meadow fescue, *Festuca arundinacea* Schreb. トールフェスク

tall fleabane, *Conyza sumatrensis* (Retz.) Walker (= *Erigeron sumatrensis* Retz.) オオアレチノギク

tall goldenrod, *Solidago altissima* L. セイタカアワダチソウ

tall grass 長草

tall meadow fescue, tall fescue, *Festuca arundinacea* Schreb. トールフェスク

tall oatgrass, *Arrhenatherum elatius* (L.) K. Presl トールオートグラス

tall training 高作り

tamarind, *Tamarindus indica* L. タマリンド

Tamarindus indica L., tamarind タマリンド

tame pasture 造成草地，人工草地
tamping, packing 鎮圧【土木】
tandem selection 順繰り選抜法
tangential division 接線分裂【細胞分裂】
tannia, yautia, *Xanthosoma sagittifolium* (L.) Schott アメリカサトイモ
tannic acid タンニン酸
tannin タンニン
tannin cell タンニン細胞
tannin crop タンニン料作物
tapetal cell タペート細胞
tapete cell タペート細胞
tapetum タペート[組織]
tapioca タピオカ
tapioca plant, cassava, manioc, *Manihot esculenta* Crantz (= *M. utilissima* Pohl) キャッサバ
tapping 切付け，タッピング【ゴム・ウルシなど】
taproot 1) 直根　2) 主根
taproot system 主根型根系
Taq polymerase Taqポリメラーゼ
Taraxacum kok-saghyz L. E. Rodin, Russian dandelion, kok-saghyz ゴムタンポポ

Taraxacum officinale Weber, dandelion　セイヨウタンポポ

Taraxacum platycarpum Dahlst. ssp. *platycarpum*　カントウタンポポ

taro, dasheen, *Colocasia esculenta* (L.) Schott var. *esculenta* Hubbard & Rehder　タロイモ

tarragon, estragon, *Artemisia dracunculus* L.　タラゴン

Tartary buckwheat, Kangra buckwheat, *Fagopyrum tataricum* (L.) Gaertn.　ダッタンソバ(韃靼蕎麦)

tassel　雄穂(ゆうすい)【トウモロコシ】

taste　食味, 味

taste panelist　食味検定者

taste substance　味物質

taster, panelist　パネリスト, パネル構成員

taxis　走性

taxonomy, systematics　分類学

TCA (tricarboxylic acid) cycle　TCA(トリカルボン酸)回路

TDN (total digestible nutrients)　可消化養分総量

TDR (time domain reflectometry)　TDR法

tea, *Camellia sinensis* (L.) O. Kuntze　チャ(茶)

tea bush　茶樹

tea field　茶園

tea garden　茶園

tea manufacturing　製茶

tea plant　茶樹

tea plucker, plucking machine　摘採機【チャ】

tea season of first crop　一番茶期

tea seed oil　茶実油

teasel, *Dipsacus fullonum* L.　チーゼル

teff [grass], *Eragrostis abyssinica* (Jacq.) Link (= *E. tef* Trotter)　テフ

telome theory　テローム説

telophase　終期【細胞分裂】

TEM (transmission electron microscope)　透過型電子顕微鏡

temperate climate　温帯気候

temperate crop　温帯[性]作物

temperate forages　寒地型牧草

temperate humid climate　温帯湿潤気候

temperate plant　温帯[性]植物

temperate zone　温帯

temperature coefficient　温度係数

tempered straw　打わら(藁)

tempering　テンパリング

temporary grassland　転換牧草地

temporary planting, provisional planting　仮植(かしょく)

temporary storage, tentative storage　仮貯蔵

ten-are yield (10-are yield)　反当収量

ten-days　旬

tenant　小作人

tenant farmer　小作農

tenant farming　小作

tending　管理【動植物】

tendril　巻きひげ

tensiometer　テンシオメーター

tension　張力

tentative storage, temporary storage　仮貯蔵

teosinte, *Euchlaena mexicana* Schrad.　テオシント

tepary bean, *Phaseolus acutifolius* A. Gray var. *latifolius* Freem.　テパリビーン

terata, malformation, deformity, deformation　奇形

terminal　頂生の

terminal bud, apical bud　頂芽

terminal flower　頂花

terminator　ターミネーター

ternately compound leaf　三出複葉

Terra Rossa　テラロサ

terrace culture　テラス栽培, 階段耕作

terrace farming　テラス栽培, 階段耕作

terrace field　段畑

terrace paddy field, rice terrace　棚田

terrestrial stem　地上茎
test cross　検定交雑
test field　検定圃
test field for disease-tolerance [evaluation], disease garden　耐病性検定圃
test for physiological character　特性検定，特性試験
test for regional adaptability, local adaptability test　系統適応性検定試験
test of goodness of fit　適合度検定
test of homogeneity, homogeneity test　均一性の検定【分散等】
test of hypothesis　仮説検定
test of normality　正規性検定
test of significance, significance test　有意性検定
test of specific character　特性検定，特性試験
test plant, assay plant　検定植物
test statistic　検定統計量
test tube　試験管
testa, seed coat　種皮
tetrad　1) 四分子　2) 四分染色体
tetraploid [plant]　四倍体植物
tetraploidy　四倍性
textile fiber crop　紡績繊維[料]作物
texture　テクスチャー
texturometer　テクスチュロメータ
The Agricultural Basic Law　農業基本法
The Basic Law on Food, Agriculture and Rural Areas　食料・農業・農村基本法
The Crop Science Society of Japan　日本作物学会
The Environment Basic Law　環境基本法
the first ten-days [of a month]　上旬
the last ten-days [of a month]　下旬
the second ten-days [of a month]　中旬
the third ten-days [of a month]　下旬
Theobroma cacao L., cacao, cocoa　カカオ

thermal capacity, heat capacity　熱容量
thermistor thermometer　サーミスタ温度計
thermocouple　熱電対
thermocouple psychrometer　サーモカップルサイクロメータ
thermodynamics　熱力学
thermometer　温度計
thermoperiodism　温度周期性，温周性
thermophase　感温相
thermosensitive variety　感温性品種
thermosensitivity, sensitivity to temperature　感温性
thick root　太根【チャ】
thick sowing (seeding), dense sowing (seeding)　厚播き，密播 (みっぱ，みっぱん)
thick stand　密生
thickening　肥厚，肥大
thickening growth　肥大成長
thigmomorphogenesis　接触形態形成
thin layer chromatography (TLC)　薄層クロマトグラフィー
thin sowing (seeding), sparse sowing (seeding)　薄播き，疎播 (そは)
thinning　間引き
thinning-out pruning　間引きせん (剪) 定
thorn apple, Jimson weed, *Datura stramonium* L.　ヨウシュチョウセンアサガオ
thousand-grain weight (1000-grain weight), one-thousand-grain weight　千粒重
three-cornered grass, Chinese mat grass, *Cyperus malaccensis* Lam. ssp. *brevifolius* (Boeck.) T. Koyama　シチトウイ (七島藺)
three-course rotation, three-field system　三圃式農法
three-crop system, triple cropping　三毛作
three-field system, three-course rotation

三圃式農法
three major nutrients, NPK elements 肥料三要素
three phases of soil 土壌三相
three prong digging hook 三本ぐわ(鍬)
three-way cross, triple cross 三系交配, 三元交配
three-way stopcock 三方コック
three-year rotation 三年輪作
threeseeded copperleaf, *Acalypha australis* L. エノキグサ
thremmatology, breeding science 育種学
threonine (Thr) トレオニン
threshability, shattering habit, shedding habit 脱粒性
thresher 脱穀機, スレッシャ
threshing 脱穀
threshing comb, comb thresher 千歯, 千歯こ(扱)き
threshing sticks こ(扱)き箸
threshold value いき(閾)値
throw-in plowing, gathering plowing 内返し耕
thrum flower, short-styled flower 短[花]柱花
thunderstorm 雷雨
thylakoid チラコイド
thymeleaf sandwort, *Arenaria serpyllifolia* L. ノミノツヅリ
thymine (T) チミン
Thymus vulgaris L., common thyme, garden thyme タイム
Ti plasmid Ti プラスミド
TIBA (2,3,5-triiodobezoic acid) 2,3,5-トリヨード安息香酸
tiger nut, *Cyperus esculentus* L. ショクヨウガヤツリ
tiger's-tongue grass, bareet grass, southern cutgrass, *Leersia hexandra* Sw. タイワンアシカキ
tile drainage, underdrainage, tile drainage 暗きょ(渠)排水
tillage, tilling 耕うん(耘)
tiller 分げつ(蘗)【形態】
tiller bud 分げつ芽
tiller number 分げつ数
tillering 分げつ(蘗)【発育現象】
tillering node 分げつ節
tillering position on stem 分げつ節位
tillering stage 分げつ期
tillers at high nodal position, upper nodal tiller 高位分げつ
tilling, tillage 耕うん(耘)
tilth 耕起地
time constant 時定数
time domain reflectometry (TDR) TDR法
time of bud opening, sprouting time 萌芽期【チャ】
time series 時系列
time series analysis 時系列解析, 時系列分析
timopheevi wheat 1) チモフェービ系コムギ 2) チモフェービコムギ *Triticum timopheevi* Zhuk.
timothy, *Phleum pratense* L. チモシー
tiny vetch, *Vicia hirsuta* (L.) S. F. Gray スズメノエンドウ
tip, apex 頂部
tip burn 縁腐れ【レタス, ハクサイ等】
tips 天葉(てんぱ)【タバコ】
tissue 組織
tissue culture 組織培養
tissue-specific gene expression 組織特異的遺伝子発現
titration 滴定
tjereh チェレー【イネの生態型】
TLC (thin layer chromatography) 薄層クロマトグラフィー
tobacco, *Nicotiana tabacum* L. タバコ(煙草)
toddy palm, wine palm, *Caryota urens* Jacq. クジャクヤシ(孔雀椰子)
tolerance 耐性

tolerance limit　許容限界
toluidine blue　トルイジンブルー【染色】
tomato, *Lycopersicon esculentum* Mill.　トマト
tonoplast　液胞膜, トノプラスト
tooth pick, *Ammi visnaga* (L.) Lam.　アンミ
top, aboveground part, aerial part　地上部
top clipping　先刈り【イグサ】
top cross　トップ交雑
top grafting, top working　高接ぎ
top grass　上繁草
top pruning, topping, pinching　摘心
top-root ratio (T-R ratio)　TR率
top soil　1) 表[層]土　2) 作土
top working, top grafting　高接ぎ
topdressing, supplement application　追肥
topdressing after harvest　お礼肥
topdressing at panicle (ear) formation stage　穂肥
topdressing at ripening stage　実肥
topee-tambu, *Calathea allouia* (Aubl.) Lindl　トラフヒメバショウ
topographic map, relief map　地形図
topography　地形
topping　1) 心止め, 摘心【タバコ】　2) うら(梢)切り【イグサ】　3) トッピング【テンサイ】
topsoil, cultivated soil, arable soil　耕地土壌, 耕土
topsoiling　土入れ
Torilis japonica (Houtt.) DC., Japanese hedgeparsley　ヤブジラミ
torpedo grass, *Panicum repens* L.　ハイキビ
Torreya nucifera Sieb. et Zucc., Japanese torreya　カヤ(榧)
total digestible nutrients (TDN)　可消化養分総量
total nitrogen　全窒素
totalizing rain gauge　積算雨量計

totipotency　全能性, 分化全能性
towel gourd, *Luffa acutangula* (L.) Roxb.　トカドヘチマ
toxicity　毒性
toxicity to mammals　人畜毒性
toxin　毒素
trace element, microelement, minor element　微量元素
tracer　トレーサー
tracheary element　管状要素
tracheid　仮導管
Tracheophyta, vascular plant　維管束植物
Trachycarpus fortunei (Hook.) Wendl., windmill palm　シュロ(棕櫚)
tractor　トラクタ
tragacanth milkvetch, *Astragalus gummifer* Labill.　トラガカントゴムノキ
trailing　伏臥性
training　仕立て, 整枝, 誘引【園芸】
trait　1) 形質　2) 特性【統計】
trampling　踏圧
transcription　転写
transduction　形質導入
transfer cell　転送細胞, 輸送細胞
transfer function　伝達関数
transfer genes　転移遺伝子
transfer pipette　ホールピペット
transfer RNA (tRNA)　転移 RNA
transformation　形質転換, トランスフォーメーション
transgenic plant　形質転換植物
transgression breeding　超越育種法
transgressive segregation　超越分離
transient expression　一過性発現
transition probability　推移確率
transitory starch　移動デンプン
translation　翻訳
translocating herbicide　移行性除草剤
translocation　1) 転流　2) 転座【染色体】
translocation analysis　転座分析

transmission electron microscope (TEM) 透過型電子顕微鏡
transmissivity 透過率
transmittance 透過率
transparency 透明度
transpiration 蒸散
transpiration coefficient 蒸散係数
transpiration efficiency 蒸散効率
transpiration rate 蒸散速度
transpiration ratio 蒸散比
transplantation 移植
transplanter 移植機
transplanting 移植, 植替え
transplanting bed 移植床
transplanting culture 移植栽培
transplanting in nursery 苗床移植
transplanting injury 植傷み
transplanting time 移植期
transposable genetic element 転移性遺伝因子
transposon トランスポゾン
transverse division 横分裂
Trapa bispinosa Roxb. var. *iinumai* Nakano, water chestnut ヒシ (菱)
traumatotropism 傷屈性
treading 麦踏み
treatment 処理
tree, arbor tree 高木
tree p[-a-]eony, moutan, *Paeonia suffruticosa* Andr. (= *P. moutan* Sims) ボタン (牡丹)
trellis training 棚仕立て
trench method ざんごう (塹壕) 法
trencher トレンチャ, 溝掘り機
trend トレンド
trend function トレンド関数
triacontanol トリアコンタノール
trial 試行【確率論】
triangular planting 三角植え
tribe 族
tricarboxylic acid (TCA) cycle トリカルボン酸 (TCA) 回路
Trichloma matsutake Sing., matsutake fungus マツタケ
Trichosanthes cucumerina Buch.-Ham. ex Wall. (= *T. anguina* L.), snake gourd ヘビウリ
trickle irrigation, drip irrigation 点滴灌漑
Trifolium alexandrinum L., Egyptian clover, Berseem clover エジプシャンクローバ
Trifolium fragiferum L., strawberry clover ストロベリクローバ
Trifolium hybridum L., alsike clover アルサイククローバ
Trifolium incarnatum L., crimson clover クリムソンクローバ
Trifolium pratense L., red clover アカクローバ
Trifolium repens L., white clover シロクローバ
Trifolium repens L. var. *giganteum*, ladino clover ラジノクローバ
Trifolium resupinatum L. (= *T. suaveolens* Willd.), Persian clover ペルシャンクローバ
Trifolium subterraneum L., subterranean clover サブタレニアンクローバ, subclover サブクローバ
trigenomic hexaploid 三基六倍体
trigenomic species 三基種
Trigonotis peduncularis (Trevir.) Benth. ex Hemsl. キュウリグサ
trihybrid 三性雑種
2,3,5-triiodobezoic acid (TIBA) 2,3,5-トリヨード安息香酸
trimming 刈込み, 整枝
trimming cut 掃除刈り【牧草】
triose 三炭糖, トリオース
triple cropping, three-crop system 三毛作
triple cropping in two years 二年三作
triple cross, three-way cross 三系交配, 三元交配
triple hybrid 三系雑種

triple staining 三重染色
triploid [plant] 三倍体[植物]
triploidy 三倍性
tripping トリッピング【アルファルファ】
trisomic plant 三染色体植物
Trispsacum dactyloides L., gama grass ガマグラス
triticale ライコムギ
Triticum aestivum L., bread wheat, common wheat パンコムギ
Triticum araraticum Jakubz., wild timopheevi wheat アルメニアコムギ
Triticum boeoticum Boiss., wild einkorn wheat 野生一粒系コムギ
Triticum carthlicum Nevski, Persian wheat ペルシアコムギ
Triticum compactum Host, club wheat クラブコムギ
Triticum dicoccoides (Körn.) Schwein., wild emmer wheat パレスチナコムギ
Triticum dicoccum Schubl., Emmer エンマーコムギ
Triticum durum Desf., macaroni wheat マカロニコムギ, durum wheat デュラムコムギ
Triticum macha Dek. et Men., macha wheat マッハコムギ
Triticum monococcum L., cultivated einkorn wheat 栽培一粒系コムギ
Triticum paleocolchicum Men. グルジアコムギ
Triticum polonicum L., Polish wheat ポーランドコムギ
Triticum spelta L., spelt [wheat] スペルトコムギ
Triticum sphaerococcum Perc., Indian dwarf wheat インド矮性コムギ
Triticum spp., wheat コムギ(小麦)
Triticum timopheevi Zhuk., timopheevi wheat チモフェービコムギ
Triticum turanicum Jakubz. オリエントコムギ
Triticum turgidum L., rivet wheat リベットコムギ
Triticum vavilovii Jakubz. バビロフコムギ
Triticum zhukovskyi Men. et Er. ジュコブスキーコムギ
tritium 三重水素, トリチウム
trituration 粉砕, 摩砕
trivalent [chromosome] 三価染色体
tRNA (transfer RNA) 転移RNA
Tropaeolum majus L., common nasturtium, garden nasturtium キンレンカ(金蓮花)
Tropaeolum tuberosum Ruiz. et Pav., anu, jicamas アヌウ
tropic ageratum, white-weed, *Ageratum conyzoides* L. カッコウアザミ
tropical 熱帯の
tropical crop 熱帯[性]作物
tropical forages 暖地型牧草
tropical plant 熱帯[性]植物
tropical soil 熱帯土壌
tropical zone 熱帯
tropics 熱帯
tropism 屈性
truck gardening 輸送園芸
true lavender, lavender, *Lavandula angustifolia* Mill. (= *L. officinalis* Chaix, *L. vera* DC.) ラベンダー
true photosynthesis 真の光合成
true [potato] seed (TPS) 真正種子
true resistance 真正抵抗性
truffle, *Tuber* spp.【*T. melanosporum* Vittl. 他】トリュフ
truncated distribution 切断分布
truncation selection, selection by truncation 切断型選抜
trunk 幹
trusted seed production 委託採種
tryptophan (Trp) トリプトファン
tube 鏡筒【顕微鏡】
tuber いも, 塊茎
tuber bulking, bulking 塊茎肥大

tuber formation, tuberization 結しょ(薯)
Tuber spp.【*T. melanosporum* Vittl. 他】, truffle トリュフ
tuberization 1) 塊茎形成 2) 結しょ(薯)
tuberous root 塊根
tubular leaf 筒状葉
tubulin チューブリン
tumor 腫瘍
tundra climate ツンドラ気候
tung, *Aleurites montana* (Lour.) E. H. Wilson カントンアブラギリ(広東油桐)
tung, Japanese tung-oil tree, *Aleurites cordata* (Thunb.) R. Br. ex Steud. アブラギリ(油桐)
tung oil 桐油(とうゆ)
tung tree, *Aleurites fordii* Hemsl. シナアブラギリ(支那油桐)
tunica 外衣
tunica-corpus theory 外衣内体説
turbulence 乱流【気象】
turbulent flow 乱流【気象】
turf 芝地, 芝生
turgor pressure 膨圧
turka, kendyr, *Apocynum sibiricum* Jacq. ツルカ, ケンディル
turmeric, *Curcuma longa* L. (= *C. domestica* Valet.) ウコン(欝金)
turnip, *Brassica rapa* L. カブ, 飼料カブ(蕪, 蕪菁)
tussock そう(叢)生
twining plant, volubile plant 巻きつき植物
twist, convolution 撚り(より)【棉毛】
two-rowed barley, *Hordeum distichon* L. 二条オオムギ
two-rowed wild barley, *Hordeum spontaneum* K. Koch 二条野生オオムギ
two-sided test 両側検定
two-tailed test 両側検定

two-way layout 二元配置
two-year rotation 二年輪作
tying 結束【タバコ】
Typha latifolia L., cattail ガマ(蒲)
Typha orientalis Presl コガマ
typhoon 台風
typhoon damage 台風害
tyrosine (Tyr) チロシン

[U]

uchiko flour 打ち粉【ソバ】
udo, *Aralia cordata* Thunb. ウド(独活)
UDP (uridine diphosphate) ウリジンニリン酸
ulluco, papa lisas, *Ullucus tuberosus* Caldas ウルーコ
Ullucus tuberosus Caldas, ulluco, papa lisas ウルーコ
ultracentrifuge 超遠心機
ultrafiltration 限外ろ(濾)過
ultramicroscope 限外顕微鏡
ultramicrotome ウルトラミクロトーム
ultrasonication 超音波処理
ultrastructure 微細構造
ultrathin section 超薄切片
ultraviolet radiation (UV) 紫外放射
umbel 散形花序
unavailable moisture 無効水[分]
unavailable nutrient 不可給態養分
unavailable water 無効水[分]
Uncaria gambir Roxb., gambir ガンビール
uncultivated arable land, potential arable land 未耕地
uncultivated paddy field 休耕田
Undaria pinnatifida (Harvey) Suringar, wakame seaweed ワカメ
underdrain, conduit 暗きょ(渠)
underdrainage, tile drainage, pipe drainage 暗きょ(渠)排水
underground part, subterranean part 地

下部
undergrowth, bottom grass 下草, 下繁草
undertips 上葉(うわは)【タバコ】
undulate 波形の
unelongated stem part 非伸長茎部
unequal [cell] division 不等分裂
unfertilization 不受精
unfolding, emergence 抽出【発育】
unfruitfulness 不結果性, 不結実
unglazed pot, clay pot 素焼鉢
unhulled rice, rough rice, paddy 籾
unicellular organism 単細胞生物
uniconazole ウニコナゾール
uniform application of fertilizer to top soil 全層施肥
uniform distribution 一様分布
uniform random number 一様乱数
uniformity trial, homogeneity trial 一様性試験, 均一性試験【圃場等】
unimodal distribution 単峰分布
unisexual flower 単性花
unit membrane 単位膜
univalent [chromosome] 一価染色体
unloading アンローディング
unsaturated fatty acid 不飽和脂肪酸
unseasonable bolting 不時抽だい(苔)
unseasonable flowering 不時開花
unseasonable heading, premature heading 不時出穂
unusual weather, abnormal weather 異常気象
upland cotton, *Gossypium hirsutum* L. リクチメン(陸地棉)
upland farming, field crop cultivation 畑作
upland field, field 畑
upland irrigation 畑地灌漑
upland rice 陸稲
upland rice-nursery 畑苗代
upland [rice] seedling 畑苗
upland soil 畑土壌
upland-cultured paddy rice 畑[作]水稲

upper leaf 上位葉
upper nodal tiller, tillers at high nodal position 高位分げつ
upper part of tuberous root なり首, 藷梗(しょこう)【サツマイモ】
upright 立性の
upright habit 直立性
upright planting 直立植え, 直立挿し【サツマイモ】
upright stem, erect stem 直立茎
upright type, erect type 直立型
uprooting of seedling, pulling of seedling 苗取り
upside down plowing 反転耕
uptake, absorption 吸収
uracil (U) ウラシル
urd, black gram, black matpe, *Vigna mungo* (L.) Hepper (= *Phaseolus mungo* L.) ケツルアズキ(毛蔓小豆), ブラックマッペ
urea 尿素
urea nitrogen 尿素態窒素
Urena lobata L. var. *lobata*, aramina オオバボンテンカ
uridine diphosphate (UDP) ウリジン二リン酸
uridine triphosphate (UTP) ウリジン三リン酸
useful wild plant 有用野草
utilization rate of arable land 耕地利用率
UTP (uridine triphosphate) ウリジン三リン酸
UV (ultraviolet radiation) 紫外放射
uzu type 渦(うず)性【オオムギ】

[V]

vacant hill, missing plant 欠株
vacuole 液胞
vacuum infiltration method 減圧浸潤法
Valeriana fauriei Briq. カノコソウ

valine (Val)　バリン
Vandellia anagallis (Burm. f.) Yamazaki var. *verbenaefolia* (Colsm.) Yamazaki　スズメノトウガラシ
vanilla, *Vanilla planifolia* Andr.　バニラ
Vanilla planifolia Andr., vanilla　バニラ
vapor pressure　蒸気圧
vapour pressure deficit (VPD)　水蒸気圧差
variability　変異性
variable　変数
variable charge　変異荷電
variance　分散【統計】
variance component, component of variance　分散成分
variance-covariance matrix　分散共分散行列
variance ratio　分散比
variate　変量
variation　変異
variegation　1) 斑入り　2) 絞り【模様】
varietal characteristics　品種特性
varietal cross　品種間交雑
varietal difference　品種間差異
varietal differentiation　品種分化
variety　1) 品種　2) 変種
variety for processing　加工用品種
variety preservation　品種保存
variety registration　品種登録
variety test　品種比較試験
varnish tree, lacquer tree, *Rhus* spp.　ウルシ(漆)【広義】
vascular bundle　維管束
vascular bundle sheath　維管束鞘
vascular plant, Tracheophyta　維管束植物
vector　ベクター, 媒介生物
vector data　ベクトルデータ
vegetable [crop]　野菜
vegetable fat　植物脂
vegetable oil　植物油
vegetable gardening, olericulture　野菜園芸

vegetable oil and fat　植物油脂
vegetable wax　木ろう(蠟)
vegetation　植生
vegetation cover　植被
vegetation index　植生指数, 植生指標
vegetative branch, vegetative shoot　発育枝
vegetative cell　栄養細胞
vegetative growth　栄養成長
vegetative growth stage　栄養成長期
vegetative hybrid　栄養雑種
vegetative nucleus　栄養核
vegetative organ　栄養器官
vegetative phase　栄養相
vegetative propagation, cloning　栄養繁殖
vegetative shoot, vegetative branch　発育枝
vegetatively propagated plant　栄養繁殖植物
vein, nerve　葉脈
vein ending　脈端
venation　脈系, 葉脈系
veneer grafting　切接ぎ
ventilation　換気
ventral　腹側の
ventral scale　前鱗(ぜんりん)
ventral side　腹面
Verbena officinalis L., vervain　クマツヅラ
vermiculite　バーミキュライト
vernalization　春化[処理], バーナリゼーション
Veronica arvensis L., corn speedwell　タチイヌノフグリ
Veronica hederifolia L., ivy-leaf speedwell　フラサバソウ
Veronica persica Poir., Persian speedwell　オオイヌノフグリ
vertical resistance　垂直抵抗性
vervain, *Verbena officinalis* L.　クマツヅラ

very fine sand　微砂
vesicular-arbuscular mycorrhiza (VAM)　VA菌根
vessel　導管
vestige　退化痕跡
vetiver, khus khus, *Vetiveria zizanioides* Stapf (= *V. zizanioides* (L.) Nash ex Small)　ベチベル
Vetiveria zizanioides Stapf (= *V. zizanioides* (L.) Nash ex Small), vetiver, khus khus　ベチベル
viability, survival rate　生存率
viability of seed, germination ability, germinability　発芽力
Vicia angustifolia L., narrowleaf vetch　カラスノエンドウ
Vicia cracca L., cow vetch　クサフジ
Vicia faba L., broad bean　ソラマメ (蚕豆)
Vicia hirsuta (L.) S. F. Gray, tiny vetch　スズメノエンドウ
Vicia sativa L., common vetch　コモンベッチ
Vicia tetrasperma (L.) Schreb., fourseeded vetch, sparrow vetch　カスマグサ
Vicia villosa Roth, hairy vetch, wooly vetch　ヘアリベッチ
Vigna mungo (L.) Hepper (= *Phaseolus mungo* L.), black gram, urd, black matpe　ケツルアズキ(毛蔓小豆), ブラックマッペ
Vigna aconitifolia (Jacq.) Maréchal (= *Phaseolus aconitifolius* Jacq.), moth bean, mat bean　モスビーン
Vigna angularis (Willd.) Ohwi et Ohashi (= *Phaseolus angularis* L.), adzuki bean, azuki bean, small red bean　アズキ(小豆)
Vigna radiata (L.) R. Wilczek (= *Phaseolus aureus* Roxb.), green gram, mung bean　リョクトウ(緑豆)
Vigna subterranea (L.) Verdc. (= *Voandzeia subterranea* (L.) Thouars), bambara bean, bambara groundnut　バンバラマメ
Vigna umbellata (Thunb.) Ohwi et Ohashi (= *Phaseolus calcaratus* Roxb.), rice bean　タケアズキ
Vigna umbellata (Thunb.) Ohwi et Ohashi (= *Phaseolus pendulus* Makino)　ツルアズキ(蔓小豆)
Vigna unguiculata (L.) Walp. (= *V. sinensis* Endl.), cowpea, southern pea　ササゲ(豇豆, 大角豆)
Vigna unguiculata (L.) Walp. var. *catjang* (Burm. f.) H. Ohashi, catjang　ハタササゲ
Vigna unguiculata (L.) Walp. var. *sesquipedalis* (L.) H. Ohashi, asparagus bean　ジュウロクササゲ(十六豇豆)
vine　つる(蔓)[植物]
vinegar　酢
viny, pole climbing　つる(蔓)性の
vinyl film　ビニルフィルム
vinyl house　ビニルハウス
vinyl sash　ビニル障子
violet crabgrass, *Digitaria violascens* Link　アキメヒシバ
Virginia pepperweed, peppergrass, *Lepidium virginicum* L.　マメグンバイナズナ
viroid　ウイロイド
virtual reality　バーチャルリアリティー, 仮想現実
virus　ウイルス
virus disease　ウイルス病
viscoelasticity　粘弾性
viscosity　粘度, 粘性
visible light　可視光[線]
visible radiation　可視光[線]
visual selection　肉眼選抜
visual trait　可視形質
vital staining　生体染色
vitamin　ビタミン

Vitellaria paradoxa (A. DC.) C. F. Gaertn. (= *Butyrospermum parkii* Don Kotschy), shea [butter] tree　シアバターノキ (シアバターの木)
Vitis vinifera L., grape　ブドウ (葡萄)
vitreous break　ガラス状断面
vitreous grains　硬質粒
vitrification　ガラス化【培養】
viviparous seed　胎生種子
vivipary　胎生
viviparity　胎生
Voandzeia subterranea (L.) Thouars (= *Vigna subterranea* (L.) Verdc.), bambara bean, bambara groundnut　バンバラマメ
void, pore space　孔げき (隙)
volatile　揮発性の
volcanic ash soil　火山灰土壌
volubile plant, twining plant　巻きつき植物
volume　容積
volumetric flask　メスフラスコ
voluntary intake　自由採食 [量]

[W]

wadding crop　充てん (填) 料作物
Wagner pot　ワグナーポット
waiting meristem　待機分裂組織
wakame seaweed, *Undaria pinnatifida* (Harvey) Suringar　ワカメ
walking tractor　歩行用トラクタ, ハンドトラクタ
wall pressure　壁圧
walnut, *Juglans* spp.　クルミ (胡桃)
wandering cudweed, *Gnaphalium pensylvanicum* Willd. (= *G. purpureum* L. var. *stathulatum* (Lam.) Baker)　チチコグサモドキ
Warburg effect　ワールブルグ効果
Warburg's manometer　ワールブルグ検圧計
WARDA (West Africa Rice Development Association)　西アフリカ稲開発協会
Waring blender　ワーリングブレンダー
warm temperate zone　暖温帯
warmth index　温量指数, 暖かさの指数
wasabi, *Eutrema japonica* (Miq.) Koidz. (= *E. wasabi* Maxim.)　ワサビ (山葵)
waste　廃棄物
wasteland, barren land, barrens　荒廃地, 不毛地
watch glass　時計皿
water absorption　吸水
water apple, *Syzygium aqueum* (Burm. f.) Alston　ミズレンブ
water balance　水収支
water channel　水チャンネル
water chestnut, *Trapa bispinosa* Roxb. var. *iinumai* Nakano　ヒシ (菱)
water clover, pepperwort, *Marsilea quadrifolia* L.　デンジソウ
water content, moisture content　水分含量, 含水量
water control, water management　水管理
water convolvulus, water spinach, swamp cabbage, *Ipomoea aquatica* Forsk.　ヨウサイ (蕹菜)
water-cress, cresson, *Nasturtium officinale* R. Br. (= *Roripa nasturtium-aquaticum* (L.) Hayek.)　クレソン (和蘭芥)
water culture　水耕, 水耕栽培
water deficit　水分欠乏
water-dispersible powder, wettable agent (powder)　水和剤
water economy　水分経済
water erosion　水食
water foxtail, *Alopecurus aequalis* Sobol. var. *amurensis* (Komar.) Ohwi　スズメノテッポウ
water hyacinth, *Eichhornia crassipes* (Mart.) Solms-Laub.　ホテイアオイ (布袋葵)

water inlet　水口（みなくち）
water leak prevention　漏水防止
water-leaking paddy field, high permeable paddy field　漏水田
water lifting machinery　揚水機
water loss in depth, water requirement in depth　減水深
water management, water control　水管理
water outlet　水尻（みなじり）
water pepper, *Persicaria hydropiper* (L.) Spach (= *Polygonum hydropiper* L.)　ヤナギタデ
water permeability　透水性
water plantain, *Alisma canaliculatum* A. Br. et Bouch ex Sam.　ヘラオモダカ
water pollution　水質汚濁
water pore　水孔
water potential　水ポテンシャル
water relations　水分生理, 水関係
water requirement　要水量
water requirement in depth, water loss in depth　減水深
water retentivity　保水性
water retting　浸水精練【繊維作物】
water right　水利権
water saturation deficit　水欠差
water-saving culture　節水栽培
water shoot　徒長枝
water-soluble carbohydrate　水溶性炭水化物
water-soluble concentrate　水溶剤
water spinach, water convolvulus, swamp cabbage, *Ipomoea aquatica* Forsk.　ヨウサイ（蕹菜）
water sprout　徒長枝
water starwort, *Stellaria aquatica* (L.) Scop.　ウシハコベ
water starwort, *Callitriche palustris* L.　ミズハコベ
water stress　水ストレス
water temperature　水温
water transmission　水媒伝染
water transport　水輸送
water use efficiency (WUE)　水利用効率
water utilization association, irrigation association　水利組合
water warming canal　温水路
water warming paddy field　温水田
water warming pond　温水池
water yam, greater yam, winged yam, *Dioscorea alata* L.　ダイジョ（大薯）
watering, irrigation　灌水
watering pot　じょうろ（如雨露）
waterlogged soil, flooded soil, submerged soil　湛水土壌
waterlogging, flooding　湛水
watermelon, *Citrullus lanatus* (Thunb.) Matsum. et Nakai (= *C. vulgaris* Schrad.)　スイカ（西瓜）
Watson pomelo, *Citrus natsudaidai* Hayata　ナツミカン（夏蜜柑）
wax　ろう（蠟）
wax apple, Java apple, *Syzygium samarangense* (Blume) Merr. et Perry　レンブ
wax crop　ろう（蠟）料作物
waxy, glutinous　もち（糯）性の
waxy barley, glutinous barley　もち（糯）オオムギ
waxy bloom, bloom　果粉
waxy corn, *Zea mays* L. var. *amylosaccharata* Sturt.　ワキシーコーン, もち（糯）トウモロコシ
waxy rice, glutinous rice　もち（糯）米
waxy wheat, glutinous wheat　もち（糯）コムギ
weak-strawed variety　弱稈品種
weather fleck, physiological leaf spot　生理的斑点病【タバコ】
weather forecasting　天気予報
weather information system　気象情報システム
weathering　風化[作用]
weed　雑草

weed control　雑草防除
weed control program　雑草防除体系
weed control spectrum, weeding spectrum, herbicidal spectrum　殺草スペクトル
weed loss, loss from weed　雑草害
weed management　雑草管理
weed science　雑草学
weed scraping　草削り
weeder　除草機
weeding　除草
weeding spectrum, weed control spectrum, herbicidal spectrum　殺草スペクトル
weeding system　除草体系
weeping lovegrass, *Eragrostis curvula* (Schrad.) Nees　ウィーピングラブグラス
Weibull distribution　ワイブル分布
weight of a head　一穂重
weighted mean　加重平均，重み付き平均
well-drained paddy field　乾田
well-experienced farmer　篤農
Welsh onion, *Allium fistulosum* L.　ネギ(葱)
West Africa Rice Development Association (WARDA)　西アフリカ稲開発協会
West Indian arrowroot, arrowroot, *Maranta arundinacea* L.　アロールート
West Indian indigo, *Indigofera suffruticosa* Mill.　ナンバンコマツナギ
West Indian lemongrass, *Cymbopogon citratus* (D. C. ex Nees) Stapf　西インドレモングラス
western blot technique, western blotting　ウェスタンブロット法
western elodea, *Elodea nuttallii* (Planch.) St. John　コカナダモ
western mugwort, *Artemisia japonica* Thunb.　オトコヨモギ
western wheatgrass, *Pascopyrum smithii* (Rydb.) A. Löve　ウェスタン・ウィートグラス
wet-bulb temperature　湿球温度
wet endurance, excess water tolerance　耐湿性
wet injury, excess-moisture injury　湿害
wet-season cropping, rainy season cropping　雨季作, 雨期作
wettable agent　水和剤
wettable powder　水和剤
wetting agent　展着剤
wheat, *Triticum* spp.　コムギ(小麦)
wheat bran　ふすま(麩)
wheat cropping　麦作(むぎさく, ばくさく)【コムギ】
wheat culture　麦作(むぎさく, ばくさく)【コムギ】
wheat flour　小麦粉
wheat meals　コムギ粗砕粉
wheat starch　小麦デンプン
wheat straw　麦稈(ばっかん)【コムギ】
wheat threshing　麦こ(扱)き, 麦打ち【コムギ】
white-back rice　背白米
white-based rice kernel　基白米(もとじろまい)
white-belly rice　腹白米
white clover, *Trifolium repens* L.　シロクローバ
white core rice　心白米
white datura, *Datura metel* L. (= *D. alba* Nees)　チョウセンアサガオ
white guinea yam, *Dioscorea rotundata* Poir.　ギニアヤム【白肉】
white head　白穂(しらほ)
white jute, jute, *Corchorus capsularis* L.　ジュート
white lupine, *Lupinus albus* L.　シロバナルーピン(白花ルーピン)
white mustard, *Sinapis alba* L. (= *Brassica alba* (L.) Boiss.)　シロガ

ラシ(白芥子)
white popinac, ipil-ipil, *Leucaena leucocephala* (Lam.) De Wit　ギンゴウカン(銀合歓)
white sandalwood, sandalwood, *Santalum album* L.　ビャクダン(白檀)
white sweetclover, *Melilotus albus* Medik.　シロバナスィートクローバ(白花スィートクローバ)
white-weed, tropic ageratum, *Ageratum conyzoides* L.　カッコウアザミ
whiteness　白度【コムギ】
whittle grafting, crown grafting　そぎ接ぎ
whole crop silage　ホールクロップサイレージ
whole grain　整粒
whole rice grain, perfect rice grain　完全米
whole sampling　全刈り
whorled　輪生の
wide cross　遠縁交雑
wide hybridization　遠縁交雑
wide [regional] adaptability　広域適応性
wild ancestor　野生原種
wild basil, wild spikenard, *Hyptis suaveolens* Poir.　ニオイイガクサ
wild einkorn wheat, *Triticum boeoticum* Boiss.　野生一粒系コムギ
wild emmer wheat, *Triticum dicoccoides* (Körn.) Schwein.　パレスチナコムギ
wild grass　野草【イネ科】
wild herb　野草【一般】
wild oat, *Avena fatua* L.　カラスムギ【≠エンバク】
wild relatives　近縁野生種
wild soybean, *Glycine max* (L.) Merr. ssp. *soja* (Sieb. et Zucc.) H.Ohashi　ツルマメ(蔓豆)
wild species　野生種
wild spikenard, wild basil, *Hyptis suaveolens* Poir.　ニオイイガクサ
wild strain　野生株
wild timopheevi wheat, *Triticum araraticum* Jakubz.　アルメニアコムギ
wild type　野生型
wildrice, Indian rice, *Zizania aquatica* L. および *Z. palustris* L.　アメリカマコモ
wilting　1) しお(萎)れ　2) 予乾(よかん)【飼料】
wilting coefficient　しお(萎)れ係数
wilting point　しお(萎)れ点
wimmera ryegrass, *Lolium rigidium* Gaud.　ウィメラライグラス
wind and flood damage　風水害
wind-borne infection　風媒伝染
wind damage (injury)　風害
wind erosion　風食
wind machine　防霜ファン, ウインドマシン【チャ】
wind pollination, anemophily　風媒
wind protection　防風
wind selection, winnowing　風選
wind speed　風速
wind tolerance　耐風性
wind vane, anemoscope　風向計
wind vane and anemometer　風向風速計
wind velocity　風速
windbreak, hedge　風除け
windbreak forest　防風林
windbreak net　防風網
windmill palm, *Trachycarpus fortunei* (Hook.) Wendl.　シュロ(棕櫚)
windrow　地干し列
wine palm, toddy palm, *Caryota urens* Jacq.　クジャクヤシ(孔雀椰子)
winged bean, four-angled bean, goa bean, asparagus pea, *Psophocarpus tetragonolobus* (L.) DC.　シカクマメ
winged waterprimrose, *Ludwigia decurrens* Walt.　ヒレタゴボウ
winged yam, greater yam, water yam,

Dioscorea alata L. ダイジョ (大薯)
winnow み (箕)
winnowed rough rice 精籾
winnower 唐み (箕)
winnowing, wind selection 風選
winnowing machine 唐み (箕)
winter annual 冬一年草, 越年草
winter annual crop 越年生作物
winter annual weed 越年生雑草
winter bud 冬芽, 越冬芽
winter cereals 冬穀物, 麦 [類]
winter crop 冬作 [物]
winter cropping 冬作
winter cropping on drained paddy field 水田裏作
winter dressing 寒肥
winter fallow 冬期休閑, 冬季休閑
winter habit 秋播き性
winter hardiness 耐冬性
winter irrigation 冬期灌漑
winter plowing 冬耕
winter pruning 冬期せん (剪) 定
winter soaking 寒水浸【イネ種子】
winter sowing (seeding) 冬播き
winter squash, pumpkin, *Cucurbita maxima* Duch. ex Lam. セイヨウカボチャ
winter squash, pumpkin, *Cucurbita moschata* (Duch. ex Lam.) Duch. ex Poir. ニホンカボチャ
winter type 秋播き型
winter variety 秋播き品種
winter weed 冬雑草
winter wheat 冬コムギ
wiregrass, goosegrass, *Eleusine indica* (L.) Gaertn. オヒシバ
witche's broom 天狗巣病
within-group variance 群内分散
wood 材
wood-pasture, grazable forestland 混牧林, 林内草地
wooden base hoe 風呂ぐわ (鍬), 平ぐわ (鍬)
wooden base spade 風呂すき (鋤)
woodland pasture, grazable forestland 混牧林, 林内草地
woody, arboreous 木本 [性] の
woody plant 木本植物
wooly vetch, hairy vetch, *Vicia villosa* Roth ヘアリベッチ
work system 作業体系【広義】
working distance 作動距離【顕微鏡】
wormseed goosefoot, *Chenopodium ambrosioides* L. var. *anthelminticum* (L.) A. Gray アメリカアリタソウ
wound hormone ゆ (癒) 傷ホルモン
wounded rough rice 傷籾
WUE (water use efficiency) 水利用効率

[X]

X-ray diffraction X線回折
Xanthium strumarium L., common cocklebur オナモミ
xanthophyll キサントフィル
Xanthosoma sagittifolium (L.) Schott, yautia, tannia アメリカサトイモ
xenia キセニア
xeromorphism 乾生形態
xeromorphy 乾生形態
xerophyte 乾生植物
xerophytic weed 乾生雑草
xylem 木部
xylem fiber 木部繊維
xylem parenchyma 木部柔組織
xylene キシレン
xyloglucan キシログルカン

[Y]

Y-shaped tubing connector Y字管
YAC (yeast artificial chromosome) vector YACベクター, 酵母人工染色体ベクター
yacon, *Polymnia sonchifolia* Poepp. et

Endl. ヤーコン
yam ヤムイモ
yam bean, *Pachyrhizus erosus* (L.) Urban クズイモ
yard, paddock 追込み場
yautia, tannia, *Xanthosoma sagittifolium* (L.) Schott アメリカサトイモ
year-round cropping 周年栽培
year-round culture 周年栽培
year-round supply 周年供給
year-to-year correlation 年次相関
year-to-year variation, interannual variation 年次変異
yeast 酵母
yeast artificial chromosome (YAC) vector 酵母人工染色体ベクター, YACベクター
yellow fieldcress, *Rorippa sylvestris* (L.) Bess. キレハイヌガラシ
yellow foxtail, *Setaria glauca* (L.) Beauv. キンエノコロ
yellow gentian, *Gentiana lutea* L. ゲンチアナ
yellow guinea yam, *Dioscorea cayenensis* Lam. ギニアヤム【黄肉】
yellow lupine, *Lupinus luteus* L. キバナルーピン(黄花ルーピン)
yellow mosaic 縞萎縮病【オオムギ】
yellow ripe 黄熟
yellow ripe stage 黄熟期
Yellow soil 黄色土
yellow sweetclover, *Melilotus officinalis* (L.) Lam. キバナスィートクローバ(黄花スィートクローバ)
yellow velvetleaf, *Limnocharis flava* (L.) Buchenau キバナオモダカ
yellowed rice 黄変米
yellowing, etiolation 黄化
yellowing stage 黄変期【タバコ】
yellows 黄化病, 萎黄[病]
yield 収量
yield component 収量構成要素
yield decrease 減収

zea (329)

yield forecast 収量予測
yield increase 増収
yield potential 生産力, 収量ポテンシャル
yield prediction 収量予測
yield survey 収量調査
yield trial 収量試験, 生産力検定[試験]
yielding ability 収量性
yielding percentage, finishing ratio 歩留り
ylang-ylang, ilang-ilang, *Canangium odoratum* Hook. f. et Thoms. (= *Cananga odorata* Baill.) イランイラン
Yokohama [velvet] bean, *Mucuna pruriens* (L.) DC. var. *utilis* (Wight) Burck (= *Stizolobium hassjoo* Piper et Tracy) ハッショウマメ(八升豆)
Yorkshire fog, common velvetgrass, *Holcus lanatus* L. ベルベットグラス
young panicle 幼穂
young tea field 幼木園【チャ】
young tree 若木, 幼木
Youngia japonica (L.) DC., Asiatic hawk's-beard オニタビラコ

[Z]

Z scheme ゼットスキーム【光合成】
Zamia floridana A. DC., Florida arrowroot, coontie フロリダアロールート
Zanthoxylum piperitum (L.) DC., Japanese pepper, Japanese prickly ash サンショウ(山椒)
Zea mays L., corn, maize, Indian corn トウモロコシ(玉蜀黍)
Zea mays L. var. *amylacea* Sturt., soft corn ソフトコーン
Zea mays L. var. *amylosaccharata* Sturt., waxy corn ワキシーコーン
Zea mays L. var. *everta* Bailey, pop corn ポップコーン

Zea mays L. var. *indentata* Bailey, dent corn　デントコーン
Zea mays L. var. *indurata* Bailey, flint corn　フリントコーン
Zea mays L. var. *saccharata* Bailey, sweet corn　スイートコーン
Zea mays L. var. *tunicata* St. Hil., pod corn　ポドコーン
zeatin　ゼアチン
zedoary, *Curcuma zedoaria* (Christm.) Roscoe　ガジュツ(莪蒁)
zero emission　ゼロエミッション
zhukovskyi wheat　ジュコブスキー系コムギ
zigzag planting　千鳥植え
zinc (Zn)　亜鉛
Zingiber mioga (Tumb. ex Murray) Roscoe, mioga ginger　ミョウガ(茗荷)
Zingiber officinale Rosc., ginger　ショウガ(生姜)
Zizania aquatica L. および *Z. palustris* L., wildrice, Indian rice　アメリカマコモ
Zizania latifolia (Griseb.) Turcz. ex Stapf　マコモ(真菰)
Zizania shoot　マコモダケ(菰筍,白筍)
Ziziphus jujuba Mill., common jujube, Chinese jujube　ナツメ(棗)
Zoysia japonica Steud., Japanese lawngrass　シバ(芝)
zygote　接合子,接合体
zygotene stage　ザイゴテン期,合糸期【減数分裂】
zygotic lethal　接合体致死
zygotic ratio　接合体比

(331)

付表1 国際単位系 (SI)
表1.1 SI基本単位,補助単位,併用単位および主要組立単位

量	固有の単位名称	単位記号	SI基本単位による表示
基本単位			
長さ	メートル (meter)	m	
質量	キログラム (kilogram)	kg	
時間	秒 (second)	s	
電流	アンペア (ampere)	A	
熱力学温度	ケルビン (kelvin)	K	
物質量	モル (mole)	mol	
光度	カンデラ (candela)	cd	
補助単位			
平面角	ラジアン (radian)	rad	
立体角	ステラジアン (steradian)	sr	
併用単位			
時間	分 (minute)	min	
	時 (hour)	h	
	日 (day)	d	
平面角	度 (degree)	°	
	分 (minute)	′	
	秒 (second)	″	
体積	リットル (liter)	l, L	
質量	トン (ton)	t	
組立単位			
周波数	ヘルツ (hertz)	Hz	s^{-1}
エネルギー	ジュール (joule)	J	$m^2\,kg\,s^{-2}$
力	ニュートン (newton)	N	$m\,kg\,s^{-2}$
圧力	パスカル (pascal)	Pa	$m^{-1}\,kg\,s^{-2}$
仕事率	ワット (watt)	W	$m^2\,kg\,s^{-3}$
温度	セルシウス度 (degree Celcius)	℃	K
電圧	ボルト (volt)	V	$m^2\,kg\,s^{-3}\,A^{-1}$
電気抵抗	オーム (ohm)	Ω	$m^2\,kg\,s^{-3}\,A^{-2}$
放射能	ベクレル (becquerel)	Bq	s^{-1}
モル濃度			$mol\,m^{-3}$

表1.2 倍数に関するSI接頭語

倍数	名称	記号	倍数	名称	記号
10^{18}	エクサ (exa)	E	10^{-1}	デシ (deci)	d
10^{15}	ペタ (peta)	P	10^{-2}	センチ (centi)	c
10^{12}	テラ (tera)	T	10^{-3}	ミリ (milli)	m
10^{9}	ギガ (giga)	G	10^{-6}	マイクロ (micro)	μ
10^{6}	メガ (mega)	M	10^{-9}	ナノ (nano)	n
10^{3}	キロ (kilo)	k	10^{-12}	ピコ (pico)	p
10^{2}	ヘクト (hecto)	h	10^{-15}	フェムト (femto)	f
10	デカ (deka)	da	10^{-18}	アト (atto)	a

付表2 単位の換算表

SI単位	非SI単位への換算係数	非SI単位	SI単位への換算係数
		長　さ	
km	0.621	mile (mi)	1.609
m	1.094	yard (yd)	0.914
m	3.28	foot (ft)	0.304
m	3.3	尺	0.303
mm	3.94×10^{-2}	inch (in)	25.4
		面　積	
ha	2.47	acre	0.405
km^2	0.386	square mile (mi^2)	2.590
m^2	0.3025	歩 (ぶ), 坪	3.3058
m^2	1.5×10^{-3}	畝 mn (ムー) (中国)	666.667
m^2	6.25×10^{-4}	rai (タイ)	1600
mm^2	1.55×10^{-3}	square inch (in^2)	645
		容　積	
m^3	35.3	cubic foot (ft^3)	2.83×10^{-2}
m^3	6.10×10^4	cubic inch (in^3)	1.64×10^{-5}
m^3	554.35	升	1.804×10^{-3}
L ($10^{-3} m^3$)	2.84×10^{-2}	bushel (bu)	35.24
L ($10^{-3} m^3$)	0.265	gallon	3.78
		質　量	
g	2.205×10^{-3}	pound (lb)	454
g	3.52×10^{-2}	ounce (oz)	28.4
g	0.2667	匁 (もんめ)	3.75
Mg(t)	1.102	ton (U.S.)	0.907
		収　量	
kg ha^{-1}	0.893	lb $acre^{-1}$	1.12
L ha^{-1}	0.107	gallon per acre	9.35
		圧　力	
MPa	9.90	atmosphere	0.101
MPa	10	bar	0.1
		温　度	
K	1.00 (K-273.15)	℃*	1.00 (℃+273.15)
℃	(9/5℃)+32	Fahrenheit (°F)	5/9 (°F-32)
		エネルギー	
J	0.239	calorie (cal)	4.19
Wm^{-2}	1.43×10^{-3}	cal $cm^{-2} min^{-1}$	698
		平面角	
rad	57.3	degrees (°)	1.75×10^{-2}
		放射能	
Bq	2.7×10^{-11}	curie (Ci)	3.7×10^{10}
Gy	100	rad (rd)	0.01

* SI組立単位である.

付表3 旧国立研究機関等の新旧名称対照表

改 称 後	改 称 前	頁
独立行政法人　国際開発機構 the Japan International Cooperation Agency(JICA)	国際開発事業団 Japan International Cooperative Agency (JICA)	p.51, 243, 245
独立行政法人　国際農林水産業研究センター Japan International Research Center for Agricultural Sciences (JIRCAS)	国際農林水産業研究センター Japan International Research Center for Agricultural Sciences (JIRCAS)	p.52,245
独立行政法人　森林総合研究所 Forestry and Forest Products Research Institute	森林総合研究所 Forestry and Forest Products Research Institute	p.79
独立行政法人　農業環境技術研究所 National Institute for Agro-Environmental Sciences	農業環境技術研究所 National Institute for Agro-Environmental Sciences	p.123
独立行政法人　農業・食品産業技術総合研究機構 果樹研究所 National Agriculture and Food Research Organization National Institute of Fruit Tree Science	果樹試験場 National Institute of Fruit Tree Science	p.24
独立行政法人　農業・食品産業技術総合研究機構 九州沖縄農業研究センター National Agriculture and Food Research Organization National Agricultural Research Center for Kyushu Okinawa Region	九州農業試験場 Kyushu National Agricultural Experiment Station	p.36
独立行政法人　農業・食品産業技術総合研究機構 近畿中国四国農業研究センター National Agriculture and Food Research Organization National Agricultural Research Center for Western Region	四国農業試験場,中国農業試験場 Shikoku National Agricultural Experiment Station, Chugoku National Agricultural Experiment Station	p.65（四）, 103（中）
独立行政法人　農業・食品産業技術総合研究機構 作物研究所 National Agriculture and Food Research Organization National Institute of Crop Science	農業研究センター National Agriculture Research Center	p.124
独立行政法人　農業・食品産業技術総合研究機構 食品総合研究所 National Agriculture and Food Research Organization National Food Research Institute	食品総合研究所 National Food Research Institute	p.75
独立行政法人　農業・食品産業技術総合研究機構 畜産草地研究所 National Agriculture and Food Research Organization National Institute of Livestock and Grassland Science	草地試験場, 畜産試験場 National Grassland Research Institute, National Institute of Animal Industry	p.91（草）, 101（畜）
独立行政法人　農業・食品産業技術総合研究機構 中央農業総合研究センター National Agriculture and Food Research Organization National Agricultural Research Center	農業研究センター National Agriculture Research Center	p.124
独立行政法人　農業・食品産業技術総合研究機構 中央農業総合研究センター　北陸研究センター National Agriculture and Food Research Organization National Agricultural Research Center Hokuriku Research Center	北陸農業試験場 Hokuriku National Agricultural Experiment Station	p.148

改　称　後	改　称　前	頁
独立行政法人　農業食品産業技術総合研究機構 動物衛生研究所 National Agriculture and Food Research Organization National Institute of Animal Health	家畜衛生試験場 National Institute of Animal Health	p.26
独立行政法人　農業・食品産業技術総合研究機構 東北農業研究センター National Agriculture and Food Research Organization National Agricultural Research Center for Tohoku Region	東北農業試験場 Tohoku National Agricultural Experiment Station	p.112
独立行政法人　農業・食品産業技術総合研究機構 農村工学研究所 National Agriculture and Food Research Organization National Institute for Rural Engineering	農業工学研究所 National Research Institute of Agricultural Engineering	p.124
独立行政法人 農業・食品産業技術総合研究機構 北海道農業研究センター National Agriculture and Food Research Organization National Agricultural Research Center for Hokkaido Region	北海道農業試験場 Hokkaido National Agricultural Experiment Station	p.150
独立行政法人　農業・食品産業技術総合研究機構 野菜茶業研究所 National Agriculture and Food Research Organization National Institute of Vegetable and Tea Science	野菜・茶業試験場 National Research Institute of Vegetable, Ornamental Plants and Tea	p.159
独立行政法人　農業生物資源研究所 National Institute of Agrobiological Sciences	蚕糸・昆虫農業技術研究所，農業生物資源研究所 National Institute of Sericultural and Entomological Sciences, National Institute of Agrobiological Resources	p.62 (蚕), 124 (農)
農林水産省　農林水産政策研究所 Policy Research Institute, Ministry of Agriculture, Forestry and Fisheries	農業総合研究所 National Research Institute of Agricultural Economics	p.124

ℝ〈学術著作権協会委託〉		
2009 新編 作物学用語集	2000年 6 月20日 第 1 版発行 2009年11月20日 訂正第 2 版	
著者との申し合せにより検印省略	編 著 者	日本作物学会
©著作権所有	発 行 者	株式会社 養賢堂 代表者 及川 清
定価3990円 (本体3800円) 税 5％	印 刷 者	株式会社 丸井工文社 責任者 今井晋太郎

〒113-0033 東京都文京区本郷5丁目30番15号
発 行 所 株式会社 養賢堂
TEL 東京(03)3814-0911 振替00120
FAX 東京(03)3812-2615 7-25700
URL http://www.yokendo.com/
ISBN978-4-8425-0063-8 C3061

PRINTED IN JAPAN　　　製本所　株式会社三水舎
本書の無断複写は、著作権法上での例外を除き、禁じられています。
本書からの複写許諾は、学術著作権協会（〒107-0052 東京都港区赤坂9-6-41 乃木坂ビル、電話 03-3475-5618・ＦＡＸ03-3475-5619）から得てください。